Understanding Complex Systems

For further volumes:
http://www.springer.com/series/5394

Understanding Complex Systems

Future scientific and technological developments in many fields will necessarily depend upon coming to grips with complex systems. Such systems are complex in both their composition - typically many different kinds of components interacting simultaneously and nonlinearly with each other and their environments on multiple levels - and in the rich diversity of behavior of which they are capable.

The Springer Series in Understanding Complex Systems series (UCS) promotes new strategies and paradigms for understanding and realizing applications of complex systems research in a wide variety of fields and endeavors. UCS is explicitly transdisciplinary. It has three main goals: First, to elaborate the concepts, methods and tools of complex systems at all levels of description and in all scientific fields, especially newly emerging areas within the life, social, behavioral, economic, neuroand cognitive sciences (and derivatives thereof); second, to encourage novel applications of these ideas in various fields of engineering and computation such as robotics, nano-technology and informatics; third, to provide a single forum within which commonalities and differences in the workings of complex systems may be discerned, hence leading to deeper insight and understanding.

UCS will publish monographs, lecture notes and selected edited contributions aimed at communicating new findings to a large multidisciplinary audience.

Springer Complexity

Springer Complexity is an interdisciplinary program publishing the best research and academic-level teaching on both fundamental and applied aspects of complex systems - cutting across all traditional disciplines of the natural and life sciences, engineering, economics, medicine, neuroscience, social and computer science.

Complex Systems are systems that comprise many interacting parts with the ability to generate a new quality of macroscopic collective behavior the manifestations of which are the spontaneous formation of distinctive temporal, spatial or functional structures. Models of such systems can be successfully mapped onto quite diverse "real-life" situations like the climate, the coherent emission of light from lasers, chemical reaction-diffusion systems, biological cellular networks, the dynamics of stock markets and of the internet, earthquake statistics and prediction, freeway traffic, the human brain, or the formation of opinions in social systems, to name just some of the popular applications.

Although their scope and methodologies overlap somewhat, one can distinguish the following main concepts and tools: self-organization, nonlinear dynamics, synergetics, turbulence, dynamical systems, catastrophes, instabilities, stochastic processes, chaos, graphs and networks, cellular automata, adaptive systems, genetic algorithms and computational intelligence.

The two major book publication platforms of the Springer Complexity program are the monograph series "Understanding Complex Systems" focusing on the various applications of complexity, and the "Springer Series in Synergetics", which is devoted to the quantitative theoretical and methodological foundations. In addition to the books in these two core series, the program also incorporates individual titles ranging from textbooks to major reference works.

Massimiliano Daniele Rosini

Macroscopic Models for Vehicular Flows and Crowd Dynamics: Theory and Applications

Classical and Non-classical Advanced Mathematics for Real Life Applications

Springer

Massimiliano Daniele Rosini
University of Warsaw
ICM
Warsaw
Poland

ISSN 1860-0832 ISSN 1860-0840 (electronic)
ISBN 978-3-319-03348-8 ISBN 978-3-319-00155-5 (eBook)
DOI 10.1007/978-3-319-00155-5
Springer Heidelberg New York Dordrecht London

Printed on acid-free paper

Springer is part of Springer Science+Business Media (www.springer.com)

Foreword

This book is devoted to new approaches in modelling and mathematical analysis of a wide range of phenomena arising from the dynamics of traffic flows. The study is focused on addressing various types of crowd effects. Crowding is a phenomenon that results from a non-equilibrium development in a mobile population, hence often may contribute to a destabilization of an involved system and results in singular dynamic behaviour such as congestion effects, in particular. On the mathematical modelling side, there is a range of approaches to those classes of processes, differing in scales of resolution (both in space and time) and formalisms applied. The finer is the scale, the more natural it becomes to apply discrete forms of the relevant mathematical models. The discrete models prove to be a preferable choice when describing the processes in particular in biological populations at cellular level or higher resolutions. Unlike in the latter context, the scope of this book is primarily focused on macroscopic situations, with continuum framework exploited. The macroscopic continuum models treated in the book are governed by nonlinear hyperbolic balance laws and their systems. The nonlinearities accounted for in the hyperbolic balance equations enable treatment of extreme developments related to crowding at the scale of populations that are large enough to admit a representation in terms of density distributions. Two classes of dynamical processes are studied, referring to vehicular traffic and pedestrian flows, with a wide variety of mathematical models introduced and discussed. The classes of models explored in the book refer to specific applied contexts, with special focus on real-time capacity of the implemented mathematical analysis techniques, equally of qualitative and quantitative types. In the context of real-world applications, the analytic treatments are required to address questions on the asymptotic large-time behaviour of the modelled systems, existence and structure of the resulting equilibrium structures, stability properties of the latter ones, to give just few characteristic references. Beyond addressing basic mathematical questions on the structure and properties of the models, a comprehensive analysis of their possible applications to real-world situations is given. The study presented in the book may be considered a systematic introduction to full-scale modelling of real phenomena, proving useful for addressing transitions from regular to singular patterns, handling emergency situations and providing procedures for system stabilization.

Warsaw, Poland, January 2013

Marek Niezgódka
Director of the ICM, University of Warsaw

Acknowledgements

For the opportunity and encouragement to develop applied mathematics, the author wishes sincerely to thank Prof. M. Niezgódka and everybody at ICM, University of Warsaw; in particular: Tamara Cetnar, Piotr Cierpiatka, Karolina Fabiszewska, Magdalena Jarosz, Joanna Jeżewska, Katarzyna Kamińska, Kerstin Kantiem, Marek Kępka, Piotr Kmieć, Magdalena Mozga, Arkadiusz Niegowski, Anna Psoda, Franciszek Rakowski, Robert Sot, Ewa Szafranek, Hanna Szymanowska, Zuzanna Szymańska, Anna Trykozko, Hubert Wojtasik and Beata Wolicka.

The author would like to thank Prof. P. Marcati, with whom will always owe a great debt of personal and scientific gratitude; Prof. R.M. Colombo for suggesting and introducing the author to the subjects of research that represent the topics of this book; Prof. F. Przytycki, Prof. B. Jakubczyk and Dr. A. Świerczewska–Gwiazda, who first gave to the author the opportunity to work in Warsaw.

My parents, Elena and Domenico, receive my deepest gratitude and love for their dedication and the many years of support. Many thanks to Natalia for the thankless task of helping in rewriting the many drafts. Having no interest in Mathematics, this was an exceptionally difficult effort.

The author was partially supported by ICM (Interdisciplinary Centre for Mathematical and Computational Modelling), University of Warsaw; Narodowe Centrum Nauki, grant 4140; and Polonium 2011 (French-Polish cooperation program) under the project "CROwd Motion Modeling and Management".

Contents

Chapter 1
Introduction

Abstract. This introductory chapter is devoted to the description and explanation of the main ideas lying behind the book. We explain the real issues that will be addressed and show how to treat them in a robust mathematical framework.

1.1 Motivations and Applications

In the recent years, the number of different elements and aspects involved in the management of urban traffic has increased enormously. Besides, many transportation problems arise from the lack of timely, easily usable and accurate information, or from the lack of appropriate coordination among decision makers. To address these issues, national, regional and urban authorities invest in modern, computerized traffic control centers and traffic management centers. Nowadays advanced management systems allow for an automatic dynamical collection of data on traffic flows and speeds, control of traffic signals based on these data, traffic forecast and provision of data for travel information services. These systems minimize the reaction times, ensure optimal performances and are becoming the norm world wide, to control entire urban networks, varying from a country to country in level of acceptance, take up and local applicability.

A development of intelligent transport systems for integrated applications of communications, control and information processing technologies to the whole transport system is mandatory to consult the wide range of interests and combine the actions to a joint problem–solving. The required skills encompass a broad array of techniques and approaches. Traffic management tools (monitoring current traffic conditions, predicting what can be expected, rerouting guidance, coordinating traffic signals in a dynamic way, detecting and managing incidents, giving green waves to give priority to public transport and emergency vehicles, road pricing, access control, speed enforcement, journey time estimate, real–time information, etc.), safety and environmental control (assistance for vulnerable road users, weather and road condition monitoring, adaptive speed control, collision detection, enhanced vehicle

M.D. Rosini: *Macroscopic Models for Vehicular Flows & Crowd Dynamics*, UCS, pp. 1–7.
DOI: 10.1007/978-3-319-00155-5_1

safety systems, hazardous load monitoring, cargo screening, intelligent evacuation route signing and priority, air quality monitoring, implementation of strategies to ease air quality problems, etc.), making public transport more attractive (giving priority to public vehicles, improving reliability and punctuality, providing real–time information at stops, electronic payments systems, etc.) are the main aspects to be considered in intelligent transport systems. The integration of these aspects certainly adds complexity, but it also provides improvements in overall system effectiveness and inter–operability ensures that the different components can function together, such as for instance advanced transportation management systems with advanced traveler information systems, transport providers and transport network operators in an intermodal transfer of passengers or freights, traffic police and rescue services in a life threatening incident, or traffic control and information centers for real–time traffic information.

The overall resultant benefits are to make better informed, safer, more coordinated and more intelligent decisions, often in real–time, by transport network controllers, such as highway authorities and agencies, and a smarter use of networks by road operators, public and commercial transport providers and individual travelers. The resulting improvements of safety, security, efficiency and comfort of the urban and inter–urban transport system save lives, money, energy and the environment.

These are evolutionary steps towards the future goal of the complete automation of the driving process, that will make the vehicle part of the highway. Mathematical models represent the fundamental bricks of this construction. In fact, Mathematics can be applied to various subjects in science and engineering, after understanding the physical context. Here problems in vehicular and pedestrian traffic flows are studied and developed starting from first principles.

1.2 Mathematical Framework

Many models for traffic flows have been developed resorting to different approaches, ranging from microscopic ones, taking into account each single individual, to kinetic and continuum ones, dealing with averaged quantities. However, recently the use of sensors imbedded in the road and cameras focused above is becoming more common and the consequent availability of on–line data allows to implement real–time strategies to avoid or mitigate congested traffics. Opposed to direct numerical simulations of large number of individual interacting subjects, as typical when dealing with microscopic models, researchers advice the use of continuum models for traffic. The main advantages of this approach with respect to the microscopic one are the following:

- The model is completely evolutive and is able to rapidly describe any traffic situation at every instant of time.
- The resulting description of queues evolution and of traveling times is accurate as the position of shock waves can be exactly computed and corresponds to queues tails.

- The theory enables the development of efficient numerical schemes also to describe very large number of individuals.
- The model can be easily calibrated, validated and implemented as the number of parameters is low.
- The theory allows to state and possibly solve optimal management problems.

Aim of this book is to present macroscopic models for vehicular and pedestrian dynamics, respectively, in the second and third part. The conservation of vehicles and pedestrians is assumed to derive continuum models based on one–dimensional nonlinear hyperbolic conservation laws with the macroscopic traffic variables as unknowns. The theory for conservation laws is carefully developed in the first part of the book. Beside the classical theory of entropy weak solutions, the concepts of non–entropy weak solutions are also introduced to describe phenomena typical of the traffic, such as those related to the presence of constraints along the paths and, in the case of crowds, to the rise of panic. The method of characteristics and the wave front tracking algorithm are detailed and applied to various typical traffic situations. This previous exposure in depth to the mathematical theory for hyperbolic conservation laws provides for the reader the tools of use in the modeling of traffic. The topics discussed supplement rather than substitute for a mathematical course on PDEs. However, no more advanced techniques are necessary, as the used results are slowly, carefully and fully explained in the first part of the book.

Although its specific subjects, the book is intended as an introduction to applied mathematics. The choice of the areas to investigate stems from similar reasons. In both, the background knowledge necessary to formulate and understand the mathematical models are relatively well known to the average reader. This makes unnecessary to refer to exceedingly technical research results. Furthermore, the described models are on one side simple enough to permit a complete understanding and, on the other side, reach enough to describe important phenomena typical of urban traffics. Finally, these two topics are strictly related to each other, and serve as introduction to more specialized investigations, such as traffic on networks, supply chains, telecommunications, blood flow, phase transitions, fluid dynamics, etc.

1.3 Book Chapters

Each part of the book is divided into many subsections of differen length. Few of them correspond to as much as a lecture. However, depending on the background of the reader, most of the sections require an amount of time less than that of a lecture.

The book is organized as follows:

Part I Mathematical Theory [5, 21, 22, 24, 29]

Chapter 2 introduces notations, terminologies and various general mathematical basic results.

In Chap. 3 we highlight some of the fundamental issues and difficulties arising in the general theory of conservation laws. We also present some basic definitions and concepts about one–dimensional scalar conservation laws.

Chapter 4 is devoted to study the Riemann problems for one–dimensional scalar conservation laws. Inter alia, we introduce the definitions of entropy and non–entropy shock waves, rarefaction waves and contact discontinuities.

Chapter 5 deals with Cauchy problems for one–dimensional scalar conservation laws. We also prove that it is well posed in the class of entropy weak solutions.

In Chap. 6 we study initial–boundary value problems [2], constrained Cauchy problems [9] and constrained initial–boundary value problems [14] for one–dimensional scalar conservation laws.

Chapter 7 deals with one–dimensional systems of conservation laws.

Chapter 8 is concerned with one–dimensional systems of balance laws weakly coupled [7]. We prove the well–posedness of the corresponding Cauchy problem in the space of functions with bounded total variation generalizing the results obtained by the author in [18, 19] for initial–boundary value problem for Temple systems of balance laws and in [15, 16] for general scalar balance laws.

Part II Models for Vehicular Traffic

Chapter 9 introduces the main theoretical definitions, concepts and mathematical models for vehicular traffic present in the literature [3, 23].

Chapter 10 deals with equilibrium traffic models for traffic flows, such as the LWR model [25, 27].

In Chap. 11 we review macroscopic models for vehicular traffic resulting from a generalization of the equilibrium traffic models presented in Chap. 10 and used to describe a road with an entrance and time depending constraint [14], or two merging roads [14], or a traffic circle [6], or a multi–population traffic [4], or a multi–lane road [7].

In Chap. 12 we propose some criteria to be optimized in an intelligent control of traffic and prove the existence of optimal management strategies [14].

In Chap. 13 the constrained initial–boundary value problems resulting from the models presented in Chap. 11 are numerically integrated [13, 14].

Chapter 14 deals with non–equilibrium traffic models for traffic flows, namely the PW model [26, 30] and the AR model [1, 31].

Part III Models for Pedestrian Traffic

In Chap. 15 we introduce the general concepts dealing with the description of crowd dynamics [12]. We also define the so called panic and highlight its dynamic effects, such as the Braess' paradox (for pedestrian flows).

In Chap. 16 we describe the CR model [12, 13, 17, 20, 28], the unique macroscopic model whose main assumption is experimentally confirmed and able to reproduce the fall in the efficiency of the exit, the rise of panic and the Braess' paradox.

In Chap. 17 we apply the CR model to describe the evacuation of two corridors [8, 10–13]. The two resulting numerical integrations display the ability of the CR model to reproduce the Braess' paradox.

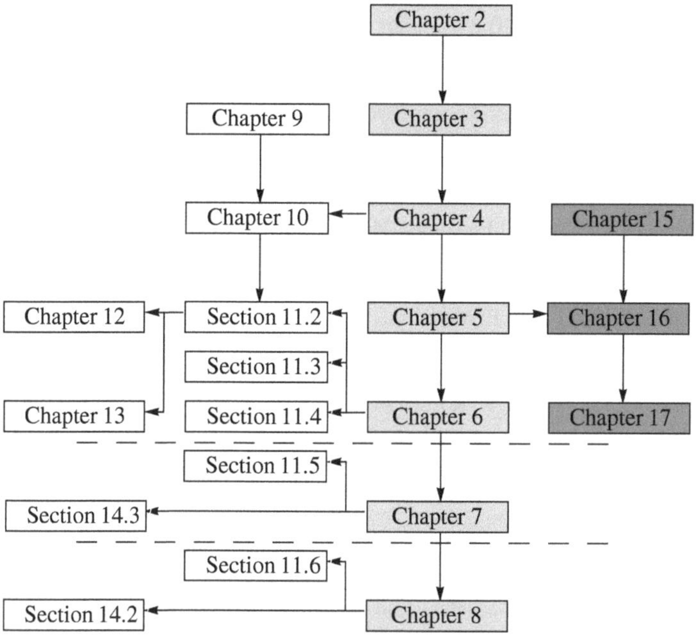

The links between the chapters and sections of the book are given in the table above. The white boxes are related to the part about vehicular traffic, the light shaded boxes to that about the mathematical theory and the dark shaded boxes to that about the pedestrian traffic.

The book can be used for undergraduate courses in mathematical modeling, physics and civil engineering. For a short course on urban traffic, the reader can focus on the sections above the first horizontal dashed line. In this way, the book can be substantially covered in one semester. However, the book provides models and techniques of fundamental interest and should motivate the reader for further studies on the subject. Longer course can reach the second horizontal dashed line or complete the entire book.

References

1. Aw, A., Rascle, M.: Resurrection of "second order" models of traffic flow. SIAM J. Appl. Math. 60(3), 916–938 (2000) (electronic)
2. Bardos, C., le Roux, A.Y., Nédélec, J.C.: First order quasilinear equations with boundary conditions. Comm. Partial Differential Equations 4(9), 1017–1034 (1979)
3. Bellomo, N., Dogbe, C.: On the Modeling of Traffic and Crowds: A Survey of Models, Speculations, and Perspectives. SIAM Rev. 53(3), 409–463 (2011)
4. Benzoni-Gavage, S., Colombo, R.M.: An *n*-populations model for traffic flow. European Journal of Applied Mathematics 14(05), 587–612 (2003)
5. Bressan, A.: Hyperbolic systems of conservation laws. Oxford Lecture Series in Mathematics and its Applications, vol. 20. Oxford University Press, Oxford (2000)

6. Chitour, Y., Piccoli, B.: Traffic circles and timing of traffic lights for cars flow. Discrete and Continuous Dynamical Systems Series B 5(3), 599–630 (2005)
7. Colombo, R.M., Corli, A., Rosini, M.D.: Non local balance laws in traffic models and crystal growth. ZAMM Z. Angew. Math. Mech. 87(6), 449–461 (2007)
8. Colombo, R.M., Facchi, G., Maternini, G., Rosini, M.D.: On the continuum modeling of crowds. American Mathematical Society (AMS), Providence (2009)
9. Colombo, R.M., Goatin, P.: A well posed conservation law with a variable unilateral constraint. J. Differential Equations 234(2), 654–675 (2007)
10. Colombo, R.M., Goatin, P., Maternini, G., Rosini, M.D.: Using conservation Laws in Pedestrian Modeling, pp. 73–79 (2009)
11. Colombo, R.M., Goatin, P., Maternini, G., Rosini, M.D.: Macroscopic Models for Pedestrian Flows. In: Big Events and Transport: the Transportation Requirements for the Management of Large Scale Events, pp. 11–22. IUAV – TTL Research Unit (2010)
12. Colombo, R.M., Goatin, P., Rosini, M.D.: A macroscopic model for pedestrian flows in panic situations. In: Proceedings of the 4th Polish-Japanese Days. GAKUTO International Series. Mathematical Sciences and Applications, vol. 32, pp. 255–272 (2010)
13. Colombo, R.M., Goatin, P., Rosini, M.D.: Conservation laws with unilateral constraints in traffic modeling. In: Mussone, L., Crisalli, U. (eds.) Transport Management and Land-Use Effects in Presence of Unusual Demand, Atti del Convegno SIDT 2009 (June 2009)
14. Colombo, R.M., Goatin, P., Rosini, M.D.: On the modelling and management of traffic. ESAIM: Mathematical Modelling and Numerical Analysis 45(05), 853–872 (2011)
15. Colombo, R.M., Mercier, M., Rosini, M.D.: Stability and total variation estimates on general scalar balance laws. Commun. Math. Sci. 7(1), 37–65 (2009)
16. Colombo, R.M., Mercier, M., Rosini, M.D.: Stability estimates on general scalar balance laws. Comptes Rendus Mathematique 347(1-2), 45–48 (2009)
17. Colombo, R.M., Rosini, M.D.: Pedestrian flows and non-classical shocks. Math. Methods Appl. Sci. 28(13), 1553–1567 (2005)
18. Colombo, R.M., Rosini, M.D.: Well posedness of balance laws with boundary. J. Math. Anal. Appl. 311(2), 683–702 (2005)
19. Colombo, R.M., Rosini, M.D.: Well posedness of balance laws with non-characteristic boundary. Bollettino Bollettino UMI 10-B(8), 875–894 (2007)
20. Colombo, R.M., Rosini, M.D.: Existence of nonclassical solutions in a Pedestrian flow model. Nonlinear Analysis: Real World Applications 10(5), 2716–2728 (2009)
21. Dafermos, C.M.: Hyperbolic systems of conservation laws. In: Systems of Nonlinear Partial Differential Equations, Oxford (1982); NATO Adv. Sci. Inst. Ser. C. Math. Phys. Sci., vol. 111, pp. 25–70. Reidel, Dordrecht (1983)
22. Godlewski, E., Raviart, P.A.: Numerical approximation of hyperbolic systems of conservation laws. Applied Mathematical Sciences, vol. 118. Springer, New York (1996)
23. Hoogendoorn, S.P., Bovy, P.H.L.: State-of-the-art of Vehicular Traffic Flow Modelling, pp. 283–303. Delft University of Technology, Delft (2001)
24. Lefloch, P.G.: Hyperbolic systems of conservation laws. The theory of classical and nonclassical shock waves. Lectures in Mathematics ETH Zürich. Birkhäuser Verlag, Basel (2002)
25. Lighthill, M.J., Whitham, G.B.: On kinematic waves. II. A theory of traffic flow on long crowded roads. Proc. Roy. Soc. London. Ser. A 229, 317–345 (1955)
26. Payne, H.J.: Models of freeway traffic and control. Math. Models Publ. Sys. Simul. Council Proc. (28), 51–61 (1971)
27. Richards, P.I.: Shock waves on the highway. Operations Res. 4, 42–51 (1956)

28. Rosini, M.D.: Nonclassical interactions portrait in a macroscopic pedestrian flow model. J. Differential Equations 246(1), 408–427 (2009)
29. Serre, D.: Systems of conservation laws. 1 & 2. Cambridge University Press, Cambridge (1999); Translated from the 1996 French original by I. N. Sneddon
30. Whitham, G.B.: Linear and nonlinear waves. Pure and Applied Mathematics. Wiley-Interscience [John Wiley & Sons], New York (1974)
31. Zhang, H.M.: A non-equilibrium traffic model devoid of gas-like behavior. Transportation Research Part B: Methodological 36(3), 275–290 (2002)

Part I
Mathematical Theory

Chapter 2
Mathematical Preliminaries

Abstract. This chapter presents notations, terminologies and various mathematical basic results, which will be used in later chapters.

2.1 Introduction

We introduce the notation that is used in this book.

Acronym List

a.e. : almost everywhere
iff : if and only if

Notations in $\mathbb{R}^n$

In the case $n = 1$, special subsets of the set of **real numbers** $\mathbb{R} = \mathbb{R}^1$ are

$\mathbb{N} = \{1, 2, \ldots\}$ the set of **positive natural numbers**
$\mathbb{N}^* = \mathbb{N} \cup \{0\}$ the set of **natural numbers**
$\mathbb{Z} = \mathbb{N}^* \cup \{-\mathbb{N}\}$ the set of **integer numbers**
$\mathbb{Q} = \left\{ \frac{p}{q} \, : \, p \in \mathbb{N}^* \text{ and } q \in \mathbb{N} \right\}$ the set of **rational numbers**
$\mathbb{R}_+^* = [0, +\infty[\,, \quad \mathbb{R}_+ =]0, +\infty[\,, \quad \mathbb{R}_-^* =]-\infty, 0]\,, \quad \mathbb{R}_- =]-\infty, 0[\,.$

For $n \in \mathbb{N}$, $\mathbb{R}^n$ denotes the n–dimensional real **Euclidean space**. A typical point of $\mathbb{R}^n$ is $u = (u_1, \ldots, u_n)^T$. By $0 = 0_n$ we denote the element $(0, \ldots, 0)^T$. The **Euclidean**

M.D. Rosini: *Macroscopic Models for Vehicular Flows & Crowd Dynamics*, UCS, pp. 11–21.
DOI: 10.1007/978-3-319-00155-5_2

norm of $u \in \mathbb{R}^n$ is $\|u\| = \sqrt{\sum_{i=1}^n u_i^2}$. The **Euclidean product** of $u, v \in \mathbb{R}^n$ is $u \cdot v = \sum_{i=1}^n u_i \, v_i$. If $U \subset \mathbb{R}^n$, then ∂U is the **boundary** of U. We say that ∂U is $\mathbf{C^k}$ if for each point $\bar{u} \in \partial U$ there exist $r \in \mathbb{R}_+$ and a $\mathbf{C^k}$ function $\gamma : \mathbb{R}^{n-1} \to \mathbb{R}$ such that (upon relabeling and reorienting the coordinate axes) we have

$$U \cap \left\{ u \in U \, : \, \|u - \bar{u}\| < r \right\} = \left\{ u \in \mathbb{R}^n \, : \, \|u - \bar{u}\| < r \text{ and } u_1 > \gamma(u_2, \ldots, u_n) \right\} .$$

The **measure** of a Lebesgue measurable subset U of $\mathbb{R}^n$ is denoted by meas U. The **counting measure** of a set V with a finite number of elements is denoted by $\#(V)$ and gives the number of elements in V.

Notation for Matrices

$A = (a_{ij})_{1 \le i \le n, 1 \le j \le m}$	: an $n \times m$ **matrix** with a_{ij} as the (i,j)–th entry
$A^T = (a_{ji})_{1 \le j \le m, 1 \le i \le n}$	: **transpose** of the matrix A
$\mathrm{Id}_n = (\delta_{ij}^K)_{1 \le ij \le n}$	: the $n \times n$ **identity matrix**
$AB = \left(\sum_{j=1}^k a_{ij} b_{jk} \right)_{1 \le i \le n, 1 \le k \le l}$	: **matrix product** of A and $B = (b_{jk})_{1 \le j \le m, 1 \le k \le l}$

Notation for Functions

Let U be an open subset of $\mathbb{R}^m$ and W be an open subset of $\mathbb{R}^n$. The **set of functions** from U to W is denoted by W^U. If $f \in W^U$, we write $f(u) = \left(f_1(u), \ldots, f_n(u) \right)^T$ for all $u \in U$. The **support** of a function f is denoted by $\mathrm{spt}\{f\}$ and is defined as the closure of the set of points where f is not zero. We will consider the following subspaces of W^U:

- $\mathbf{BV}(U;W)$ is the space of functions with bounded total variation, see Sect. 2.5.
- $\mathbf{C^k}(U;W)$ is the space of functions that are k times continuously differentiable, with uniformly bounded derivatives up to order k. If $f = (f_1, \ldots, f_n)^T \in \mathbf{C^1}(U;W)$, we denote by $Df(u)$ the $n \times m$ **Jacobian matrix** of first–order **partial derivatives** $(\partial_{u_j} f_i)$, $1 \le i \le n$, $1 \le j \le m$, computed at the point $u \in U$. The $\mathbf{C^k}$**–norm** of $f \in \mathbf{C^k}(U;W)$ is $\|f\|_{\mathbf{C^k}(U;W)} = \sup_{\substack{j \in \{0,\ldots,k\} \\ u \in U}} \left\| D^j f(u) \right\|$ where $D^j(f) = \{ \partial_{u_1}^{\alpha_1} \ldots \partial_{u_m}^{\alpha_m} f(u) \, : \, \alpha_1 + \ldots + \alpha_m = j \}$. If $m = n$, the **divergence** of $f \in \mathbf{C^1}(U;W)$ is $\mathrm{div}(f)(u) = \sum_{i=1}^n \partial_{u_i} f_i(u)$.
- $\mathbf{C_c^k}(U;W)$ is the space of functions in $\mathbf{C^k}(U;W)$ with compact support.
- $\mathbf{C^\infty}(U;W) = \cap_{k \in \mathbb{N}} \mathbf{C^k}(U;W)$.
- $\mathbf{C_c^\infty}(U;W) = \cap_{k \in \mathbb{N}} \mathbf{C_c^k}(U;W)$.
- $\mathbf{Lip}(U;W)$ denotes the space of **Lipschitz functions** f whose **Lipschitz constant** $\mathscr{L}ip(f) = \sup_{\substack{u,v \in U \\ u \neq v}} \frac{\|f(u) - f(v)\|}{\|u - v\|}$ is finite.

- $\mathbf{C^{0,1}}(U;W)$ is the space of **locally Lipschitz functions**, i.e. $f \in \mathbf{C^{0,1}}(U;W)$ iff for all $u \in U$ there exists a neighborhood of u in which f is Lipschitz.
- $\mathbf{L^k}(U;W)$ is the space of Lebesgue measurable functions f whose $\mathbf{L^k}$**–norm** $\|f\|_{\mathbf{L^k}(U;W)} = (\int_U \|f(u)\|^k \,\mathrm{d}u)^{1/k}$ is finite.
- $\mathbf{L^k_{loc}}(U;W)$ is the space of functions that are in $\mathbf{L^k}(V;W)$ for all compact subset V of U.
- $\mathbf{L^\infty}(U;W)$ is the space of Lebesgue measurable functions f whose $\mathbf{L^\infty}$**–norm** $\|f\|_{\mathbf{L^\infty}(U;W)} = \operatorname{ess\,sup}_U \|f\|$ is finite.
- $\mathbf{L^\infty_{loc}}(U;W)$ is the space of functions that are in $\mathbf{L^\infty}(V;W)$ for all compact subset V of U.
- $\mathbf{PC}(U;W)$ is the space of **piecewise constant functions** taking a finite number of values and having a finite number of discontinuities.
- $\mathbf{PC}_c(U;W)$ is the space of functions in $\mathbf{PC}(U;W)$ with compact support.
- $\mathbf{PLC}(U;W)$ is the space of **piecewise linear continuous functions** f such that f' has a finite number of discontinuities.

We introduce the following special functions:

- The **Kronecker delta** $\delta^K : \mathbb{N}^2 \to \{0,1\}$ is defined by

$$\delta^K_{ij} = \begin{cases} 1 & \text{if } i = j \\ 0 & \text{if } i \neq j \end{cases}, \qquad i,j \in \mathbb{N}\,. \tag{2.1.1}$$

- The **Dirac delta function** centered in $v \in U$ is δ^D_v and is defined by

$$\int_U f(u)\ \delta^D_v(u)\ \mathrm{d}u = f(v)\,, \qquad f \in \mathbf{C^0}(U,W)\,. \tag{2.1.2}$$

- The **characteristic function** of a set $V \subset \mathbb{R}^m$ is χ_V and is defined by

$$\chi_V(u) = \begin{cases} 1 & \text{if } u \in V \\ 0 & \text{if } u \notin V \end{cases}, \qquad u \in \mathbb{R}^m\,. \tag{2.1.3}$$

- The **sign function** $\mathrm{sgn} : \mathbb{R} \to \{-1,0,1\}$ is defined by

$$\mathrm{sgn}(u) = \begin{cases} -1 & \text{if } u \in \mathbb{R}_- \\ 0 & \text{if } u = 0 \\ 1 & \text{if } u \in \mathbb{R}_+ \end{cases}, \qquad u \in \mathbb{R}\,. \tag{2.1.4}$$

- The **translation operator** $\mathscr{T} : U \times W^U \to W^U$ is defined by

$$(\mathscr{T}_v f)(u) = f(u+v)\,, \qquad f \in W^U \text{ and } u,v \in U\,. \tag{2.1.5}$$

2.2 Preliminary Lemmas

Lemma 2.1 (Gronwall's inequality [2, page 14]). *If $t_o < T \leq +\infty$ and*

$$f(t) \leq a(t) + \int_{t_o}^{t} b(s)\, f(s)\, \mathrm{d}s\,, \qquad t \in [t_o, T[\,,$$

where all the involved functions are continuous on $[t_o, T[$ and $b(t) \in \mathbb{R}_+^*$ a.e. in $[t_o, T[$, then

$$f(t) \leq a(t) + \int_{t_o}^{t} a(s)\, b(s)\, \exp\left(\int_{s}^{t} b(r)\, \mathrm{d}r\right) \mathrm{d}s\,, \qquad t \in [t_o, T[\,.$$

If, in addition, $[t \mapsto a(t)]$ is non–decreasing, then

$$f(t) \leq a(t)\, \exp\left(\int_{t_o}^{t} b(s)\, \mathrm{d}s\right), \qquad t \in [t_o, T[\,.$$

Assume that U is a bounded open subset of $\mathbb{R}^n$ and ∂U is $\mathbf{C}^1$. Let $\nu = (\nu_1, \ldots, \nu_n)^T$ be the outward pointing unit normal vector field.

Lemma 2.2 (Green's formula). *If $f \in \mathbf{C}^1(\overline{U}; \mathbb{R})$, then*

$$\int_U \partial_{u_i} f(u)\, \mathrm{d}u = \int_{\partial U} f(u)\, \nu_i\, \mathrm{d}\gamma \qquad i = 1, \ldots, n\,. \tag{2.2.1}$$

Lemma 2.3 (Integration by parts formula). *If $f, g \in \mathbf{C}^1(\overline{U}; \mathbb{R})$, then*

$$\int_U \partial_{u_i} f(u)\, g(u)\, \mathrm{d}u = -\int_U f(u)\, \partial_{u_i} g(u)\, \mathrm{d}u + \int_{\partial U} f(u)\, g(u)\, \nu_i\, \mathrm{d}\gamma \qquad i = 1, \ldots, n\,. \tag{2.2.2}$$

2.3 Implicit Function Theorems

Theorem 2.1. *Let $f = (f_1, \ldots, f_n)^T : \mathbb{R}^m \times \mathbb{R}^n \to \mathbb{R}^n$ be a $\mathbf{C}^\mathbf{k}$ function, with $k \geq 1$, see Fig. 2.1. If $(\overline{u}, \overline{v}) \in \mathbb{R}^m \times \mathbb{R}^n$ is such that $f(\overline{u}, \overline{v}) = \overline{z}$ and the Jacobian matrix $D_v f(\overline{u}, \overline{v}) = (\partial_{v_j} f_i)_{1 \leq ij \leq n}(\overline{u}, \overline{v})$ is invertible in a neighborhood of $(\overline{u}, \overline{v})$, then there exist a neighborhood U of $\overline{u}$, a neighborhood V of $(\overline{u}, \overline{v})$ and a $\mathbf{C}^\mathbf{k}$ function $g : U \to \mathbb{R}^n$ such that*

$$g(\overline{u}) = \overline{v}$$
$$f\big(u, g(u)\big) = \overline{z} \quad \text{for all } u \in U$$
$$\text{if } (u, v) \in V \text{ and } f(u, v) = \overline{z}, \text{ then } v = g(u)\,.$$

If the k–th derivatives of f are Lipschitz continuous, then the same is true of the k–th derivatives of g. The derivative of g at the point $\overline{u}$ is the $n \times m$ Jacobian matrix

$$Dg(\overline{u}) = -\left(D_v f(\overline{u}, \overline{v})\right)^{-1} D_u f(\overline{u}, \overline{v}) \,,$$

where $D_u f = (\partial_{u_j} f_i)_{1 \le i \le n,\ 1 \le j \le m}$.

For a proof, see for instance [3, 5].

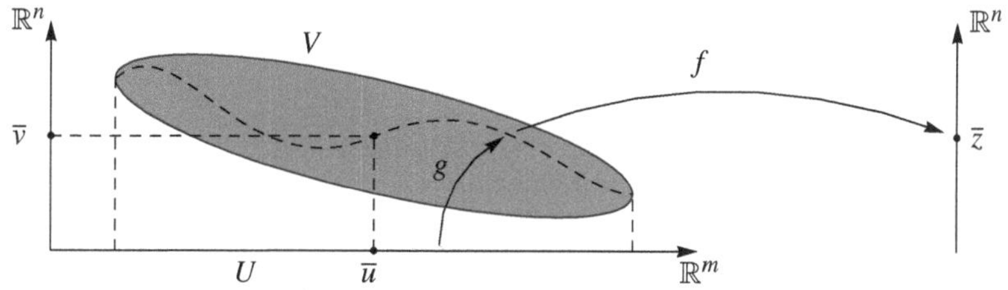

Fig. 2.1 The geometrical meaning of Theorem 2.1

We present also a parameterized version of the implicit function theorem.

Theorem 2.2. *Let $\Gamma \subseteq \mathbb{R}^n$ be an open set and $f : \mathbb{R}^m \times \mathbb{R}^n \times \Gamma \to \mathbb{R}^n$ be a $\mathbf{C^k}$ function, with $k \ge 1$. If $[\gamma \mapsto (\overline{u}_\gamma, \overline{v}_\gamma)]$ is a $\mathbf{C^k}$ map from Γ into $\mathbb{R}^m \times \mathbb{R}^n$ such that $f(\overline{u}_\gamma, \overline{v}_\gamma; \gamma) = \overline{z}$ and the Jacobian matrix $D_v f(\overline{u}_\gamma, \overline{v}_\gamma; \gamma)$ is invertible for every γ in a compact set $K \subset \Gamma$, then there exists a neighborhood U of $\{\overline{u}_\gamma : \gamma \in K\}$ and a $\mathbf{C^k}$ function $g : U \times K \to \mathbb{R}^n$ such that*

$$g(\overline{u}_\gamma, \gamma) = \overline{v}_\gamma \,, \qquad f\left(u, g(u, \gamma); \gamma\right) = \overline{z} \quad \textit{for all } \gamma \in K,\ u \in U \,.$$

If the k–th derivatives of f are Lipschitz continuous, then the same is true of the k–th derivatives of g.

2.4 Linear Algebra

Let A be an $n \times n$ square matrix. We say that a scalar λ is an **eigenvalue** of A if it satisfies the **characteristic polynomial**

$$\det(A - \lambda\ \mathrm{Id}_n) = 0 \,. \tag{2.4.1}$$

Equation (2.4.1) is equivalent to require the existence of a **right eigenvector**, see Fig. 2.2, namely of a non–zero vector $r \in \mathbb{R}^n$ such that

$$A\, r = \lambda\, r \,, \tag{2.4.2}$$

and of a **left eigenvector**, namely of a non–zero vector $l \in \mathbb{R}^n$ such that

$$l^T A = l^T \lambda \,. \tag{2.4.3}$$

Observe that any scalar multiple of an eigenvector is also an eigenvector of A with

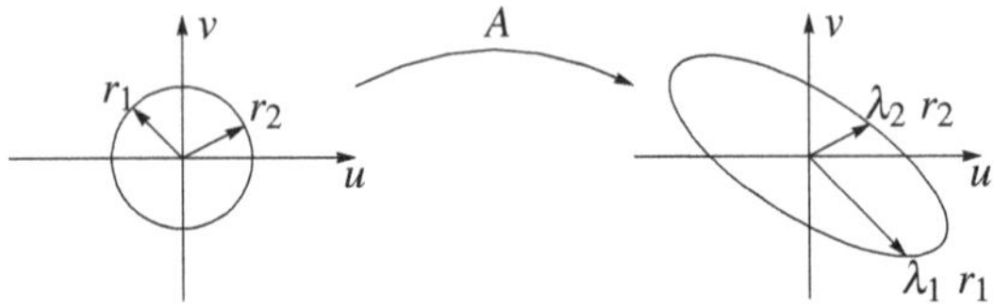

Fig. 2.2 The geometrical meaning of the condition (2.4.2) for the right eigenvectors in the case $n = 2$ and $\lambda_1 < 0 < \lambda_2$.

the same eigenvalue, namely $\alpha\, r$ and $\alpha\, l$ are also, respectively, right and left eigenvectors of A corresponding to the eigenvalue λ.

The **algebraic multiplicity** of an eigenvalue is the multiplicity of the corresponding root of the characteristic polynomial (2.4.1). The **geometric multiplicity** of an eigenvalue is defined as the dimension of the associated **eigenspace**, i.e. the number of linearly independent eigenvectors with that eigenvalue.

If A has n real distinct eigenvalues, say $\lambda_1 < \lambda_2 < \ldots < \lambda_n$, then A admits n linearly independent right eigenvectors $r_1, r_2, \ldots, r_n$ and n linearly independent left eigenvectors $l_1, l_2, \ldots, l_n$. These eigenvectors are uniquely determined, except for the orientation, by imposing the relations

$$\|r_i\| = 1, \qquad l_i \cdot r_j = \delta_{ij}^K \,. \tag{2.4.4}$$

Then, the **characteristic decomposition** of a vector $v \in \mathbb{R}^n$ is

$$v = \sum_{i=1}^{n} (l_i \cdot v)\, r_i \,. \tag{2.4.5}$$

Proposition 2.1. *Assume that the entries of the matrix* $A = (a_{ij})_{1 \le ij \le n}$ *are* $\mathbf{C^k}$ *functions of a parameter* $\gamma \in \mathbb{R}^n$*, with* $k \ge 1$*. If* $A(\overline{\gamma})$ *has* n *real distinct eigenvalues* $\lambda_1(\overline{\gamma}) < \ldots < \lambda_n(\overline{\gamma})$*, then there exists* $r \in \mathbb{R}_+$ *such that, for* $|\gamma - \overline{\gamma}| < r$*, the matrix* $A(\gamma)$ *has* n *distinct eigenvalues* $\lambda_1(\gamma) < \ldots < \lambda_n(\gamma)$*. Moreover, one can choose bases of left and right eigenvectors* $l_i(\gamma)$*,* $r_i(\gamma)$ *satisfying* (2.4.4) *such that the maps* $[\gamma \mapsto \lambda_i(\gamma)]$*,* $[\gamma \mapsto l_i(\gamma)]$*,* $[\gamma \mapsto r_i(\gamma)]$ *are all* $\mathbf{C^k}$.

For the proof, see for instance [1, 6].

2.5 Functions with Bounded Variation

In this section we give some basic results about functions with bounded variation, see [4, 7] for more details.

Definition 2.1. The **total variation** of a function $u:\ \mathbb{R}\to\mathbb{R}^n$ is defined as

$$\mathrm{TV}(u)=\sup_{\eta\in\mathbb{N}}\left\{\sum_{j=1}^{\eta}\left\|u(x_j)-u(x_{j-1})\right\|\ :\ x_0<\ldots<x_\eta\right\}. \tag{2.5.1}$$

We say that u has **bounded variation** when $\mathrm{TV}(u)$ is finite, in this case we write $u\in\mathbf{BV}(\mathbb{R};\mathbb{R}^n)$.

Remark 2.1. Being a function in $\mathbf{BV}(\mathbb{R};\mathbb{R})$ means that the distance along the direction of the y-axis, neglecting the contribution of motion along x-axis, traveled by a point moving along the graph has a finite value. As a consequence, a function with bounded total variation is always also a bounded function, $\mathbf{BV}(\mathbb{R};\mathbb{R}^n)\subseteq\mathbf{L}^\infty(\mathbb{R};\mathbb{R}^n)$, and $u(-\infty),u(+\infty)$ are well defined. The converse is false. In fact, every non–constant periodic and bounded function has unbounded total variation. □

Another important property of functions with bounded total variation is the existence of left and right limits.

Lemma 2.4. *Let $u\in\mathbf{BV}(\mathbb{R};\mathbb{R}^n)$. Then u has at most countably many points of discontinuity $\mathscr{D}$ and for every $x\in\mathscr{D}$, the following two limits are well defined*

$$u(x-)=\lim_{y\nearrow x}u(y)\,,\qquad u(x+)=\lim_{y\searrow x}u(y)\,. \tag{2.5.2}$$

Proof. Let $x\in\mathbb{R}$ and consider an increasing sequence x_j converging to x. Since

$$\sum_{j\in\mathbb{N}}\left\|u(x_j)-u(x_{j-1})\right\|\leq\mathrm{TV}(u)\,,$$

$u(x_j)$ is a Cauchy sequence and converges to some limit $u(x-)$. Observe that if also x'_j is a second increasing sequence tending to x, then

$$\sum_{j\in\mathbb{N}}\left\|u(x_j)-u(x'_j)\right\|\leq\sum_{j\in\mathbb{N}}\left(\left\|u(x_j)-u(x_{j-1})\right\|+\left\|u(x'_j)-u(x'_{j-1})\right\|\right)\leq 2\ \mathrm{TV}(u)\,.$$

Therefore also $u(x'_j)$ tends to $u(x-)$ and the first limit in (2.5.2) is well defined.

Let $j\in\mathbb{N}$ and consider the set

$$\mathscr{D}_j=\left\{x\in\mathbb{R}\ :\ \left\|u(x-)-u(x)\right\|+\left\|u(x)-u(x+)\right\|>j^{-1}\right\}.$$

The number of points contained in $\mathscr{D}_j$ can not be bigger than $j\ \mathrm{TV}(u)$. Indeed,

$$\mathrm{TV}(u)\geq\sum_{x\in\mathscr{D}_j}\left(\left\|u(x+)-u(x)\right\|+\left\|u(x)-u(x-)\right\|\right)\geq j^{-1}\,\#\left(\mathscr{D}_j\right)$$

and $\mathscr{D}\subseteq\cup_{j\in\mathbb{N}}\mathscr{D}_j$. □

Remark 2.2. By Lemma 2.4, we can assume that any $u \in \mathbf{BV} \cap \mathbf{L^k}(\mathbb{R};\mathbb{R}^n)$ is right continuous by redefying $u(x) = u(x+)$ for each $x \in \mathscr{D}$. □

Lemma 2.5. *Let $u \in \mathbf{BV}(\mathbb{R};\mathbb{R}^n)$ be right continuous. Then, for every $\varepsilon \in \mathbb{R}_+$, with $\varepsilon \ll 1$, there exists a piecewise constant function v such that*

$$\mathrm{TV}(v) \leq \mathrm{TV}(u)\,, \quad \|v\|_{\mathbf{L}^\infty(\mathbb{R};\mathbb{R}^n)} \leq \|u\|_{\mathbf{L}^\infty(\mathbb{R};\mathbb{R}^n)}\,, \quad \|v-u\|_{\mathbf{L}^\infty(\mathbb{R};\mathbb{R}^n)} \leq \varepsilon\,. \tag{2.5.3}$$

Proof. The case $\mathrm{TV}(u) = 0$ is obvious. Assume that $\mathrm{TV}(u) \in \mathbb{R}_+$ and introduce

$$\Upsilon u(x) = \sup_{\eta \in \mathbb{N}} \left\{ \sum_{j=1}^{\eta} \left\| u(x_j) - u(x_{j-1}) \right\| \,:\, x_0 < \ldots < x_\eta = x \right\}. \tag{2.5.4}$$

Observe that $\Upsilon u(x)$ is the total variation of u on the interval $]-\infty, x]$ and satisfies

$$\begin{array}{l} \Upsilon u \text{ is a right continuous non–decreasing function} \\ \Upsilon u(-\infty) = 0 \quad , \qquad \Upsilon u(+\infty) = \mathrm{TV}(u) \\ \left\| u(y) - u(x) \right\| \leq \Upsilon u(y) - \Upsilon u(x) \text{ for all } x < y\,. \end{array} \tag{2.5.5}$$

Let $N \in \mathbb{N}$ be such that $N < \mathrm{TV}(u)\,\varepsilon^{-1} \leq N+1$ and consider the points

$$x_j = \min\left\{x \in \mathbb{R} \,:\, \Upsilon(x) \geq j\,\varepsilon\right\}\,, \quad j = 1, \ldots, N\,, \tag{2.5.6}$$

see Fig. 2.3. Then

$$v(x) = \begin{cases} u(-\infty) & \text{if } x \in \left]-\infty, x_1\right[\\ u(x_j) & \text{if } x \in \left[x_j, x_{j+1}\right[\,, \quad j = 1, \ldots, N-1 \\ u(x_N) & \text{if } x \in \left[x_N, +\infty\right[\end{cases}$$

satisfies the conditions required in (2.5.3). □

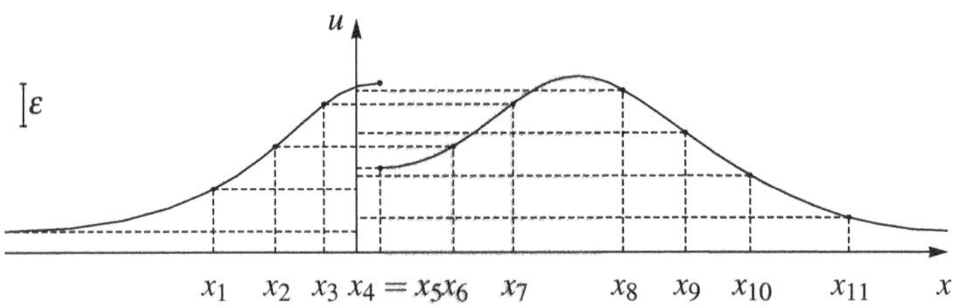

Fig. 2.3 The points x_j defined by (2.5.6). In particular, in the case considered in the figure, we have $N = 11$, $\Upsilon(x_4) = \Upsilon(x_5) = 5.5\,\varepsilon$ while for the other cases $\Upsilon(x_j) = j\,\varepsilon$.

Lemma 2.6. *If $u \in \mathbf{BV}(\mathbb{R};\mathbb{R}^n)$ is right continuous, then for every $\varepsilon \in \mathbb{R}_+$ one has*

$$\frac{1}{\varepsilon} \int_{\mathbb{R}} \left\| u(x+\varepsilon) - u(x) \right\| \,\mathrm{d}x \leq \mathrm{TV}(u)\,. \tag{2.5.7}$$

Proof. Consider the operator Υ defined by (2.5.4). By (2.5.5) we have

$$\begin{aligned}
\int_{\mathbb{R}} \left\| u(x+\varepsilon) - u(x) \right\| \, dx &\leq \int_{\mathbb{R}} \left(\Upsilon u(x+\varepsilon) - \Upsilon u(x) \right) \, dx \\
&= \int_{\mathbb{R}} \text{meas} \left\{ y \in \mathbb{R} : \Upsilon u(x) < y < \Upsilon u(x+\varepsilon) \right\} \, dx \\
&= \int_{\Upsilon u(-\infty)}^{\Upsilon u(+\infty)} \text{meas} \left\{ x \in \mathbb{R} : \Upsilon u(x) < y < \Upsilon u(x+\varepsilon) \right\} \, dy \\
&= \int_0^{\text{TV}(u)} \varepsilon \, dy = \varepsilon \, \text{TV}(u) .
\end{aligned}$$

Therefore (2.5.7) is established. □

Theorem 2.3 (Helly). *Consider a sequence of functions $u_j \in \mathbf{BV}(\mathbb{R};\mathbb{R}^n)$. Assume that there exist two constants C, M such that for any $j \in \mathbb{N}$*

$$\text{TV}(u_j) \leq C , \qquad \left\| u_j \right\|_{\mathbf{L}^\infty(\mathbb{R};\mathbb{R}^n)} \leq M .$$

Then there exists a function $u \in \mathbf{BV}(\mathbb{R};\mathbb{R}^n)$ and a subsequence u_m such that

$$\lim_{m \to +\infty} u_m(x) = u(x) \text{ for any } x \in \mathbb{R} , \quad \text{TV}(u) \leq C , \qquad \|u\|_{\mathbf{L}^\infty(\mathbb{R};\mathbb{R}^n)} \leq M . \quad (2.5.8)$$

Proof. By a diagonal procedure and (2.5.5), there exists a subsequence Υu_m of Υu_j and a non–decreasing function $q : \mathbb{Q} \to [0, C]$ such that $\Upsilon u_m(x) \to q(x)$ for every $x \in \mathbb{Q}$ as $m \to +\infty$. For each $m \in \mathbb{N}$, define the set of jump points for q

$$J_m = \left\{ x \in \mathbb{R} : \lim_{\substack{y \searrow x \\ y \in \mathbb{Q}}} q(y) - \lim_{\substack{y \nearrow x \\ y \in \mathbb{Q}}} q(y) \geq \frac{1}{m} \right\} .$$

The set J_m is finite and $\#(J_m) \leq C\, m$. Therefore, the set $J = \cup_{m \in \mathbb{N}} J_m$ of points where the right and left limits of q are distinct is at most countable. Furthermore, we can choose a subsequence u_l such that $u_l(x) \to u(x)$ for every $x \in J \cup \mathbb{Q}$ as $l \to +\infty$. We claim that the limit holds for every $x \in \mathbb{R}$. Indeed, if $x \notin J$, then for each $m \in \mathbb{N}$, there exist rational points $p_1 < x < p_2$ such that $q(p_2) - q(p_1) < 2/m$. By using (2.5.5), the triangular inequality and the fact that $u_l(p_1) \to u(p_1)$, we obtain

$$\begin{aligned}
&\limsup_{l, \tilde{l} \to +\infty} \left\| u_l(x) - u_{\tilde{l}}(x) \right\| \leq 2 \limsup_{l \to +\infty} \left\| u_l(x) - u_l(p_1) \right\| \\
&\leq 2 \limsup_{l \to +\infty} \left(\Upsilon u_l(p_2) - \Upsilon u_l(p_1) \right) = 2 \left(q(p_2) - q(p_1) \right) < \frac{4}{m} .
\end{aligned}$$

Since m is arbitrary, our claim is proved and the first part of the theorem is established. Finally, for any given points $x_0 < x_1 < \ldots < x_\eta$, we have

$$\sum_{j=1}^{\eta} \left\| u(x_j) - u(x_{j-1}) \right\| = \lim_{l \to +\infty} \left(\sum_{j=1}^{\eta} \left\| u_l(x_j) - u_l(x_{j-1}) \right\| \right) \leq \limsup_{l \to +\infty} \mathrm{TV}(u_l) \leq C .$$

This proves the second inequality in (2.5.8). The third one is obvious. □

Theorem 2.4. *Consider a sequence of functions $u_j : \mathbb{R}_+^* \to \mathbf{BV}(\mathbb{R};\mathbb{R}^n)$. Assume that there exist three constants C,M,L such that for all $j \in \mathbb{N}$ and $t,s \in \mathbb{R}_+^*$*

$$\mathrm{TV}\left(u_j(t)\right) \leq C , \quad \left\| u_j(t) \right\|_{\mathbf{L}^\infty(\mathbb{R};\mathbb{R}^n)} \leq M , \quad \left\| u_j(t) - u_j(s) \right\|_{\mathbf{L}^1(\mathbb{R};\mathbb{R}^n)} \leq L \left| t - s \right| . \tag{2.5.9}$$

Then there exists a function $u \in \mathbf{L}^1_{\mathbf{loc}}(\mathbb{R}_+^* \times \mathbb{R};\mathbb{R}^n)$ and a subsequence u_m such that

$$\begin{aligned} & u_m \to u \text{ in } \mathbf{L}^1_{\mathbf{loc}}(\mathbb{R}_+^* \times \mathbb{R};\mathbb{R}^n) \text{ as } m \to +\infty \\ \text{and } & \left\| u(t) - u(s) \right\|_{\mathbf{L}^1(\mathbb{R};\mathbb{R}^n)} \leq L \left| t - s \right| \text{ for all } t,s \in \mathbb{R}_+^* . \end{aligned} \tag{2.5.10a}$$

The point values of the limit function u can be uniquely determined by requiring its right semi–continuity. In this case, one has for all $t \in \mathbb{R}_+^*$

$$\mathrm{TV}\left(u(t)\right) \leq C , \quad \left\| u(t) \right\|_{\mathbf{L}^\infty(\mathbb{R};\mathbb{R}^n)} \leq M . \tag{2.5.10b}$$

Proof. By Helly's theorem, it is possible to construct a subsequence u_m, such that, for every $t \in \mathbb{Q} \cap \mathbb{R}_+^*$ the limit $u_m(t) \to u(t)$ pointwise and hence also in $\mathbf{L}^1_{\mathbf{loc}}(\mathbb{R};\mathbb{R}^n)$. This limit function clearly satisfies (2.5.10) for every $t,s \in \mathbb{Q} \cap \mathbb{R}_+^*$. Fix now $t \in \mathbb{R}_+^*$ and consider a sequence of rational times $t_l \to t$. By (2.5.9) we can define $u(t) = \lim_{l \to +\infty} u(t_l)$. Observing that the bounds (2.5.10b) are satisfied for $t = t_l$, by eventually modifying the function u on a set of measure zero, we achieve the estimates (2.5.10b).

Consider now, for $\varepsilon \in \mathbb{R}_+$ sufficiently small, $u_\varepsilon(t,x) = \frac{1}{\varepsilon} \int_x^{x+\varepsilon} u(t,y)\, dy$ and $\tilde{u}(t,x) = \lim_{\varepsilon \searrow 0} u_\varepsilon(t,x) = u(t,x+)$. Observe that u_ε is uniformly Lipschitz continuous with respect to both variables t,x. Therefore $\tilde{u}$ is Borel measurable. For each $t \in \mathbb{R}_+^*$ the identity $u(t) = \tilde{u}(t)$ holds a.e. in $\mathbb{R}$. By replacing u with $\tilde{u}$, all estimates (2.5.10) are clearly satisfied. □

References

1. Bressan, A.: Hyperbolic systems of conservation laws. Oxford Lecture Series in Mathematics and its Applications, vol. 20. Oxford University Press, Oxford (2000)
2. Corduneanu, C.: Principles of Differential and Integral Equations. AMS Chelsea Publishing Series. American Mathematical Society (2008)
3. Dieudonne, J.: Foundations of Modern Analysis. Pure and Applied Mathematics. Hesperides Press (2008)
4. Evans, L.C., Gariepy, R.F.: Measure theory and fine properties of functions. In: Studies in Advanced Mathematics. CRC Press, Boca Raton (1992)

5. Rudin, W.: Principles of mathematical analysis. International series in pure and applied mathematics. McGraw-Hill (1976)
6. Strang, G.: Introduction to linear algebra. Wellesley-Cambridge Press (2003)
7. Ziemer, W.P.: Weakly Differentiable Functions: Sobolev Spaces and Functions of Bounded Variation. Graduate Texts in Mathematics, vol. 120 (1989)

Chapter 3
One–Dimensional Scalar Conservation Laws

Abstract. In this chapter we give some basic preliminaries about one–dimensional scalar conservation laws to highlight some of the fundamental issues and difficulties arising in the general theory of conservation laws. We also present some basic definitions and concepts which will be of constant use in this book.

3.1 Introduction

A **scalar conservation law** in one space dimension written in **conservative form** is a first–order homogeneous partial differential equation of the form

$$\partial_t u + \partial_x f(u) = 0\,, \tag{3.1.1}$$

where t and x are, respectively, time and space coordinates, ∂_t and ∂_x are, respectively, the partial time derivative and the partial space derivative, $u = u(t,x) \in \mathbb{R}$ is the unknown function, while $f : \mathbb{R} \to \mathbb{R}$ is a given smooth function.

Definition 3.1. Let $f : \mathbb{R} \to \mathbb{R}$ be a given $\mathbf{C^1}$ function. A function $u : \mathbb{R}_+^* \times \mathbb{R} \to \mathbb{R}$ is a **smooth solution** of (3.1.1) if is $\mathbf{C^1}$ and satisfies (3.1.1) pointwise.

To better understand the physical meaning of Eq. (3.1.1), we consider a smooth solution u to (3.1.1). Then, the application of the Green's formula (2.2.1) on some interval $[a,b]$ immediately yields the following identities

$$\begin{aligned}\frac{\mathrm{d}}{\mathrm{d}t}\left(\int_a^b u(t,x)\ \mathrm{d}x\right) &= \int_a^b \partial_t u(t,x)\ \mathrm{d}x \\ &= -\int_a^b \partial_x f\left(u(t,x)\right)\ \mathrm{d}x = f\left(u(t,a)\right) - f\left(u(t,b)\right)\,.\end{aligned} \tag{3.1.2}$$

As a consequence, the total amount of the quantity measured by u contained in $[a,b]$, namely the quantity $\int_a^b u(t,x)\ \mathrm{d}x$, can change only due to the flow $f\left(u(t,a)\right)$ of u

M.D. Rosini: *Macroscopic Models for Vehicular Flows & Crowd Dynamics*, UCS, pp. 23–42.
DOI: 10.1007/978-3-319-00155-5_3

across the point $x = a$ and to the flow $f\left(u(t,b)\right)$ of u across the point $x = b$. In other words, the quantity u is neither created nor destroyed. For this reason, it is natural to refer to u as the **conserved quantity** and to f as the **flux function** related to the Eq. (3.1.1). In particular, when $\lim_{|x|\to+\infty} u(t,x) = 0$ for all $t \in \mathbb{R}_+$ and the flux is normalized so that $f(0) = 0$, we can let a go to $-\infty$ and b go to $+\infty$ in (3.1.2) and obtain that the integral of the conserved quantity on the whole space is independent of time.

Remark 3.1. A natural generalization of the conservation law (3.1.1) writes

$$\partial_t u + \partial_x f(t,x,u) = g(t,x,u) . \tag{3.1.3}$$

Equation (3.1.3) is referred to as **balance law**, where g represents the **source term**, see Sect. 8 or, for instance, [5, 13] and the references therein for more details. □

If $f \in \mathbf{C^1}(\mathbb{R};\mathbb{R})$ and u is a smooth solution of (3.1.1), then is possible to apply the chain rule and (3.1.1) can be rewritten in the so called **quasilinear form**

$$\partial_t u + a(u)\, \partial_x u = 0 , \tag{3.1.4}$$

where $a = f' \in \mathbf{C^0}(\mathbb{R};\mathbb{R})$ is the derivative of f.

Remark 3.2. For smooth solutions, the two equations in (3.1.1) and (3.1.4) are entirely equivalent. However, if u has a jump, the quasilinear equation (3.1.4) is in general not well defined, because a product of a discontinuous function, $a(u)$, with a Dirac measure, $\partial_x u$, is involved. Hence (3.1.4) is meaningful only within the class of continuous functions, while (3.1.1) can be interpreted in a distributional sense, see Sect. 3.4. □

Remark 3.3. The generalization of (3.1.4) is represented by the **transport equation**

$$\partial_t u + a(t,x,u)\, \partial_x u = g(t,x,u) , \tag{3.1.5}$$

see for instance [1, 14] for more details. □

For a given **initial datum** $\overline{u} : \mathbb{R} \to \mathbb{R}$, we aim to study the **Cauchy problem**, also called **initial value problem (IVP)**, for the Eq. (3.1.1)

$$\partial_t u + \partial_x f(u) = 0 \tag{3.1.6a}$$

$$u(t=0) = \overline{u} , \tag{3.1.6b}$$

see Chap. 5 for more details. The simplest Cauchy problem is the so called Riemann problem, which corresponds to a Heaviside initial datum, see Chap. 4.

Definition 3.2. Let $f, \overline{u} : \mathbb{R} \to \mathbb{R}$ be given $\mathbf{C^1}$ functions. Then $u : \mathbb{R}_+^* \times \mathbb{R} \to \mathbb{R}$ is a **global smooth solution** of the Cauchy problem (3.1.6) if is $\mathbf{C^1}$ and satisfies (3.1.6) pointwise.

In the next section we introduce the method of characteristics, a technique for solving the Cauchy problem (3.1.6) in the space of smooth functions. As we will see in Sect. 3.3, in general global smooth solutions to the Cauchy problem (3.1.6) do not exist beyond some finite time interval, even when the initial function $\bar{u}$ is very smooth. For this reason Sect. 3.4 addresses the necessity to consider (3.1.6) in the distributional sense and to introduce the definition of discontinuous weak solutions.

3.2 Method of Characteristics

Definition 3.3. Let u be a smooth solution to the Cauchy problem (3.1.6) and $x_o \in \mathbb{R}$. The **characteristic curve** $[t \mapsto x(t;x_o)]$ associated to u and starting from the point $(t,x) = (0,x_o)$ is the integral curve of the following Cauchy problem for an ordinary differential equation

$$x'(t;x_o) = a\Big(u\big(t,x(t;x_o)\big)\Big) \quad , \qquad x(0;x_o) = x_o \, . \tag{3.2.1}$$

Proposition 3.1. *Let u be a smooth solution of* (3.1.6) *and $x_o \in \mathbb{R}$. Then u is constant along the characteristic curve* (3.2.1)*, which results to be a straight line with the following expression*

$$x(t;x_o) = a\big(\bar{u}(x_o)\big)\; t + x_o \, . \tag{3.2.2}$$

Proof. Let $x = x(t;x_o)$ be the characteristic curve starting from x_o given by (3.2.1). It exists at least on a small time interval $[0,t_o[$. Along such a curve, u is constant since for any $t \in [0,t_o[$

$$\begin{aligned}\frac{\mathrm{d}}{\mathrm{d}t} u\big(t,x(t;x_o)\big) &= \partial_t u\big(t,x(t;x_o)\big) + x'(t;x_o)\;\partial_x u\big(t,x(t;x_o)\big)\\ &= \partial_t u\big(t,x(t;x_o)\big) + a\Big(u\big(x(t;x_o),t\big)\Big)\;\partial_x u\big(t,x(t;x_o)\big) = 0\,,\end{aligned}$$

the last equality being the conservation law in the quasilinear form (3.1.4). Thus, u is constant along the characteristic $x = x(t;x_o)$, taking the value $\bar{u}(x_o)$. As a consequence, (3.2.1) implies (3.2.2). □

This important property gives a way to construct smooth solutions. Indeed, if we can invert the relation (3.2.2) and write $x_o = x_o(t,x)$, then

$$u(t,x) = \bar{u}\big(x_o(t,x)\big) \tag{3.2.3}$$

is a smooth solution to the Cauchy problem (3.1.6). This is the so called **method of characteristics** used to compute smooth solutions to Cauchy problems for scalar conservation laws, as it will be done, for instance, in Example 3.1 and Example 3.2.

There are many ifs in the above construction. Indeed, the solution to (3.2.1) may only be local and it may be not possible to invert (3.2.2). With the next proposition we give sufficient conditions for the existence and uniqueness of a smooth solution globally in time.

Proposition 3.2. *Let $\overline{u} \in \mathbf{C}^1(\mathbb{R};\mathbb{R})$ be bounded together with its derivative. If $a \circ \overline{u} : \mathbb{R} \to \mathbb{R}$ is increasing, then the function* (3.2.3) *is well defined for all $(t,x) \in \mathbb{R}_+^* \times \mathbb{R}$ and is the only one global smooth solution to the Cauchy problem* (3.1.6).

Proof. Let $s = a \circ \overline{u}$. Making the derivative of (3.2.2) with respect to x_o we have $\partial_{x_o} x(t;x_o) = s'(x_o)\ t + 1 > 0$ for all $t \in \mathbb{R}_+^*$. Therefore, the relation (3.2.2) can be inverted. In addition, the implicit function theorem, Theorem 2.1, ensures that $[(t,x) \mapsto x_o(t,x)]$ is a $\mathbf{C}^1$ function. Let us then verify that (3.2.3) is indeed a solution to (3.1.6). By the definition (3.2.1), we have $x(0,x_o) = x_o$, and therefore (3.1.6b) is satisfied. By the equality $s\left(x_o\left(t,y\right)\right)\ t + x_o\left(t,y\right) = x\left(t;x_o\left(t,y\right)\right) = y$ we get

$$\begin{aligned}\left(s'\left(x_o\left(t,y\right)\right)\ t+1\right)\partial_t x_o(t,y) &= -s\left(x_o\left(t,y\right)\right)\\ \left(s'\left(x_o\left(t,y\right)\right)\ t+1\right)\partial_y x_o(t,y) &= 1\,.\end{aligned}$$

Then, the function (3.2.3) satisfies also (3.1.6a), because from the above equalities we obtain

$$\partial_t u + \partial_y f(u) = \left(\partial_t x_o + s\left(x_o\left(t,y\right)\right)\partial_y x_o\right)\ \overline{u}' = 0$$

and this ends the proof. □

Example 3.1. When $f(u) = a\ u$ for a constant $a \in \mathbb{R}$, (3.1.1) reduces to the **linear advection equation** and the corresponding Cauchy problem writes

$$\partial_t u + a\ \partial_x u = 0 \quad , \qquad u(t=0) = \overline{u}\,. \tag{3.2.4}$$

In this case, it is possible to invert the relation (3.2.2) and to obtain $x_o(t) = x - a\,t$. If $\overline{u} \in \mathbf{C}^1(\mathbb{R};\mathbb{R})$, then the traveling wave $u(t,x) = \overline{u}(x - a\,t)$ is a global smooth solution to (3.2.4). □

In the general nonlinear case (3.1.4), where a is a function of u, the shape of the solution may vary in time. This may possibly lead to the formation of discontinuities in a finite time. This is an essential feature of the Cauchy problems for conservation laws: in general there exist no smooth solutions beyond some finite time interval, even when the initial condition $\overline{u}$ is very smooth, see for instance the Example 3.2.

3.3 Loss of Regularity

Compared to linear ones, nonlinear problems display new features such as the occurrence of discontinuous solutions. Indeed, the nonlinearity implies that the speed of propagation of a wave is not constant but depends on the state. Hence, in general, a solution may experience a wave overtaking, which results in the creation of discontinuities in a finite time, even for a smooth initial datum. We illustrate this aspect by studying the next example (for more general situations see [11]).

Example 3.2. Consider the Cauchy problem for the **Hopf** or **inviscid Burgers' equation**

$$\partial_t u + \partial_x (u^2/2) = 0 \tag{3.3.1a}$$

$$u(0,x) = (1+x^2)^{-1} \tag{3.3.1b}$$

used in physics to describe, for instance, a medium consisting of non–interacting particles, see [3, 4]. If u is a smooth solution, (3.3.1a) is equivalent to the equation

$$\partial_t u + u\, \partial_x u = 0\,. \tag{3.3.2}$$

Consider the characteristic curve $[t \mapsto x(t)]$ in the (t,x)–plane starting from the point $(t,x) = (0,x_o)$, i.e. a solution of the Cauchy problem

$$x'(t) = u\big(t, x(t)\big) \quad , \qquad x(0) = x_o\,. \tag{3.3.3}$$

By Proposition 3.1, the solution to (3.3.3) is

$$x(t) = x_o + \frac{t}{1+x_o^2} \tag{3.3.4}$$

and the solution to the Cauchy problem (3.3.1) is implicitly given by

$$u\left(t, x + \frac{t}{1+x^2}\right) = \frac{1}{1+x^2}\,. \tag{3.3.5}$$

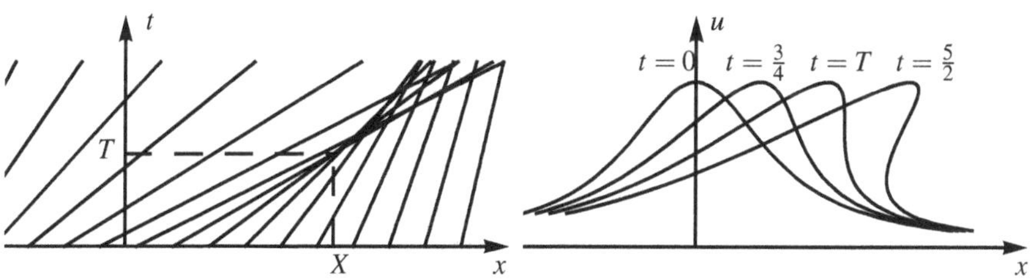

Fig. 3.1 Left: Representation in the (x,t)–plane of the characteristic lines defined by (3.3.4). Right: Representation in the (x,u)–plane of possibly multi–valued functions implicitly given by (3.3.5) and corresponding to $t = 0$, $t = 3/4$, $t = T$ and $t = 5/2$.

In Fig. 3.1, left, are represented characteristic lines (3.3.4) corresponding to different values of x_o. This figure points out that the characteristic lines start to intersect at time $t = T$. To compute the exact value of T, it is sufficient to consider the possible intersection of two characteristics starting from x and y, with $x \neq y$. As a result, one get for the t–coordinate of the intersection the value

$$t = \frac{(1+x^2)(1+y^2)}{x+y}, \tag{3.3.6}$$

which represents a surface in the (x,y,t)–coordinates. Its unique minimum point is $(x,y,t) = (1/\sqrt{3}, 1/\sqrt{3}, 8/\sqrt{27})$, and therefore $T = 8/\sqrt{27}$.

On the other hand, when $t > T$, the characteristic lines start to intersect, the map

$$\left[x \mapsto x + \frac{t}{1+x^2}\right]$$

is not one-to-one and (3.3.5) no longer defines a single valued solution of (3.3.1), see Fig. 3.2. Accordingly, no smooth solution exists beyond time $t = T$. The only possibility is then to prolong it for all times $t \geq 0$ within a class of discontinuous functions and to consider (3.3.1) in the sense of distributions.

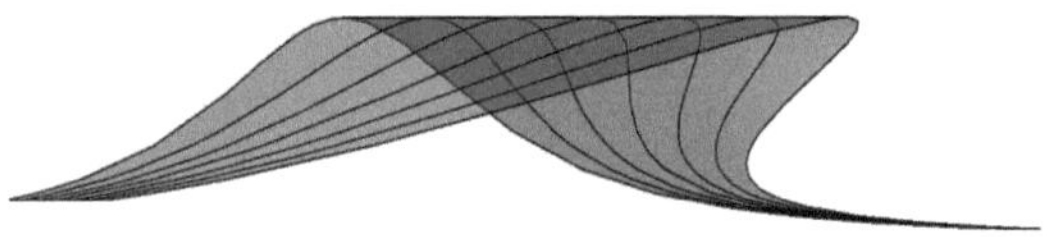

Fig. 3.2 Representation in the (t,x,u)–coordinates of the multi–valued function u implicitly given by (3.3.5).

An alternative point of view is that described in Fig. 3.1, right. The points on the graph of $u = u(t)$ move horizontally with speed equal to their distance from the x–axis. As a consequence, as t approaches the critical time T one has $\lim_{t \nearrow T} \left\{\inf_{x \in \mathbb{R}} \partial_x u(t,x)\right\} = -\infty$, for $t = T$ the graph "folds over" and for $t > T$ there are some x that have associated three values for u. To prolong the solution beyond $t = T$ we have to choose among these three values of u. Anyhow, it is impossible to continue the solution and at the same time keep it continuous. □

Proposition 3.3. *Let $\bar{u} \in \mathbf{C}^1(\mathbb{R};\mathbb{R})$ be bounded together with its derivative. Then there exists one and only one smooth solution $u \in \mathbf{C}^1([0, T_{max}[\times\mathbb{R};\mathbb{R})$ to (3.1.6), where*

$$T_{max} = \begin{cases} +\infty & \text{if } a \circ \bar{u} \text{ is increasing} \\ -\left(\inf_{x\in\mathbb{R}} \left\{\frac{\mathrm{d}}{\mathrm{d}x} a\left(\bar{u}(x)\right)\right\}\right)^{-1} & \text{otherwise}\,, \end{cases} \tag{3.3.7}$$

and does not exist any smooth solution on a greater time interval.

Proof. Let $s = a \circ \bar{u}$. By Proposition 3.2, if s is increasing then the function (3.2.3) is the unique smooth solution to (3.1.6) and is defined globally in time. If s is not increasing, then the smooth solution provided by the method of characteristics is the only one, at least for $t \in [0, T_{max}[$. Hence it is sufficient to prove that it can not be prolonged beyond $t = T_{max}$. Let $T \in \mathbb{R}_+$ be such that there exists a smooth solution u on $[0,T] \times \mathbb{R}$. Observe that by (3.1.6a)

$$0 = \partial_x\left(\partial_t u + a(u)\ \partial_x u\right) = \partial_x \partial_t u + a(u)\ \partial_x^2 u + a'(u)\ (\partial_x u)^2\,. \tag{3.3.8}$$

Introduce $v = a'(u)\,\partial_x u$, which is well defined until u is $\mathbf{C^1}$. Differentiating along the characteristic $x = x(t;x_o)$ and recalling that $u\left(t, x(t;x_o)\right)$ is constant, by (3.2.1) and (3.3.8) we obtain

$$\frac{\mathrm{d}}{\mathrm{d}t} v\left(t, x(t;x_o)\right) = a'(u)\left(\partial_t \partial_x u + a(u)\,\partial_x^2 u\right) = -\left(a'(u)\,\partial_x u\right)^2 = -v\left(t, x(t;x_o)\right)^2 .$$

The solution of the above equation of Ricatti type is

$$v\left(t, x(t;x_o)\right) = \frac{v(0,x_o)}{1 + v(0,x_o)\,t} .$$

By hypothesis, there are values of y for which $v(0,y) = \frac{\mathrm{d}}{\mathrm{d}y}\left(a(\overline{u}(y)\right) = s'(y) \in \mathbb{R}_-$. For these values $v\left(t, x(t;y)\right)$ blows up at time $t = -s'(y)^{-1}$. Therefore, v exists finite only for $t < \inf_{y\in\mathbb{R}}\{-s'(y)^{-1}\} = T_{max}$. □

In short, by using the method of characteristics one can construct a smooth solution to (3.1.6) at least on a small time interval. On the other hand, we have seen that in the nonlinear case $a'(u) \neq 0$ discontinuities may develop after a finite time. The above considerations lead us to introduce in the next section weak solutions which are indeed weaker than the smooth ones.

3.4 Weak Solutions

The example in Sect. 3.3 points out that no matter how smooth the initial datum is, in general we can not expect to be able to define smooth solutions of nonlinear conservation laws for all times. In order to construct solutions globally in time, we are forced to work in a space of discontinuous functions, and to interpret the derivatives appearing in the conservation law (3.1.1) in a distributional sense.

Definition 3.4. Let $f \in \mathbf{C^1}(\mathbb{R};\mathbb{R})$ and $\Omega \subseteq \mathbb{R}^2$ be an open set. A function $u \in \mathbf{L^\infty_{loc}}(\Omega;\mathbb{R})$ satisfies (3.1.1) in the distributional sense if for every $\varphi \in \mathbf{C^\infty_c}(\Omega;\mathbb{R})$

$$\int\!\!\int_\Omega \left(u\,\partial_t\varphi + f(u)\,\partial_x\varphi\right)\,\mathrm{d}x\;\mathrm{d}t = 0 . \tag{3.4.1}$$

If $\overline{u} \in \mathbf{L^\infty_{loc}}(\mathbb{R};\mathbb{R})$ and $T \in \mathbb{R}_+$, then a function $u \in \mathbf{L^\infty_{loc}}([0,T]\times\mathbb{R};\mathbb{R})$ satisfies the Cauchy problem (3.1.6) in the distributional sense if for every $\varphi \in \mathbf{C^\infty_c}([0,T[\times\mathbb{R};\mathbb{R})$

$$\int_0^T\!\!\int_{\mathbb{R}} \left(u\,\partial_t\varphi + f(u)\,\partial_x\varphi\right)\,\mathrm{d}x\;\mathrm{d}t + \int_{\mathbb{R}} \overline{u}(x)\,\varphi(0,x)\,\mathrm{d}x = 0 . \tag{3.4.2}$$

A **weak solution** to the Cauchy problem (3.1.6) is a function that satisfies (3.1.6) in the distributional sense and is in $\mathbf{C^0}\left([0,T];\mathbf{L^\infty_{loc}}(\mathbb{R};\mathbb{R})\right)$.

Remark 3.4. Equivalently, a function $u \in \mathbf{C^0}\left([0,T];\mathbf{L^\infty_{loc}}(\mathbb{R};\mathbb{R})\right)$ is a weak solution to the Cauchy problem (3.1.6) iff its restriction to $\Omega =]0,T[\times \mathbb{R}$ satisfies (3.1.6a) in the distributional sense and (3.1.6b) holds a.e. in $\mathbb{R}$. □

By definition, a smooth solution is also a weak solution. Viceversa, a $\mathbf{C^1}$ weak solution is also a smooth solution, as proved by the next proposition.

Proposition 3.4. *Let $\mathscr{W}$ and $\mathscr{S}$ be, respectively, the sets of weak solutions and of smooth solutions to* (3.1.6). *Then* $\mathscr{S} = \mathscr{W} \cap \mathbf{C^1}(\mathbb{R}_+^* \times \mathbb{R};\mathbb{R})$.

Proof. Let $u \in \mathscr{S}$. If $\varphi \in \mathbf{C_c^\infty}(\mathbb{R}_+^* \times \mathbb{R};\mathbb{R})$, then we obtain by applying the integration by parts Lemma 2.3

$$
\begin{aligned}
0 &= -\int_{\mathbb{R}_+}\int_{\mathbb{R}} \left(\partial_t u + \partial_x f(u)\right) \varphi \; \mathrm{d}x \; \mathrm{d}t \\
&= \int_{\mathbb{R}_+}\int_{\mathbb{R}} \left(u\, \partial_t \varphi + f(u)\, \partial_x \varphi\right) \; \mathrm{d}x \; \mathrm{d}t + \int_{\mathbb{R}} \bar{u}(x)\; \varphi(0,x) \; \mathrm{d}x \,,
\end{aligned}
$$

and $u \in \mathscr{W}$. Viceversa, let $u \in \mathscr{W} \cap \mathbf{C^1}(\mathbb{R}_+^* \times \mathbb{R};\mathbb{R})$. If $\varphi \in \mathbf{C_c^\infty}(\mathbb{R}_+ \times \mathbb{R};\mathbb{R})$, then integrating by parts (3.4.1)

$$
\int_{\mathbb{R}_+}\int_{\mathbb{R}} \left(\partial_t u + \partial_x f(u)\right) \varphi \; \mathrm{d}x \; \mathrm{d}t = 0 \,.
$$

Since the above identity is satisfied for all test functions in $\mathbf{C_c^\infty}(\mathbb{R}_+ \times \mathbb{R};\mathbb{R})$, (3.1.6a) must hold pointwise. Next, if $\varphi \in \mathbf{C_c^\infty}(\mathbb{R}_+^* \times \mathbb{R};\mathbb{R})$, then integrating by parts (3.4.2) we obtain

$$
\int_{\mathbb{R}} \left(u(0,x) - \bar{u}(x)\right) \; \varphi(0,x) \; \mathrm{d}x = 0 \,,
$$

which yields (3.1.6b) a.e. in $\mathbb{R}$, because of the arbitrariness of φ. □

The above proposition proves that any distributional solution is a smooth solution in any domain where it is $\mathbf{C^1}$. We restrict the study to a particular type of discontinuous weak solutions. In fact, we shall be mainly interested in piecewise $\mathbf{C^1}$ weak solutions to (3.1.6).

Definition 3.5. $u : \mathbb{R}_+^* \times \mathbb{R} \to \mathbb{R}$ is **piecewise $\mathbf{C^1}$** if there exists a finite number of smooth curves $\Gamma_i : x = x_i(t)$, $i = 1,\ldots,d$, in the (t,x)–plane, across which u may experience a jump of discontinuity and such that $u \in \mathbf{C^1}\left((\mathbb{R}_+^* \times \mathbb{R}) \setminus \cup_{i=1}^d \Gamma_i;\mathbb{R}\right)$.

We now show that, even in the frame of piecewise $\mathbf{C^1}$ functions, not every discontinuity is admissible. In fact, the values of u and $f(u)$ on each side of the discontinuity have to satisfy a relation that is hidden in (3.1.6a), as results from the next theorem. For notational convenience, see [8–10, 15], let us introduce the following notation

$$
u_\pm(t,x) = \lim_{\varepsilon \searrow 0} u(t, x \pm \varepsilon) \,. \tag{3.4.3}
$$

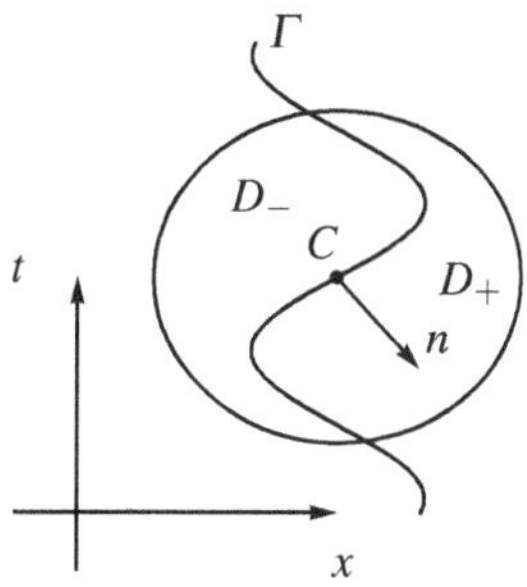

Fig. 3.3 The curve of discontinuity Γ divides the disc D centered in $C \in \Gamma$ in two parts: the left one is D_- and the right one is D_+. By n is denoted the normal vector orthogonal to Γ in C and directed to the right. Note that Γ is traditionally represented in the (x,t)–plane.

Theorem 3.1. *A piecewise* $\mathbf{C^1}$ *function u is a weak solution of* (3.1.6) *iff the following two conditions are satisfied:*

(i) u is a smooth solution of (3.1.6) *in the domains where it is* $\mathbf{C^1}$.
(ii) If $\Gamma\ :\ x = x(t)$ *is a curve of discontinuity for u and* $\sigma(t) = x'(t)$, *then u satisfies along* Γ *the* ***Rankine–Hugoniot (jump) condition***

$$(u_+ - u_-)\ \sigma = f(u_+) - f(u_-)\ . \tag{3.4.4}$$

Proof. Let u be a piecewise $\mathbf{C^1}$ weak solution to (3.1.6). We have already observed in Proposition 3.4 that u satisfies property (i). Let $C = (t_C, x_C) \in \Gamma$, with $t_C \in \mathbb{R}_+$ and $x_C = x(t_C)$, be a point of discontinuity for u and consider a disc D centered in C and small enough that $\Gamma \cap D \subset \mathbb{R}_+ \times \mathbb{R}$ is the only curve of discontinuity of u in D, see Fig. 3.3. Denote by $D_\pm$ the two open components of D on each side of Γ. Let $n = (t_n, x_n)$ be the unit normal vector to the curve Γ that points in the direction of D_+ and let $\varphi \in \mathbf{C_c^\infty}(D;\mathbb{R})$. By the integration by parts formula (2.2.2), we write

$$\begin{aligned} 0 &= \int\!\!\int_D \big(u\ \partial_t\varphi + f(u)\ \partial_x\varphi\big)\ \mathrm{d}x\ \mathrm{d}t \\ &= -\int\!\!\int_{D_+} \big(\partial_t u + \partial_x f(u)\big)\ \varphi\ \mathrm{d}x\ \mathrm{d}t - \int_{\Gamma\cap D} \big(t_n\ u_+ + x_n\ f(u_+)\big)\ \varphi\ \mathrm{d}\gamma \\ &\quad -\int\!\!\int_{D_-} \big(\partial_t u + \partial_x f(u)\big)\ \varphi\ \mathrm{d}x\ \mathrm{d}t + \int_{\Gamma\cap D} \big(t_n\ u_- + x_n\ f(u_-)\big)\ \varphi\ \mathrm{d}\gamma\ . \end{aligned}$$

The first and third integrals vanish because u is a smooth solution of (3.1.6a) in $D_\pm$. Furthermore, n is parallel to the vector $(-\sigma, 1)$, and therefore

$$\int_{\Gamma\cap D} \big(\sigma\ (u_+ - u_-) - f(u_+) + f(u_-)\big)\ \varphi\ \mathrm{d}\gamma = 0\ .$$

Since this is to hold for arbitrary φ and C, we obtain the jump relation (3.4.4) along Γ.

Viceversa, if u is a piecewise $\mathbf{C^1}$ function that satisfies properties (i) and (ii), it is a simple matter to check that u is indeed a weak solution to the Cauchy problem (3.1.6). □

Remark 3.5. The Rankine–Hugoniot condition (3.4.4) expresses conservation of u across jump discontinuities.

Clearly, any continuous solution automatically satisfies the Rankine–Hugoniot condition. On the other hand, if the solution u is not continuous in $C = (t_C, x_C)$, then the Rankine–Hugoniot condition gives a geometrical interpretation to the speed of the discontinuity. Indeed, in this case it is possible to solve (3.4.4) for σ and to obtain

$$\sigma = \frac{f(u_+) - f(u_-)}{u_+ - u_-} . \tag{3.4.5}$$

Therefore, the speed of the discontinuity is the slope of the secant line through the points $\big(u_-, f(u_-)\big)$ and $\big(u_+, f(u_+)\big)$. Furthermore, if f is a Lipschitz function, then the speed of propagation of the discontinuity satisfies the upper bound

$$|\sigma| \leq \mathscr{L}ip(f) .$$

The Rankine–Hugoniot condition has also a mathematical interpretation. Indeed, recalling that $a = f'$, (3.4.5) can be rewritten as

$$\sigma = \frac{1}{u_+ - u_-} \int_{u_-}^{u_+} a(v)\, dv . \tag{3.4.6}$$

In other words, the propagation speed of the discontinuity connecting u_- and u_+ is the average of the wave speeds $a(v)$, as v ranges in the interval between u_- and u_+.

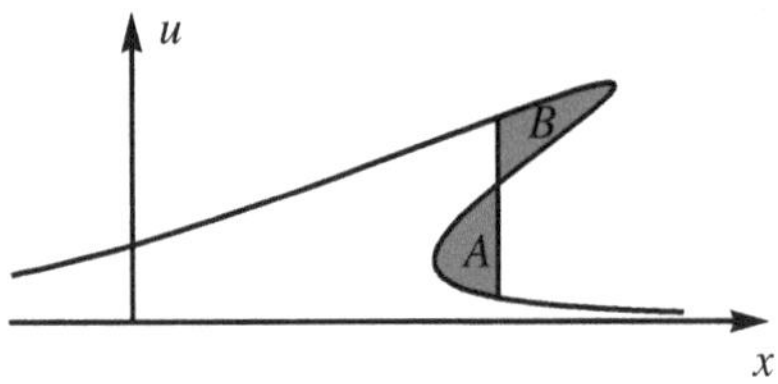

Fig. 3.4 Application of the Whitham "equal area" rule to construct the correct position for the discontinuity.

Finally, it turns out that the Rankine–Hugoniot condition corresponds to cutting off from the multi–valued profile two equal area lobes A and B as described in Fig. 3.4; this is the so called Whitham "equal area" rule, see [17]. □

The next example points out that the notion of weak solution is not invariant under nonlinear transformations of the independent variables.

Example 3.3. Consider the Hopf equation (3.3.1a). By (3.4.5), the speed of propagation of a discontinuity between u_- and u_+ is

$$\sigma_u = \frac{1}{2}(u_+ + u_-) . \tag{3.4.7}$$

Making the change of variables $[u \mapsto v(u) = u^3]$, Eq. (3.3.1a) becomes

$$\partial_t v + \partial_x \left(\frac{3}{4} v^{4/3} \right) = 0 \tag{3.4.8}$$

and the corresponding speed of propagation of the discontinuity is

$$\sigma_v = \frac{3}{4}\frac{v_+^{4/3} - v_-^{4/3}}{v_+ - v_-} = \frac{3}{4}\frac{u_+^4 - u_-^4}{u_+^3 - u_-^3},$$

which is clearly not equivalent to (3.4.7). Therefore, the (additional) conservation law (3.4.8) is satisfied by smooth solutions of (3.3.1a) but, in general, is not satisfied by discontinuous weak solutions. □

The above example points out that the choice of the variables affects the speeds of propagation of the discontinuities. Nevertheless, the structural properties of the solutions are usually invariant, in the sense that the solutions corresponding to different form of the same conservation law have similar features. In general, the choice of the variables is somewhat arbitrary. However, this choice is greatly simplified if the partial differential equation is derived from a conservation principle, indeed in this case we can rely on the directly correspondent conservation law to pick the correct conservative form. Whether this is not the case, the selection of a conservative form can be left to experimental validation, i.e. choosing the one that produces closest speeds of propagation of the discontinuities to the field measurements.

With the next example we want to point out that a weak solution of the Cauchy problem (3.1.6) is not necessarily unique, so the necessity to add an additional admissibility condition to individuate a unique solution.

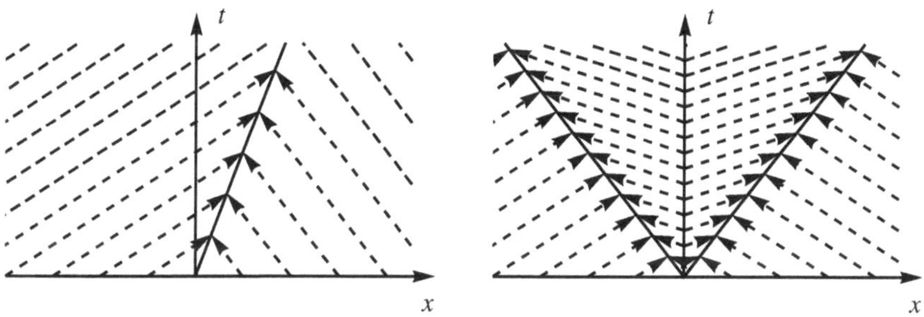

Fig. 3.5 Left: Representation of the solution (3.4.10) in the (x,t)–plane in the case $u^l > u^r$. The solid line is the discontinuity, which propagates with velocity $\sigma = (u^l + u^r)/2$. The dashed lines are the characteristic lines, which propagates with velocity $a(u^l) = u^l$ on the left of the discontinuity, and with velocity $a(u^r) = u^r$ on the right of the discontinuity. Right: Representation of the solution (3.4.11) in the (x,t)–plane in the case $\alpha = 2u^l = -2u^r > 0$. The three solid lines are the discontinuities, which propagates with velocity $s_1 = (u^l - \alpha)/2$, 0, $s_2 = (u^r + \alpha)/2$. The dashed lines are the characteristic lines, which propagates with velocity $a(u^l) = u^l$ on the left of the first discontinuity, $a(-\alpha) = -\alpha$ between the first and the second discontinuity lines, $a(\alpha) = \alpha$ between the second and the third discontinuity lines, and with velocity $a(u^r) = u^r$ on the right of the last discontinuity.

Example 3.4. We consider the Riemann problem for the Hopf equation, i.e. the Cauchy problem for (3.3.1a) with Heaviside initial datum

$$\partial_t u + \partial_x \left(\frac{u^2}{2}\right) = 0 \quad , \qquad u(0,x) = \begin{cases} u^l & \text{if } x < 0 \\ u^r & \text{if } x \geq 0 \, , \end{cases} \tag{3.4.9}$$

with $u^l \neq u^r$. The simplest weak solution is, see Fig. 3.5, left, and Fig. 4.1, left,

$$u(t,x) = \begin{cases} u^l & \text{if } x < \sigma\, t \\ u^r & \text{if } x \geq \sigma\, t \, , \end{cases} \tag{3.4.10}$$

where σ is given by (3.4.7). Furthermore, for any $\alpha > \max\{u^l, -u^r\}$, the function

$$u(t,x) = \begin{cases} u^l & \text{if } x < s_1\, t \\ -\alpha & \text{if } s_1\, t \leq x < 0 \\ \alpha & \text{if } 0 \leq x < s_2\, t \\ u^r & \text{if } x \geq s_2\, t \end{cases} \tag{3.4.11}$$

is also a weak solution if $s_1 = (u^l - \alpha)/2$ and $s_2 = (u^r + \alpha)/2$, see Fig. 3.5, right. Finally, if $u^l < u^r$, then also

$$u(t,x) = \begin{cases} u^l & \text{if } x < u^l\, t \\ x/t & \text{if } u^l\, t \leq x < u^r\, t \\ u^r & \text{if } x \geq u^r\, t \end{cases} \tag{3.4.12}$$

is a weak (continuous) solution, see Fig. 4.1, right. □

3.5 Entropy Weak Solutions

Example 3.4 points out that a weak solution of the Cauchy problem (3.1.6) is not necessarily unique, even in the case of a smooth flux. Hence, we need to add to the Rankine–Hugoniot condition (3.4.4) some "admissibility condition" that enables us to select the "physically relevant" solution among all the weak solutions. A first admissibility criterion, coming from physical considerations, was proposed by Dafermos [6]. It is the so called entropy condition or entropy inequality and will be presently discussed. In Sect. 4.3, we introduce alternative conditions motivated by the models under consideration.

Definition 3.6. A function u in $\mathbf{C^0}\left([0,T];\mathbf{L^\infty_{loc}}(\mathbb{R};\mathbb{R})\right)$ is an **entropy weak solution** of the Cauchy problem (3.1.6) if for any convex continuous function $U : \mathbb{R} \to \mathbb{R}$

$$\int_0^T \int_{\mathbb{R}} \left(U(u)\, \partial_t \varphi + F(u)\, \partial_x \varphi\right)\, \mathrm{d}x\, \mathrm{d}t + \int_{\mathbb{R}} U\left(\bar{u}(x)\right)\, \varphi(0,x)\, \mathrm{d}x \geq 0 \tag{3.5.1}$$

holds for all $\varphi \in \mathbf{C}_{\mathbf{c}}^{\infty}(]-\infty,T[\times\mathbb{R};\mathbb{R}_+^*)$, where $F(u) = F(a) + \int_a^u U'(v)\, f'(v)\, dv$ with $a, F(a) \in \mathbb{R}$ chosen arbitrarily.

The inequality (3.5.1) is referred to as **entropy condition** or **entropy inequality** associated to the **entropy pair** (U,F), where U is the **entropy** with **entropy flux** F.

Remark 3.6. Any smooth solution of the Cauchy problem (3.1.6) is an entropy weak solution. More precisely, any smooth solution u of (3.1.6) satisfies (3.5.1) with the equality and also the additional conservation law

$$\partial_t U(u) + \partial_x F(u) = 0$$

for any entropy pair (U,F). On the other hand, in general this is not true for weak solutions, since the Rankine–Hugoniot relations corresponding to the above equation and to (3.1.1) are in general not equivalent. □

Remark 3.7. A weak solution u to (3.1.6) is **reversible**, in the sense that the function $v(t,x) = u(s-t,-x)$ is also a weak solution to (3.1.6a) in $]0,s[\times\mathbb{R}$ corresponding to the initial datum $\overline{v}(x) = u(s,-x)$. On the other hand, the entropy inequality (3.5.1) changes when we pass from u to v, with the result that an entropy weak solution is in general not reversible. □

To make the Definition 3.6 clear, we do the **viscous regularization** of the conservation law (3.1.6a) by introducing a **viscosity term** $\varepsilon\, \Delta u$, where the small parameter $\varepsilon \in \mathbb{R}_+$ is a **diffusion coefficient** and $\Delta = \partial_x\partial_x$. As a result, we obtain the Cauchy problem for a scalar parabolic equation

$$\partial_t u_\varepsilon + \partial_x f(u_\varepsilon) = \varepsilon\, \Delta u_\varepsilon\,, \qquad u_\varepsilon(t=0) = \overline{u}\,.$$

Under sufficiently general hypotheses [16], the above Cauchy problem has one and only one smooth solution u_ε, which satisfies the maximum principle. Let us assume also that the sequence u_ε converges a.e. to a function u when $\varepsilon \searrow 0$. Then, Proposition 3.5 proves in fact that u is an entropy weak solution of the Cauchy problem (3.1.6). This approach to compute an entropy weak solution of (3.1.6) is the so called **vanishing viscosity method**, see [2, Sect. 4.4] or [7, Example 3.13].

Proposition 3.5. *Suppose that* $\overline{u} \in \mathbf{C}_{\mathbf{c}}^{\mathbf{0}}(\mathbb{R};\mathbb{R})$. *If* $u_\varepsilon \to u$ *a.e. in* $[0,T]\times\mathbb{R}$ *as* $\varepsilon \searrow 0$, *then* u *is an entropy weak solution to* (3.1.6).

Proof. By assumption $u_\varepsilon \in \mathbf{C}^\infty(]0,T]\times\mathbb{R};\mathbb{R}) \cap \mathbf{C}^{\mathbf{0}}([0,T]\times\mathbb{R};\mathbb{R})$. If $U\,:\,\mathbb{R}\to\mathbb{R}$ is a $\mathbf{C}^{\mathbf{2}}$ convex function with entropy flux F, then u_ε satisfies

$$\begin{aligned} \partial_t U(u_\varepsilon) + \partial_x F(u_\varepsilon) &= U'(u_\varepsilon)\, \big(\partial_t u_\varepsilon + \partial_x f(u_\varepsilon)\big) = \varepsilon\, U'(u_\varepsilon)\, \Delta u_\varepsilon \\ &= \varepsilon\, \Delta U(u_\varepsilon) - \varepsilon\, U''(u_\varepsilon)\, (\partial_x u_\varepsilon)^2 \le \varepsilon\, \Delta U(u_\varepsilon)\,. \end{aligned}$$

If we integrate over $[0,T[\times\mathbb{R}$ the above inequality multiplied by a test function $\varphi \in \mathbf{C}^\infty([0,T[\times\mathbb{R};\mathbb{R}_+^*)$, integrate by parts and pass to the limit as $\varepsilon \searrow 0$, then we obtain

$$\begin{aligned}
0 &\le \int_0^T \int_{\mathbb{R}} \left(\varepsilon\, \Delta U(u_\varepsilon) - \partial_t U(u_\varepsilon) - \partial_x F(u_\varepsilon)\right)\, \varphi \,\mathrm{d}x\ \mathrm{d}t \\
&= \int_0^T \int_{\mathbb{R}} \left(U(u_\varepsilon)\,(\varepsilon\, \Delta\varphi + \partial_t \varphi) + F(u_\varepsilon)\, \partial_x \varphi\right) \mathrm{d}x\ \mathrm{d}t + \int_{\mathbb{R}} U\left(\overline{u}(x)\right)\ \varphi(0,x)\ \mathrm{d}x \\
&\to \int_0^T \int_{\mathbb{R}} \left(U(u)\, \partial_t \varphi + F(u)\, \partial_x \varphi\right) \mathrm{d}x\ \mathrm{d}t + \int_{\mathbb{R}} U\left(\overline{u}(x)\right)\ \varphi(0,x)\ \mathrm{d}x\ .
\end{aligned}$$

It only remains to pass from $\mathbf{C}^2$ convex entropies to continuous convex entropies. Let U be a convex function. Then U is locally a uniform limit of $\mathbf{C}^\infty$ convex functions U_n. Let F and F_n be the entropy fluxes associated, respectively, to U and U_n. By definition and the integration by parts Lemma 2.3, we have

$$\begin{aligned}
F_n(u) &= \int_0^u U_n'(v)\, f'(v)\ \mathrm{d}v = U_n(u)\, f'(u) - U_n(0)\, f'(0) - \int_0^u U_n(v)\, f''(v)\ \mathrm{d}v \\
&\to U(u)\, f'(u) - U(0)\, f'(0) - \int_0^u U(v)\, f''(v)\,\mathrm{d}v = \int_0^u U'(v)\, f'(v)\ \mathrm{d}v = F(u).
\end{aligned}$$

The entropy inequality (3.5.1) is true for the entropy pairs (U_n, F_n). A new passage to the limit when $n \to +\infty$ proves that it still holds true for (U,F). □

Remark 3.8. A generalization of the theory for entropy weak solutions to the case of non–continuous fluxes can be found in [12]. □

A condition equivalent to the entropy inequality (3.5.1) is given in the following theorem:

Theorem 3.2 (Kružkov [13]). *A function* $u \in \mathbf{C}^0\left([0,T];\mathbf{L}^\infty_{\mathbf{loc}}(\mathbb{R};\mathbb{R})\right)$ *is an entropy weak solution of the Cauchy problem* (3.1.6) *on* $\Omega = [0,T] \times \mathbb{R}$ *iff*

$$\begin{aligned}
\int\int_\Omega \left(|u-k|\ \partial_t \varphi + \mathrm{sgn}\,(u-k)\ \left(f(u) - f(k)\right)\ \partial_x \varphi\right)\ \mathrm{d}x\ \mathrm{d}t & \\
+ \int_{\mathbb{R}} \left|\overline{u}(x) - k\right|\ \varphi(0,x)\ \mathrm{d}x \ge 0 &
\end{aligned} \tag{3.5.2}$$

for every constant $k \in \mathbb{R}$ *and every* $\varphi \in \mathbf{C}^\infty_\mathbf{c}\left([0,T[\times\mathbb{R};\mathbb{R}_+^*\right)$.

Proof. Let $k \in \mathbb{R}$. Then $[u \mapsto |u-k|]$ is a convex continuous function and its entropy flux is equal (up to a sum of a constant) to $[u \mapsto \mathrm{sgn}\,(u-k)\ \left(f(u)-f(k)\right)]$. Hence, (3.5.1) implies (3.5.2).

Conversely, let us assume that (3.5.2) holds and consider a given entropy pair (U,F). By assumption $\overline{u}$ and u take values in a bounded interval $[k_1,k_2]$ and for any fixed $\varepsilon \in \mathbb{R}_+$ there exists an entropy pair $(U_\varepsilon, F_\varepsilon)$ which satisfies

$U(k) \le U_\varepsilon(k) \le U(k) + \varepsilon$ for $k \in [k_1,k_2]$,
U_ε is piecewise linear convex continuous function.

Then, there exist $\alpha_0, \alpha_1 \in \mathbb{R}$ and $\beta_j \in \mathbb{R}_+$ such that

$$U_\varepsilon(k) = \alpha_0 + \alpha_1\ k + \sum_j \beta_j\ |k-k_j|$$

$$F_\varepsilon(k) = \alpha_1\ f(k) + \sum_j \beta_j\ \mathrm{sgn}(k-k_j)\ \left(f(k)-f(k_j)\right)\ .$$

By (3.5.2), it results that (3.5.1) holds for $(U_\varepsilon, F_\varepsilon)$. As U_ε and F_ε converge uniformly, respectively, to U and F on $[k_1,k_2]$, we can pass to the limit in the integrals, with the result that (3.5.1) is valid also for (U,F). □

Corollary 3.1. *An entropy weak solution of the Cauchy problem* (3.1.6) *is also a weak solution in the sense of Definition 3.4.*

Proof. Let u be a locally bounded entropy weak solution of the Cauchy problem (3.1.6) on $\Omega = [0,T]\times\mathbb{R}$ and $\varphi \in \mathbf{C}^\infty_\mathbf{c}\left([0,T[\times\mathbb{R};\mathbb{R}^*_+\right)$. Then $u|_{\mathrm{spt}\{\varphi\}}$ and $\overline{u}|_{\mathrm{spt}\{\varphi\}}$ attain values in a bounded interval $]k_1,k_2[$. For $k=k_1$, the inequality (3.5.2) implies that

$$\begin{aligned}&\int\!\!\int_\Omega \left(u\ \partial_t\varphi + f(u)\ \partial_x\varphi\right)\ \mathrm{d}x\ \mathrm{d}t + \int_\mathbb{R} \overline{u}(x)\ \varphi(0,x)\ \mathrm{d}x\\ &\geq k_1\left(\int\!\!\int_\Omega \partial_t\varphi\ \mathrm{d}x\ \mathrm{d}t + \int_\mathbb{R}\varphi(0,x)\ \mathrm{d}x\right) + f(k_1)\int\!\!\int_\Omega \partial_x\varphi\ \mathrm{d}x\ \mathrm{d}t = 0\end{aligned}$$

on deducing the last equality from the Green's formula (2.2.1). Similarly, taking $k=k_2$, the inequality (3.5.2) implies the inequality opposite to the preceding one. Thus (3.4.2) holds and u is a weak solution of (3.1.6). □

Analogously to Theorem 3.1, in the case of piecewise $\mathbf{C}^1$ entropy weak solutions we have the following result:

Theorem 3.3. *A piecewise* $\mathbf{C}^1$ *function u is an entropy weak solution of* (3.1.6) *iff the following three conditions are satisfied:*

(i) u is a smooth solution of (3.1.6) *in the domains where it is* $\mathbf{C}^1$.
(ii) u satisfies the Rankine–Hugoniot jump condition (3.4.4) *along its curves of discontinuity.*
(iii) If $\Gamma\ :\ x = x(t)$ *is a curve of discontinuity for u,* $\sigma(t) = x'(t)$ *and* $u_\pm$ *is defined by* (3.4.3)*, then one of the following equivalent three conditions is satisfied:*

(iii.a) For any entropy pair (U,F)*, u satisfies along* Γ *the* ***entropy jump condition***

$$\left(U(u_+)-U(u_-)\right)\ \sigma \geq F\left(u_+\right) - F\left(u_-\right)\ . \tag{3.5.3}$$

(iii.b) For any $\alpha\in[0,1]$

$$\begin{aligned}f\left(\alpha\ u_+ + (1-\alpha)\ u_-\right) &\geq \alpha\ f(u_+) + (1-\alpha)\ f(u_-) \quad \textit{if}\ u_- < u_+\\ f\left(\alpha\ u_+ + (1-\alpha)\ u_-\right) &\leq \alpha\ f(u_+) + (1-\alpha)\ f(u_-) \quad \textit{if}\ u_- > u_+\ .\end{aligned} \tag{3.5.4}$$

(iii.c) For any v between u_- *and* u_+

$$\frac{f(v)-f(u_+)}{v-u_+} \leq \frac{f(u_-)-f(u_+)}{u_- - u_+}\ . \tag{3.5.5}$$

Proof. By Theorem 3.1 and Corollary 3.1, a piecewise $\mathbf{C}^1$ entropy weak solution of (3.1.6) is also a piecewise $\mathbf{C}^1$ weak solution or, equivalently, conditions (i) and (ii) are satisfied.

The proof that condition (iii.a) is equivalent to (3.5.1) is analogous to the proof that condition (ii) of Theorem 3.1 is equivalent to (3.4.2) and therefore is omitted.

About condition (iii.b), by Theorem 3.2 it is sufficient to prove that (3.5.4) is equivalent to (3.5.2). Let u be a piecewise $\mathbf{C}^1$ entropy weak solution to (3.1.6) and $k \in \mathbb{R}$. With the same notations of Theorem 3.1, see Fig. 3.3, by using the integration by parts formula (2.2.2), (3.5.2) can be written

$$\begin{aligned}
0 &\le \int\!\!\int_D \Big(|u-k|\ \partial_t\varphi + \operatorname{sgn}(u-k)\ \big(f(u)-f(k)\big)\ \partial_x\varphi\Big)\ \mathrm{d}x\ \mathrm{d}t \\
&= -\int\!\!\int_{D_-} \Big(\partial_t|u-k| + \partial_x\Big(\operatorname{sgn}(u-k)\ \big(f(u)-f(k)\big)\Big)\Big)\ \varphi\ \mathrm{d}x\ \mathrm{d}t \\
&\quad + \int_{\Gamma\cap D} \Big(\operatorname{sgn}(u_- - k)\ \big(f(u_-)-f(k)\big) - \sigma\ |u_- - k|\Big)\ \varphi\ \mathrm{d}\gamma \\
&\quad - \int\!\!\int_{D_+} \Big(\partial_t|u-k| + \partial_x\Big(\operatorname{sgn}(u-k)\ \big(f(u)-f(k)\big)\Big)\Big)\ \varphi\ \mathrm{d}x\ \mathrm{d}t \\
&\quad - \int_{\Gamma\cap D} \Big(\operatorname{sgn}(u_+ - k)\ \big(f(u_+)-f(k)\big) - \sigma\ |u_+ - k|\Big)\ \varphi\ \mathrm{d}\gamma
\end{aligned}$$

for any $\varphi \in \mathbf{C}^1(\mathbb{R}_+^* \times \mathbb{R};\mathbb{R}) \cap \mathbf{C}_c^1(D;\mathbb{R})$. Since u is a smooth solution of (3.1.6a) in $D_\pm$, it results that

$$\partial_t|u-k| + \partial_x\Big(\operatorname{sgn}(u-k)\ \big(f(u)-f(k)\big)\Big) = \operatorname{sgn}(u-k)\,\big(\partial_t u + \partial_x f(u)\big) = 0$$

a.e. in $D_\pm$. Furthermore, for the arbitrariness of φ, we obtain

$$\begin{aligned}
\sigma\ \big(|u_+ - k| - |u_- - k|\big) \ge \operatorname{sgn}(u_+ - k)\ \big(f(u_+) - f(k)\big) \\
- \operatorname{sgn}(u_- - k)\ \big(f(u_-) - f(k)\big)\ .
\end{aligned} \tag{3.5.6}$$

By choosing $k = \alpha\, u_+ + (1-\alpha)\, u_-$, $\alpha \in [0,1]$, in (3.5.6) we obtain

$$\sigma\ (u_+ + u_- - 2k)\ \operatorname{sgn}(u_+ - u_-) \ge \big(f(u_+) + f(u_-) - 2f(k)\big)\ \operatorname{sgn}(u_+ - u_-)\ .$$

The above estimate and (3.4.4) give

$$\begin{aligned}
&\big(f(u_+) - f(u_-)\big)\ (1-2\alpha)\ \operatorname{sgn}(u_+ - u_-) \ge \\
&\qquad \ge \big(f(u_+) + f(u_-) - 2f(\alpha\, u_+ + (1-\alpha)\, u_-)\big)\ \operatorname{sgn}(u_+ - u_-)\ ,
\end{aligned}$$

that is equivalent to

$$\begin{aligned}
&\big(\alpha\ f(u_+) + (1-\alpha)\ f(u_-)\big)\ \operatorname{sgn}(u_+ - u_-) \le \\
&\qquad \le f(\alpha\, u_+ + (1-\alpha)\, u_-)\ \operatorname{sgn}(u_+ - u_-)\ ,
\end{aligned} \tag{3.5.7}$$

and therefore also to (3.5.4).

Viceversa, if u is a piecewise $\mathbf{C}^1$ function that satisfies properties (i)–(iii), it is a simple matter to check that u is indeed an entropy weak solution to (3.1.6). Indeed, condition (3.5.6) is equivalent to (3.5.4) when k is a value between u_+ and u_-, otherwise is equivalent to the Rankine–Hugoniot condition (3.4.4) since, when $k \leq \min\{u_-, u_+\}$ or $k \geq \max\{u_-, u_+\}$, (3.5.6) becomes respectively

$$\sigma\ (u_+ - u_-) \geq f(u_+) - f(u_-)\quad , \qquad \sigma\ (u_+ - u_-) \leq f(u_+) - f(u_-)\,.$$

Finally, let $v = \alpha\ u_+ + (1-\alpha)\ u_-$, $\alpha \in \,]0,1[$, in (3.5.7) and observe that

$$\operatorname{sgn}(u_+ - u_-) = \frac{|v - u_-|}{v - u_-} = \frac{|u_+ - v|}{u_+ - v}\,.$$

On subtracting $f(u_-)\ \operatorname{sgn}(u_+ - u_-)$ from (3.5.7) and dividing by $|v - u_-|$ we get

$$\alpha\ \frac{f(u_+) - f(u_-)}{v - u_-} \leq \frac{f(v) - f(u_-)}{v - u_-}\ ;$$

and, on subtracting $f(u_+)\ \operatorname{sgn}(u_+ - u_-)$ from (3.5.7) and dividing by $|u_+ - v|$, we get

$$(1-\alpha)\ \frac{f(u_+) - f(u_-)}{u_+ - v} \geq \frac{f(u_+) - f(v)}{u_+ - v}\,.$$

Since $\frac{1-\alpha}{u_+ - v} = \frac{\alpha}{v - u_-}$, the two obtained inequalities are equivalent to (3.5.5). □

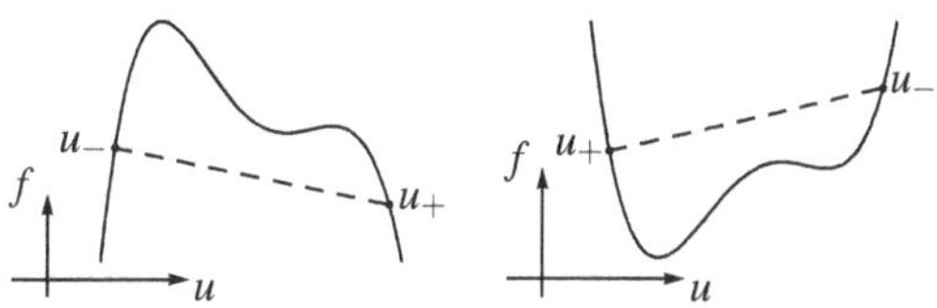

Fig. 3.6 Geometrical meaning of the Oleinik entropy inequality (3.5.5), or equivalently (3.5.4).

Condition (iii.c) is the **Oleinik entropy inequality** and have a simple geometrical interpretation, see Fig. 3.6. When $u_- < u_+$, the graph of $f|_{]u_-, u_+[}$ remains above the segment between $\big(u_-, f(u_-)\big)$ and $\big(u_+, f(u_+)\big)$. On the other hand, when $u_+ < u_-$, the graph of $f|_{]u_-, u_+[}$ remains below the segment between $\big(u_+, f(u_+)\big)$ and $\big(u_-, f(u_-)\big)$.

3.6 Lax Inequality

An important consequence of the Oleinik entropy inequality (3.5.5) is the Lax (shock) condition, also called Lax inequality, which compares the speeds of the characteristics to the right and to the left of a discontinuity with the speed of propagation of the discontinuity itself.

Definition 3.7. A piecewise $\mathbf{C}^1$ weak solution of the Cauchy problem (3.1.6) satisfies the **Lax condition** if along its curves of discontinuity holds

$$a(u_-) \geq \sigma \geq a(u_+)\ , \tag{3.6.1}$$

where $u_\pm$ are defined by (3.4.3), $a = f'$ and σ is the speed of propagation of the discontinuity defined by the Rankine–Hugoniot relation (3.4.4).

Remark 3.9. The major inconvenience of the Lax condition (3.6.1) is that in general it can not be expressed for weak solutions, but only for piecewise smooth solutions, contrary to the entropy inequality (3.5.1). □

The geometrical meaning of condition (3.6.1) is represented in the (u, f)–plane by Fig. 3.7. An alternative point of view is that described in Fig. 3.5, left. If u satisfies the Lax condition (3.6.1), then the characteristics impinge on the curve of discontinuity from both sides and can not emerge from it. In this case, the discontinuity is said to be **compressive** and will be referred to as **(entropic) shock**, see Sect. 4.2.

Definition 3.8. We shall say that a discontinuity is a **shock wave** if the inequalities of (3.6.1) are strict. We shall say that it is a **semi–characteristic shock wave** if one is strict and the other is an equality. We shall say that it is a **characteristic shock wave** if the two are equalities without f being affine between u_- and u_+. Finally, if f is affine from u_- to u_+, it is a **contact discontinuity**, see Sect. 4.5.

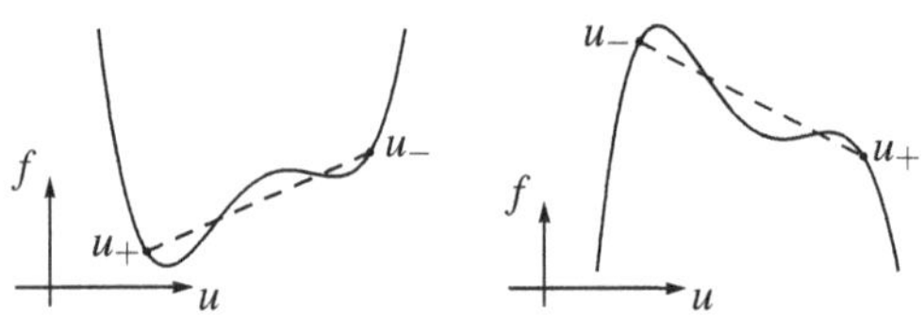

Fig. 3.7 Geometrical meaning of the Lax condition (3.6.1). In particular, these states satisfy the Lax condition (3.6.1), but not the Oleinik entropy inequality (3.5.5).

Proposition 3.6. *The Oleinik entropy inequality* (3.5.5) *implies the Lax conditions* (3.6.1).

Proof. It is geometrically obvious that (3.5.5) is equivalent to

$$\frac{f(u) - f(u_-)}{u - u_-} \geq \frac{f(u_-) - f(u_+)}{u_- - u_+} \quad \text{for all } u \text{ between } u_- \text{ and } u_+.$$

To derive the two inequalities in (3.6.1) we simply let $u \to u_-$ in the above equation and $u \to u_+$ in (3.5.5). □

Definition 3.9. The characteristic field of (3.1.1) is **genuinely nonlinear** if $f''(u) \neq 0$ for all u and is **linearly degenerate** if $f''(u) = 0$ for all u.

Remark 3.10. If the characteristic filed is genuinely nonlinear, then a is strictly monotone and f is strictly convex, or strictly concave. On the other hand, if it is linearly degenerate, then f is affine and a does not depend on u. Furthermore, if the characteristic filed is genuinely nonlinear, then the inequalities of (3.6.1) are strict, while, if it is linearly degenerate, then (3.6.1) holds with the equalities. □

Example 3.5. The linear advection equation of Example 3.1 is the prototype of a scalar conservation law with linearly degenerate characteristic field, while the Hopf equation (3.3.1a) is the prototype of conservation law with genuinely nonlinear characteristic field. Finally, the equation $\partial_t u + \partial_x u^3 = 0$ is an example of a scalar conservation law with a characteristic field which is neither linearly degenerate nor genuinely nonlinear. More precisely, it is genuinely nonlinear in $\mathbb{R}_-$ and in $\mathbb{R}_+$, but fails to be so in a neighborhood of $u = 0$. □

Proposition 3.7. *If the characteristic filed is genuinely nonlinear or linearly degenerate, then any piecewise* $\mathbf{C}^1$ *weak solution of the Cauchy problem* (3.1.6) *satisfying the Lax condition* (3.6.1) *is an entropy weak solution. Furthermore, if condition* (3.5.1) *holds for a single entropy pair, then it holds for all entropy pairs.*

Proof. The linearly degenerate case is obvious. Consider a genuinely nonlinear characteristic field. By Remark 3.10, it is not limitative to assume that f is strictly convex, being the strictly concave case similar. In this case, (3.6.1) is equivalent to require that $u_- \geq u_+$ and, from a geometrical point of view, (3.5.5) is clearly satisfied. Therefore, by Theorem 3.3, we deduce (3.5.1). Finally, assume that the condition (3.5.1), or equivalently, by Theorem 3.3, that (3.5.3) holds for a single entropy pair (U,F). Integrating by part we obtain the estimate

$$
\begin{aligned}
0 \geq F(u_+) - F(u_-) - \big(U(u_+) - U(u_-)\big)\ \sigma &= \int_{u_+}^{u_-} U'(v)\big(\sigma - a(v)\big)\ \mathrm{d}v \\
&= -\int_{u_+}^{u_-} U''(v)\big(\sigma\,(v-u_+) - f(v) + f(u_+)\big)\ \mathrm{d}v \\
&\quad + \Big[U'(v)\big(\sigma\,(v-u_+) - f(v) + f(u_+)\big)\Big]_{u_+}^{u_-} \\
&= -\int_{u_+}^{u_-} U''(v)\,(v-u_+)\left(\frac{f(u_+)-f(u_-)}{u_+-u_-} - \frac{f(v)-f(u_+)}{v-u_+}\right)\mathrm{d}v
\end{aligned}
$$

where (3.4.4) was used to cancel the boundary terms after applying the integration by parts formula (2.2.2). The integrand in the above inequality has a constant sign when f is strictly convex and is the favorable sign iff $u_- \geq u_+$. □

Example 3.6. Let us analyze whether the weak solutions proposed in Example 3.4 are entropic. First, we observe that the characteristic field is genuinely nonlinear. Therefore, Lax and entropy conditions coincide. The Lax condition for the solution (3.4.10) is equivalent to require $u^l \geq u^r$. Therefore, the weak solution (3.4.10) is entropic iff $u^l > u^r$, see Fig. 3.5, left. For solution (3.4.11), the Lax conditions corresponding to its three jumps of discontinuity are equivalent to require that

$$2\,u^l \geq u^l - \alpha \geq -2\,\alpha \quad , \qquad -\alpha \geq 0 \geq \alpha \quad , \qquad 2\,\alpha \geq u^r + \alpha \geq 2\,u^r\,.$$

Assumption $\alpha \geq \max\{u^l, -u^r\}$ ensures that $s_1 \leq 0 \leq s_2$ and, together with the above estimates, implies $(u^l, u^r, \alpha) = 0_3$. Thus, solution (3.4.11) is not entropic. □

References

1. Ambrosio, L., Caffarelli, L., Crandall, M., Evans, L., Fusco, N., Ambrosio, L.: Transport Equation and Cauchy Problem for Non-Smooth Vector Fields. In: *Calculus of Variations and Nonlinear Partial Differential Equations*. Lecture Notes in Mathematics, vol. 1927, pp. 1–41. Springer, Berlin (2008)
2. Bressan, A.: Hyperbolic systems of conservation laws. Oxford Lecture Series in Mathematics and its Applications, vol. 20. Oxford University Press, Oxford (2000)
3. Burgers, J.M.: Application of a model system to illustrate some points of the statistical theory of free turbolence. Proc. Roy. Neth. Acad. Sci. Amsterdam 43, 2–12 (1940)
4. Burgers, J.M.: The nonlinear diffusion equation. Dordrecht-Holland. D. Reidel Pub. Co., Boston (1974)
5. Colombo, R.M., Mercier, M., Rosini, M.D.: Stability and total variation estimates on general scalar balance laws. Commun. Math. Sci. 7(1), 37–65 (2009)
6. Dafermos, C.M.: Hyperbolic conservation laws in continuum physics. Grundlehren der Mathematischen Wissenschaften [Fundamental Principles of Mathematical Sciences], vol. 325. Springer, Berlin (2000)
7. Holden, H., Risebro, N.H.: Front tracking for hyperbolic conservation laws. Applied Mathematical Sciences, vol. 152. Springer, New York (2002)
8. Hugoniot, H.: Sur un théorème général relatif à la propagation du mouvement dans le corps. C. R. Acad. Sci. Paris Sér. I Math. 102, 858–860 (1886)
9. Hugoniot, H.: Mémoire sur la propagation du mouvement dans le corps et spécialement dans le gaz parfaits. J. l'Ecoles Polytechn. 57, 3–97 (1887)
10. Hugoniot, H.: Mémoire sur la propagation du mouvement dans un fluid indéfini. J. Math. Pures Appl. 3, 477–492 (1887)
11. John, F.: Formation of singularities in one-dimensional nonlinear wave propagation. Communications on Pure and Applied Mathematics 27(3), 377–405 (1974)
12. Karlsen, K.H., Towers, J.D.: Convergence of the Lax-Friedrichs scheme and stability for conservation laws with a discontinuous space-time dependent flux (2004)
13. Kružhkov, S.N.: First order quasilinear equations with several independent variables. Mat. Sb. (N.S.) 81(123), 228–255 (1970)
14. Perthame, B.: Transport equations in biology. Frontiers in mathematics. Birkhäuser (2007)
15. Rankine, W.J.M.: On the Thermodynamic Theory of Waves of Finite Longitudinal Disturbance. Phil. Trans. Roy. Soc. 160, 277–288 (1870)
16. Ritchmyer, R.D., Morton, K.W.: Difference methods for initial-value problems, 2nd edn. Interscience Tracts in Pure and Applied Mathematics, vol. 3. Interscience, New York (1967)
17. Whitham, G.B.: Linear and nonlinear waves. Pure and Applied Mathematics. Wiley-Interscience [John Wiley & Sons], New York (1974)

Chapter 4
The Riemann Problem

Abstract. This chapter is devoted to study the Riemann problems for scalar conservation laws in one space dimension. In particular, we introduce the definitions of entropy and non–entropy shock waves, rarefaction waves and contact discontinuities by first considering genuinely nonlinear characteristic fields, linearly degenerate characteristic fields and then passing to general characteristic fields.

4.1 Introduction

Let $f : \mathbb{R} \to \mathbb{R}$ be $\mathbf{C}^2$, $a = f'$ and $u^l, u^r \in \mathbb{R}$ with $u^l \neq u^r$. In this section we study the **Riemann problem** for the scalar conservation law (3.1.1), see [3, 4], namely the following special Cauchy problem with Heaviside initial datum

$$\partial_t u + \partial_x f(u) = 0 \tag{4.1.1a}$$

$$u(0,x) = \begin{cases} u^l & \text{if } x < 0 \\ u^r & \text{if } x \geq 0 \,. \end{cases} \tag{4.1.1b}$$

The invariance of both equation and datum in (4.1.1) under the change of variables

$$\left[(t,x) \mapsto (\lambda\, t, \lambda\, x)\right] \quad , \qquad \lambda \neq 0 \,, \tag{4.1.2}$$

suggests to look for solutions also invariant to (4.1.2). For this reason, at first, we restrict ourself to **self similar solutions** of (4.1.1), i.e., solutions of the form

$$u(t,x) = \phi(x/t) \,. \tag{4.1.3}$$

M.D. Rosini: *Macroscopic Models for Vehicular Flows & Crowd Dynamics*, UCS, pp. 43–50.
DOI: 10.1007/978-3-319-00155-5_4

4.2 Shock Waves

Assume that the characteristic field of (4.1.1a) is genuinely nonlinear. Given the left state $u^l \in \mathbb{R}$, we wish to describe the family of all right states $u^r \in \mathbb{R} \setminus \{u^l\}$ for which the Riemann problem (4.1.1) has a self similar piecewise constant entropy weak solution of the form

$$u(t,x) = \begin{cases} u^l & \text{if } x < \sigma\, t \\ u^r & \text{if } x \geq \sigma\, t \end{cases} \qquad (4.2.1a)$$

for some $\sigma \in \mathbb{R}$.

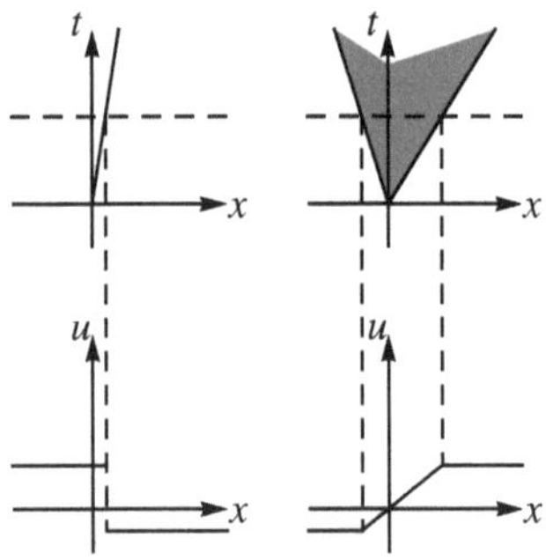

Fig. 4.1 Left: The shock wave (3.4.10), entropy weak solution of the Riemann problem (3.4.9) with $u^l > u^r$. Right: The rarefaction wave (3.4.12), entropy weak solution of the Riemann problem (3.4.9) with $u^l < u^r$.

Definition 4.1. An entropy weak solution to (4.1.1) of the form (4.2.1a) is called a **(entropy) shock wave** connecting the state u^l to the state u^r.

By Theorem 3.3, (4.2.1a) is an entropy weak solution to (4.1.1) if it satisfies the Rankine–Hugoniot jump condition (3.4.4), namely

$$\sigma = \frac{f(u^l) - f(u^r)}{u^l - u^r}, \qquad (4.2.1b)$$

and the Oleinik entropy inequality (3.5.5), namely

$$u^r = u^l - \alpha\, f''(u^l) \quad , \qquad \alpha \in \mathbb{R}_+ . \qquad (4.2.2)$$

Hence, we have proved the following proposition:

Proposition 4.1. *Assume that the characteristic field is genuinely nonlinear. Given a state $u^l \in \mathbb{R}$, for all $\alpha \in \mathbb{R}_+$, the state $u^r = u^l - \alpha\, f''(u^l)$ can be connected to u^l on the right by a shock wave* (4.2.1), *solution of* (4.1.1).

Example 4.1. The solution (3.4.10) of the Riemann problem for the Hopf equation (3.4.9) is a shock wave if $u^l > u^r$, see Fig. 3.5, left, and Fig. 4.1, left. □

On the other hand, if f is strictly convex, respectively strictly concave, and $u^l < u^r$, respectively $u^l > u^r$, then the above construction breaks down. These cases will be considered in Sect. 4.4.

4.3 Non–entropy Shock Waves

In this section we generalize the definition of non–entropic shocks introduced by LeFloch [2].

Definition 4.2. A weak solution to (4.1.1) of the form (4.2.1) is a **non–entropy shock wave** connecting the state u^l to the state u^r if the entropy condition (3.5.1) is not satisfied.

By Theorem 3.1 and Theorem 3.3, a non–entropic shock satisfies the Rankine–Hugoniot condition (3.4.4), but it does not satisfy the Oleinik entropy inequality (3.5.5), see Fig. 3.7. Further, non–entropic shock are **slow under–compressive**, in the sense that characteristics on both side of the discontinuity line pass through it.

As pointed out in Sect. 3.5, one of the aims of the Oleinik entropy inequality (3.5.5) is to ensure the uniqueness of the solution. In the framework of non–entropic shock, the uniqueness of the solution is recovered by imposing a further condition, called **kinetic condition**, which is derived by the situation at hand. For further details we defer to [2, Chap. III].

Examples of non–entropic shocks are presented in chapters 6, 11, 12, 13, 16, 17.

4.4 Rarefaction Waves

Consider continuous piecewise $\mathbf{C}^1$ self similar weak solutions of (4.1.1). These solutions satisfy a.e. the quasilinear equation

$$\partial_t u + a(u)\, \partial_x u = 0\,. \tag{4.4.1}$$

Substituting (4.1.3) in (4.4.1) and (4.1.1b) gives

$$\left[a\left(\phi\left(\xi\right)\right) - \xi\right]\, \phi'\left(\xi\right) = 0 \tag{4.4.2a}$$

$$\phi\left(\pm\infty\right) = u_{r,l}\,. \tag{4.4.2b}$$

By (4.4.2a), ϕ must be constant or has to satisfy

$$a\left(\phi\left(\xi\right)\right) = \xi\,. \tag{4.4.3}$$

Since $u^l \neq u^r$, (4.4.2b) implies that condition (4.4.3) holds on a spatial interval with non–zero Lebesgue measure. Equation (4.4.3) can not be solved if the characteristic field is linearly degenerate. Thus, assume that the vector field is genuinely nonlinear. In this case, we get from (4.4.3) that

$$a\left(\phi\left(a(u)\right)\right) = a(u)$$

and, since a is injective, the above equality implies

$$\phi\left(a(u)\right) = u\,. \tag{4.4.4}$$

If f is strictly convex (respectively strictly concave), then (4.4.3) implies that ϕ is increasing (respectively decreasing) and therefore, by (4.4.2b), it is necessary that $u^l < u^r$ (respectively $u^l > u^r$). These considerations, together with the equality $\phi\left(a(u_{l,r})\right) = u_{l,r}$ derived by (4.4.4), imply that the formula

$$u(t,x) = \begin{cases} u^l & \text{if } x < a(u^l)\,t \\ a^{-1}\left(x/t\right) & \text{if } a(u^l)\,t \le x < a(u^r)\,t \\ u^r & \text{if } x \ge a(u^r)\,t \end{cases} \tag{4.4.5}$$

defines a continuous, piecewise $\mathbf{C^1}$ and monotone solution of the Riemann problem (4.1.1).

Definition 4.3. An entropy weak solution to (4.1.1) of the form (4.4.5) is called a **centered rarefaction wave** connecting the state u^l to the state u^r, and the point at which the discontinuity occurs is called **center of the rarefaction wave**.

Proposition 4.2. *Assume that the characteristic field is genuinely nonlinear. Given a state $u^l \in \mathbb{R}$, for all $\alpha \in \mathbb{R}_+$, the state $u^r = u^l + \alpha\, f''(u^l)$ can be connected to u^l on the right by a rarefaction wave* (4.4.5) *solution of* (4.1.1).

Proof. The function defined by (4.4.5) is clearly piecewise $\mathbf{C^1}$. Therefore it is sufficient to prove the conditions (i)–(iii) of Theorem 3.3. Condition (i) is obvious in $\{(t,x) \in \mathbb{R}_+ \times \mathbb{R}\,:\, x < a(u^l)\,t \text{ or } x > a(u^r)\,t\}$, while for all $(t,x) \in \{(t,x) \in \mathbb{R}_+ \times \mathbb{R}\,:\, a(u^l)\,t < x < a(u^r)\,t\}$ it results

$$\partial_t u + \partial_x f(u) = -\frac{x}{t^2}\,(a^{-1})'\left(\frac{x}{t}\right) + \frac{1}{t}\,f'\left(a^{-1}\left(\frac{x}{t}\right)\right)\,(a^{-1})'\left(\frac{x}{t}\right) = 0\,.$$

Since (4.4.5) is continuous, conditions (ii) and (iii) are automatically satisfied. □

Example 4.2. The solution (3.4.12) of the Riemann problem for the Hopf equation (3.4.9), with $u^l < u^r$, is an example of rarefaction wave, see Fig. 3.5, right, and Fig. 4.1, right. □

4.5 Contact Waves

Assume that (4.1.1a) has a linearly degenerate characteristic field, namely that $a(u) \equiv \sigma$ for a constant $\sigma \in \mathbb{R}$. Let $u^l, u^r \in \mathbb{R}$ with $u^l \neq u^r$. By applying the method of characteristics described in Sect. 3.2, see also Example 3.1, we construct the piecewise constant function

$$u(t,x) = \begin{cases} u^l & \text{if } x < \sigma\, t \\ u^r & \text{if } x \geq \sigma\, t\,, \end{cases} \tag{4.5.1}$$

which obviously satisfies the hypotheses of Theorem 3.3. Thus (4.5.1) results to be an entropy weak solution of (4.1.1) and is called **contact discontinuity**.

4.6 The General Case

In the previous sections, we show how to construct an elementary weak solution to a Riemann problem characterized by a characteristic field, which was either genuinely nonlinear or linearly degenerate. In this section we extend our analysis to more general fluxes. We will take into account only entropy weak solutions, which result to be combinations of elementary entropy weak solutions, namely shocks, rarefaction waves and contact discontinuities.

Let $f\,:\,\mathbb{R} \to \mathbb{R}$ be a $\mathbf{C}^2$ function with finitely many inflection points. By taking the lower convex envelope and the upper concave envelope of f we obtain new fluxes, which are piecewise of the same type considered in the previous sections.

Definition 4.4. Let $f \in \mathbf{C}^2(\mathbb{R};\mathbb{R})$ be a function with finitely many inflection points and I be an interval. The **lower convex envelope** of f in the interval I is

$$f_{\smile}(u) = \sup\left\{g(u)\,:\, g(x) \leq f(x) \text{ for all } x \in I \text{ and } g \text{ convex}\right\} \tag{4.6.1a}$$

and its **upper concave envelope** in the interval I is

$$f_{\frown}(u) = \inf\left\{g(u)\,:\, g(x) \geq f(x) \text{ for all } x \in I \text{ and } g \text{ concave}\right\}\,. \tag{4.6.1b}$$

Remark 4.1. The functions $f_{\smile}$ and $f_{\frown}$ depend on the interval I, and thus are non–local functions of f. Furthermore, $f_{\smile}$ is the largest convex function that is smaller than or equal to f in the interval I, while $f_{\frown}$ is the smallest concave function that is greater than or equal to f in the interval I, see Fig. 4.3, left and center. □

Since by definition $f''_{\smile} \geq 0$, we have that $f'_{\smile}$ is non–decreasing, and hence we can define its inverse, denoted by $(f'_{\smile})^{-1}$, permitting jump discontinuities where $f'_{\smile}$ is constant, see Fig. 4.2.

In the previous sections we saw that it was important wether $u^l > u^r$ or viceversa to decide whether the solution was a shock or a rarefaction wave. Also in the general case these two cases correspond to two different constructions of the solution.

Theorem 4.1. *Let $f\,:\,\mathbb{R} \to \mathbb{R}$ be a $\mathbf{C}^2$ function with finitely many inflection points, $u^l, u^r \in \mathbb{R}$ with $u^l \neq u^r$, and I be the closed interval between u^l and u^r. Define $f_{\smile}$ as the lower convex envelope of $f|_I$, and $f_{\frown}$ as the upper concave envelope of $f|_I$. Then an entropy weak solution to the Riemann problem* (4.1.1) *is*

Fig. 4.2 Left: The dashed line is the function f and the solid line is $f_{\smile}$. Center: The dashed line is the function f' and the solid line is $f'_{\smile}$. Right: $(f'_{\smile})^{-1}$.

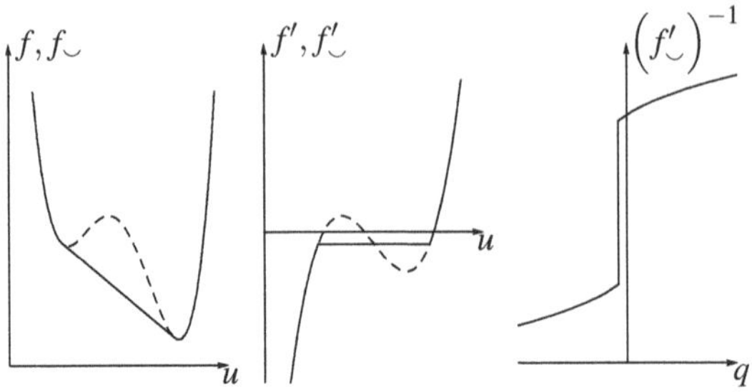

$$u(t,x) = \begin{cases} u^l & \textit{if } x < f'_{\smile}(u^l)\,t \\ (f'_{\smile})^{-1}(x/t) & \textit{if } f'_{\smile}(u^l)\,t \leq x < f'_{\smile}(u^r)\,t \\ u^r & \textit{if } x \geq f'_{\smile}(u^r)\,t \end{cases} \quad \textit{if } u^l < u^r \tag{4.6.2a}$$

$$u(t,x) = \begin{cases} u^l & \textit{if } x < f'_{\frown}(u^l)\,t \\ (f'_{\frown})^{-1}(x/t) & \textit{if } f'_{\frown}(u^l)\,t \leq x < f'_{\frown}(u^r)\,t \\ u^r & \textit{if } x \geq f'_{\frown}(u^r)\,t \end{cases} \quad \textit{if } u^l > u^r, \tag{4.6.2b}$$

where $\left(f'_*\right)^{-1}$ *denotes the inverse of the derivative of* f_*.

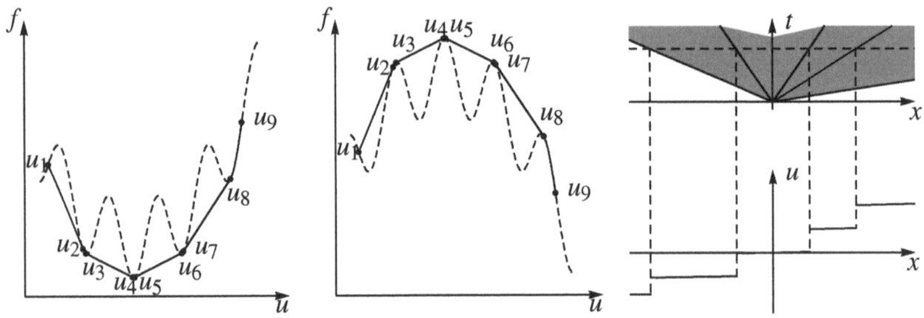

Fig. 4.3 Left: The dashed line is the function f and the solid line is the lower convex envelope $f_{\smile}$ of $f|_{[u_1,u_9]}$ defined by (4.6.1a). Center: The dashed line is the function f and the solid line is the upper concave envelope $f_{\frown}$ of $f|_{[u_1,u_9]}$ defined by (4.6.1b). Right: The solution corresponding to the case considered in the figure on the left. The solid lines in the upper picture represent the discontinuities of the solution. The solution takes the constant value u^l for $x < f'(u^l)\,t$, at $x = f'(u^l)\,t$ there is a shock, and for $x > f'(u^l)\,t$ it consists of four rarefactions separated by three shocks.

Proof. We consider only the case $u^l < u^r$, being the other case analogous. Let $I = [u^l, u^r]$. By hypothesis, there exist a finite number of intervals with endpoints $u^l = u_1 < u_2 < \ldots < u_{n-1} < u_n = u^r$, such that, if $f_{\smile} < f$ on $]u_{i-1}, u_i[\ \cup\ ]u_{i+1}, u_{i+2}[$, then $f_{\smile} = f$ on $[u_i, u_{i+1}]$, see Fig. 4.3, left. In this case, the function u defined by (4.6.2a) is piecewise $\mathbf{C^1}$ in the interval $x \in\ ]f'_{\smile}(u_{i-2})\,t, f'_{\smile}(u_{i+3})\,t[$, with discontinuities along the lines $x = f'_{\smile}(u_{i-1})\,t = f'_{\smile}(u_i)\,t$ and $x = f'_{\smile}(u_{i+1})\,t = f'_{\smile}(u_{i+2})\,t$. It results $u(t,x) = (f')^{-1}(x/t)$ for any $x/t \in \,]f'_{\smile}(u_{i-2}), f'_{\smile}(u_{i+3})[\setminus\{f'_{\smile}(u_i), f'_{\smile}(u_{i+1})\}$, see Fig. 4.3, right. By construction, $f'_{\smile}(u_{i-1}) = \big(f(u_{i-1}) - f(u_i)\big)/(u_{i-1} - u_i) = f'_{\smile}(u_i)$ and therefore the Rankine–Hugoniot conditions (3.4.4) are satisfied. By Theorem 3.1, the function (4.6.2a) is a weak solution to the Riemann problem (4.1.1), since it is made up of shocks, see Definition 4.1, and rarefactions, see Definition 4.3, which satisfy the conditions of Proposition 4.1 and Proposition 4.2. Finally, also the condition (3.5.4) is satisfied, and therefore, by Theorem 3.3, (4.6.2a) is an entropy weak solution of the Riemann problem (4.1.1). □

Observe that the solutions (4.6.2) are possible combinations of rarefactions, shocks and contact discontinuities.

Corollary 4.1. *Let $f \in \mathbf{PLC}(\mathbb{R};\mathbb{R})$ and $\mathscr{U}$ be the set of discontinuity points of f'. Then, for any $u^l, u^r \in \mathbb{R}$, the Riemann problem*

$$\partial_t u + \partial_x f(u) = 0 \quad , \qquad u(0,x) = \begin{cases} u^l & \text{if } x < 0 \\ u^r & \text{if } x \geq 0 \end{cases} \tag{4.6.3}$$

has a self similar piecewise constant weak solution with values in $\{u^l, u^r\} \cup \mathscr{U}$.

Proof. If $u^l < u^r$, let $v_1 < v_2 < \ldots < v_m$ denote the points of discontinuity of $f'_{\smile}$ in $]u^l, u^r[$. Then the weak solution to the Riemann problem (4.6.3) is

$$u(t,x) = \begin{cases} v_0 & \text{if } x < s_1\, t \\ v_i & \text{if } s_{i-1}\, t \leq x < s_i\, t\,, \quad i = 2, \ldots, m-1\,, \\ v_{m+1} & \text{if } s_{m-1}\, t \leq x\,, \end{cases} \tag{4.6.4a}$$

where $v_0 = u^l$, $v_{m+1} = u^r$ and the speeds are given by

$$s_i = \frac{f(v_{i+1}) - f(v_i)}{v_{i+1} - v_i}\,. \tag{4.6.4b}$$

If $u^l > u^r$ and $v_m < \ldots < v_1$ are the points of discontinuity of $f'_{\frown}$ in $]u^r, u^l[$, then it is sufficient to take $v_{m+1} = u_r$, $v_0 = u^l$ to have (4.6.4) as still a weak solution of the Riemann problem (4.6.3). □

4.7 Riemann Solver

As a customary in the contest of Riemann problems for conservation laws, we introduce the definition of Riemann solver.

Definition 4.5. The **Riemann solver** of the Riemann problem (4.1.1) is an operator $\mathscr{R}\,:\,\mathbb{R}^2 \to \mathbf{L}^\infty(\mathbb{R};\mathbb{R})$ such that for any pair $u^l, u^r \in \mathbb{R}$ the map $[(t,x) \mapsto \mathscr{R}[u^l, u^r](x/t)]$ is a self similar weak solution to (4.1.1). In particular, the Riemann solver selecting an entropy weak solution is called **entropic Riemann solver** and is denoted by $\mathscr{R}_e$.

In Theorem 5.2 we will prove that if $f \in \mathbf{C}^{0,1}(\mathbb{R};\mathbb{R})$, then $\mathscr{R}_e$ is well defined.

A minimal requirement for the continuity in $\mathbf{L}^1_{\mathbf{loc}}$ of a Riemann problem is given by its consistency [1, Sect. 9].

Definition 4.6. A Riemann solver $\mathscr{R}$ is **consistent**, see Fig. 4.4, if the following two conditions hold

$$\textbf{(C1)} \quad \left.\begin{array}{l} \mathscr{R}[u^l,u^m](\bar{x}) = u^m \\ \mathscr{R}[u^m,u^r](\bar{x}) = u^m \end{array}\right\} \Rightarrow \mathscr{R}[u^l,u^r] = \begin{cases} \mathscr{R}[u^l,u^m] & \text{if } x < \bar{x} \\ \mathscr{R}[u^m,u^r] & \text{if } x \geq \bar{x} \end{cases}$$

$$\textbf{(C2)} \quad \mathscr{R}[u^l,u^r](\bar{x}) = u^m \quad \Rightarrow \begin{cases} \mathscr{R}[u^l,u^m] = \begin{cases} \mathscr{R}[u^l,u^r] & \text{if } x < \bar{x} \\ u^m & \text{if } x \geq \bar{x} \end{cases} \\ \\ \mathscr{R}[u^m,u^r] = \begin{cases} u^m & \text{if } x < \bar{x} \\ \mathscr{R}[u^l,u^r] & \text{if } x \geq \bar{x}\,. \end{cases} \end{cases}$$

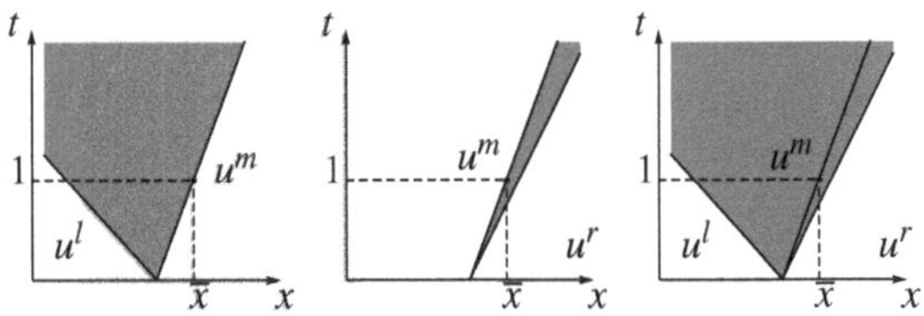

Fig. 4.4 Consistency of a Riemann solver

Essentially, **(C1)** states that whenever two solutions to two Riemann problems can be placed side by side, then their juxtaposition is again a solution to a Riemann problem, see Fig. 4.4. **(C2)** is the viceversa.

References

1. Bressan, A.: Hyperbolic systems of conservation laws. Oxford Lecture Series in Mathematics and its Applications, vol. 20. Oxford University Press, Oxford (2000)
2. Lefloch, P.G.: Hyperbolic systems of conservation laws. The theory of classical and nonclassical shock waves. Lectures in Mathematics ETH Zürich. Birkhäuser Verlag, Basel (2002)
3. Riemann, G.F.B.: Selbstanzeige: Ueber die Fortpflanzung ebener Luftwellen von endlicher Schwingungsweite. Göttinger Nachrichten 19, 192–197 (1859)
4. Riemann, G.F.B.: Ueber die Fortpflanzung ebener Luftwellen von endlicher Schwingungsweite. Abh. König. Gesell. Wiss. Göttingen 8, 43–65 (1860)

Chapter 5
The Cauchy Problem

Abstract. In this chapter we study the Cauchy problems for one–dimensional scalar conservation laws. In particular, we prove that the Cauchy problem is well posed in the class of entropy weak solutions, in the sense that it admits a unique entropy weak solution. The existence of the solutions is proved by the method of wave front tracking. The uniqueness is proved by showing the Kružkov result of the $\mathbf{L^1}$ contractiveness of the flow generated by a scalar conservation law.

5.1 Introduction

Let $f \in \mathbf{C^2}(\mathbb{R};\mathbb{R})$ and $\overline{u} \in \mathbf{L^\infty_{loc}}(\mathbb{R};\mathbb{R})$. The **Cauchy problem**, or **initial value problem (IVP)**, for a scalar conservation law is

$$\partial_t u + \partial_x f(u) = 0 \tag{5.1.1a}$$
$$u(t=0) = \overline{u}\,. \tag{5.1.1b}$$

Recalling the Definition 3.6 and Theorem 3.2, we define an **entropy weak solution** of (5.1.1) as a continuous map $u\,:\;\mathbb{R}_+^* \to \mathbf{L^\infty_{loc}}(\mathbb{R};\mathbb{R})$, which satisfies

$$\begin{aligned}\int_{\mathbb{R}_+}\int_{\mathbb{R}}\Big(|u-k|\;\partial_t\varphi + \operatorname{sgn}(u-k)\;\big(f(u)-f(k)\big)\;\partial_x\varphi\Big)\;\mathrm{d}x\;\mathrm{d}t&\\ +\int_{\mathbb{R}}\big|\overline{u}(x)-k\big|\;\varphi(0,x)\;\mathrm{d}x \geq 0&\end{aligned} \tag{5.1.2}$$

for every constant $k \in \mathbb{R}$ and every $\varphi \in \mathbf{C_c^\infty}(\mathbb{R}^2;\mathbb{R}_+^*)$.

The existence of solutions to (5.1.1) will be proved by the method of **wave front tracking**, first introduced by Dafermos [8], which we adopt in the form suited to the scalar case, see [3, Chap. 6] and [5, 9]. It has proved to be an effective and efficient tool in the analytical study of systems of conservation laws, see for instance [1, 2, 4, 6, 7, 9].

M.D. Rosini: *Macroscopic Models for Vehicular Flows & Crowd Dynamics*, UCS, pp. 51–62.
DOI: 10.1007/978-3-319-00155-5_5

As a first step we approximate the flux f by a piecewise affine continuous function f_n and the initial datum $\overline{u}$ by a piecewise constant function $\overline{u}_n$. Then we compute the exact solution u_n of the approximating Cauchy problem corresponding to the flux f_n and initial datum $\overline{u}_n$ by solving at each point of discontinuity and interaction the corresponding Riemann problem. As $n \to +\infty$, a compactness argument will yield a subsequence u_m converging in $\mathbf{L}^1_{\mathbf{loc}}$ to an exact entropy weak solution of the original Cauchy problem (5.1.1).

The uniqueness and continuous dependence of entropy weak solutions to (5.1.1) from the initial data will then be established by showing that the associated semigroup is contractive with respect to the $\mathbf{L}^1$ distance.

5.2 The Basic Case

In this section we analyze the Cauchy problems characterized by a piecewise linear continuous fluxes and piecewise constant initial data. Indeed, as pointed out in the introduction of this chapter, the approximating Cauchy problem is of this type.

As a first step, we generalize the results of Corollary 4.1.

Proposition 5.1. *Let $f \in \mathbf{PLC}(\mathbb{R};\mathbb{R})$ and $\overline{u} \in \mathbf{PC}(\mathbb{R};\mathbb{R})$. Assume that $\overline{u}$ takes only the values $u_1 < \ldots < u_n$ and that f' is continuous in $\mathbb{R} \setminus \mathscr{U}$ with $\mathscr{U} = \{u_1, \ldots, u_n\}$. Then, there exists an entropy weak solution u to the corresponding Cauchy problem* (5.1.1)*. Furthermore, u is a piecewise constant function with values in $\mathscr{U}$ and for all $t \in \mathbb{R}^*_+$ we have*

$$\mathrm{TV}\left(u(t)\right) \leq \mathrm{TV}\left(\overline{u}\right) \qquad , \qquad \left\|u(t)\right\|_{\mathbf{L}^\infty(\mathbb{R};\mathbb{R})} \leq \left\|\overline{u}\right\|_{\mathbf{L}^\infty(\mathbb{R};\mathbb{R})} .$$

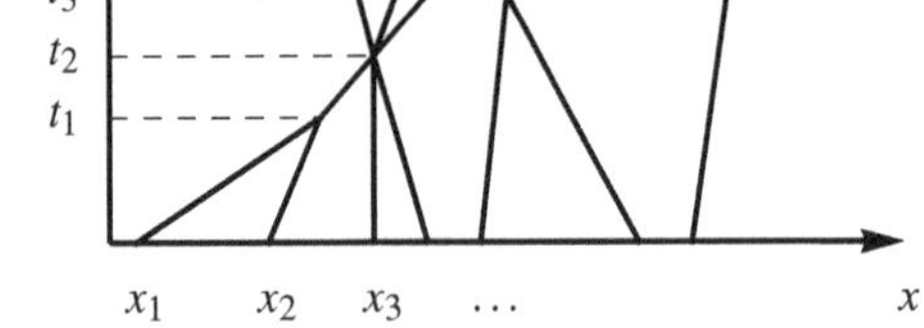

Fig. 5.1 Construction of an entropy weak solution to the Cauchy problem (5.1.1) with $f \in \mathbf{PLC}(\mathbb{R};\mathbb{R})$ and $\overline{u} \in \mathbf{PC}(\mathbb{R};\mathbb{R})$

Proof. Let $x_1 < \ldots < x_m$ be the points where $\overline{u}$ has a jump of discontinuity. At each x_i, consider the right and left limits $\overline{u}(x_i\pm) = \lim_{\varepsilon \searrow 0} \overline{u}(t, x_i \pm \varepsilon) \in \mathscr{U}$. By solving the corresponding Riemann problems as in Corollary 4.1, we obtain a local solution $u = u(t,x) \in \mathbf{PC}(\mathbb{R};\mathscr{U})$, having jumps along the discontinuity lines $\xi_i = \xi_i(t)$, $i = 1,\ldots,m$, traveling at constant speed and defined by

$$\dot{\xi}_i(t) = \frac{f(\overline{u}(x_i+)) - f(\overline{u}(x_i-))}{\overline{u}(x_i+) - \overline{u}(x_i-)} \quad , \qquad \xi_i(0) = x_i \; . \tag{5.2.1}$$

The solution can be prolonged up to a first time $t_1 \in \mathbb{R}_+$ when two or more lines of discontinuity from neighboring Riemann problems cross each other, see Fig. 5.1. In this case, let $\xi_j,\ldots,\xi_h$, with $1 \le j < h \le m$, be discontinuity lines that intersect each other at time t_1. Obviously, the other discontinuity lines ξ_k, $k \in \{1,\ldots,m\} \setminus \{j,\ldots,h\}$, continue after time t_1 with the same speed. By definition, $\xi_j(t) < \ldots < \xi_h(t)$ for $t \in [0,t_1[$ and $\xi_j(t_1) = \ldots = \xi_h(t_1)$. By Corollary 4.1, the values taken by the solution always remain within the set $\mathscr{U}$. Consequently, we can again solve the new Riemann problems generated by the interactions, according to the procedure used in the proof of Corollary 4.1.

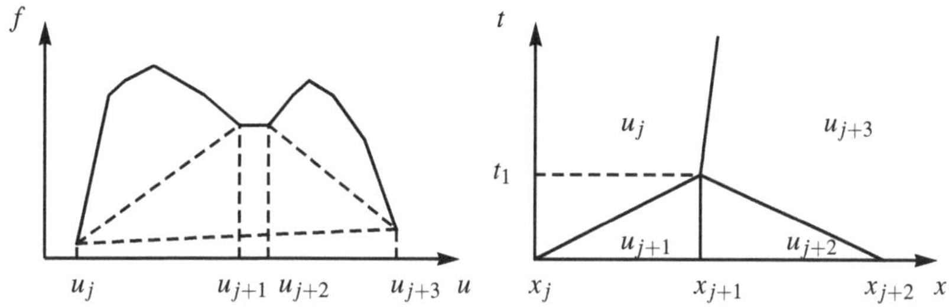

Fig. 5.2 An interaction with $h = j+2$, $u_i = \overline{u}(x_i-)$, $i = j,\ldots,h$, and $u_{j+3} = \overline{u}(x_h+)$

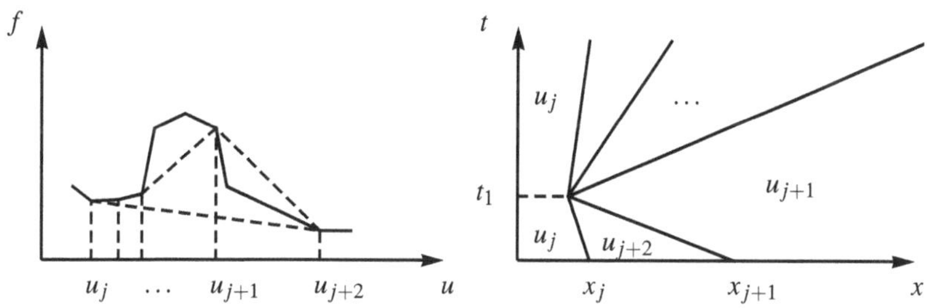

Fig. 5.3 An interaction with $h = j+1$, $u_j = \overline{u}(x_j-)$, $u_{j+1} = \overline{u}(x_{j+1}+)$ and $u_{j+2} = \overline{u}(x_{j+1}-)$

Two cases can occur.

CASE 1: Assume that all jumps $\{\overline{u}(x_i-) - \overline{u}(x_i+)\}$, $i = j,\ldots,h$, have the same sign. In this case, see Fig. 5.2, the Riemann problem at the interaction is solved by a single jump. Indeed, if for instance $\overline{u}(x_i-) < \overline{u}(x_i+)$, $i = j,\ldots,h$, by Theorem 3.3 we have for all $i = j,\ldots,h$ and for all $\alpha \in [0,1]$

$$f\left(\alpha\, \overline{u}(x_i+) + (1-\alpha)\, \overline{u}(x_i-)\right) \ge \alpha\, f\left(\overline{u}(x_i+)\right) + (1-\alpha)\, f\left(\overline{u}(x_i-)\right) \; .$$

Moreover, since the discontinuity lines meet at the same point, (5.2.1) implies that $\dot{\xi}_j > \ldots > \dot{\xi}_h$. From the above relations we deduce that for all $\alpha \in [0,1]$

$$f\left(\alpha\,\overline{u}(x_h+) + (1-\alpha)\,\overline{u}(x_j-)\right) \geq \alpha\, f\left(\overline{u}(x_h+)\right) + (1-\alpha)\, f\left(\overline{u}(x_j-)\right)\,.$$

Hence, the single shock between $\overline{u}(x_j-)$ and $\overline{u}(x_h+)$ and traveling with speed

$$\dot{\xi} = \frac{f\left(\overline{u}(x_h+)\right) - f\left(\overline{u}(x_j-)\right)}{\overline{u}(x_h+) - \overline{u}(x_j-)}$$

is an entropy weak solution. Furthermore, in this case, the total variation of the solution does not change after the interaction and the number of discontinuity lines decreases at least by one.

CASE 2: Assume that at least two jumps in $\left\{\overline{u}(x_i-) - \overline{u}(x_i+)\right\}$, $i = j,\ldots,h$, have opposite sign. In this case the total number of discontinuity lines may increase through the interaction, see Fig. 5.3. However, the total variation of the solution must decrease, owing to a cancelation effect.

Since $\mathrm{TV}\left(u(0)\right) = \mathrm{TV}\left(\overline{u}\right)$ is bounded and $\left[t \mapsto \mathrm{TV}\left(u(t)\right)\right]$ never increases, CASE 2 can occur only finitely many times, and hence also CASE 1. This implies that the total number of interactions is finite. The above algorithm thus defines a piecewise constant solution to the Cauchy problem (5.1.1), with jumps occurring along a finite number of straight segments in the (x,t)–plane. □

5.3 The General Case

In this section we show how to approximate a general Cauchy problem with one of the same type analyzed in Sect. 5.2.

5.3.1 *Approximation of the Initial Data*

Let $\overline{u} \in \mathbf{BV} \cap \mathbf{L}^1(\mathbb{R};\mathbb{R})$ and $n \in \mathbb{N}$. Then, for $r_n \gg 1$ large enough, it results

$$\int_{-\infty}^{-r_n} \left|\overline{u}(x)\right|\,\mathrm{d}x + \int_{r_n}^{+\infty} \left|\overline{u}(x)\right|\,\mathrm{d}x < \frac{1}{2\,n}\,.$$

By Lemma 2.5, there exists a function $\overline{v}_n \in \mathbf{PC}(\mathbb{R};\mathbb{R})$ such that

$$\mathrm{TV}(\overline{v}_n) \leq \mathrm{TV}(\overline{u})\,, \quad \|\overline{v}_n\|_{\mathbf{L}^\infty(\mathbb{R};\mathbb{R})} \leq \|\overline{u}\|_{\mathbf{L}^\infty(\mathbb{R};\mathbb{R})}\,, \quad \|\overline{v}_n - \overline{u}\|_{\mathbf{L}^\infty(\mathbb{R};\mathbb{R})} \leq \frac{1}{4\,n\,r_n}\,.$$

As approximation of the initial datum $\overline{u}$ we consider the piecewise constant function with compact support

$$\bar{u}_n(x) = \begin{cases} \bar{v}_n(x) & \text{if } |x| < r_n \\ 0 & \text{otherwise} . \end{cases} \tag{5.3.1}$$

We observe that $\bar{u}_n$ has the following properties:

$$\bar{u}_n(\mathbb{R}) = \left\{u_0^n, \ldots, u_N^n\right\} \text{ with } u_j^n < u_{j+1}^n \text{ and } N < 4\, n\, r_n\, \mathrm{TV}(\bar{u}) \tag{5.3.2a}$$

$$\|\bar{u}_n\|_{\mathbf{L}^\infty(\mathbb{R};\mathbb{R})} \le \|\bar{u}\|_{\mathbf{L}^\infty(\mathbb{R};\mathbb{R})} \tag{5.3.2b}$$

$$\mathrm{TV}(\bar{u}_n) \le \mathrm{TV}(\bar{u}) \tag{5.3.2c}$$

$$\|\bar{u} - \bar{u}_n\|_{\mathbf{L}^\infty(\mathbb{R};\mathbb{R})} \le \frac{1}{n} \tag{5.3.2d}$$

$$\begin{aligned} \|\bar{u} - \bar{u}_n\|_{\mathbf{L}^1(\mathbb{R};\mathbb{R})} &\le \frac{1}{2\,n} + \int_{-r_n}^{r_n} \left|\bar{u}(x) - \bar{v}_n(x)\right| \,\mathrm{d}x \\ &\le \frac{1}{2\,n} + 2\, r_n\, \|\bar{u} - \bar{v}_n\|_{\mathbf{L}^\infty(\mathbb{R};\mathbb{R})} \le \frac{1}{n} . \end{aligned} \tag{5.3.2e}$$

5.3.2 Approximation of the Flux

Let $f \in \mathbf{C}^2(\mathbb{R};\mathbb{R})$ and $n \in \mathbb{N}$. Consider $f_n \in \mathbf{PLC}(\mathbb{R};\mathbb{R})$, see Fig. 5.4, which coincides with f at all nodes u_j^n, $j = 0, \ldots, N$, given by (5.3.2a), and defined by

$$f_n(v) = \sum_{j=0}^{N-1} \left(\frac{f(u_{j+1}^n) - f(u_j^n)}{u_{j+1}^n - u_j^n} \left(v - u_j^n\right) + f(u_j^n) \right) \chi_{\left[u_j^n, u_{j+1}^n\right[}(v) . \tag{5.3.3}$$

Clearly, f_n' is continuous on $\mathbb{R} \setminus \left\{u_0^n, \ldots, u_N^n\right\}$.

Making this approximation is reasonable in many applications, since the precise form of the flux function is often the result of some measurements. These measurements are taken for a discrete set of u values, and a piecewise linear flux function is the result of a linear interpolation between these values. This method is frequently referred to as **Dafermos' method** [8].

5.4 Global Existence of BV Solutions

In this section we prove the global existence of entropy weak solutions within a class of functions with bounded variation.

Theorem 5.1 (Existence in BV). *Let $f \in \mathbf{C}^{0,1}(\mathbb{R};\mathbb{R})$ and $\bar{u} \in \mathbf{BV} \cap \mathbf{L}^1(\mathbb{R};\mathbb{R})$. Then the Cauchy problem* (5.1.1) *admits an entropy weak solution $u = u(t,x)$ defined for all $t \in \mathbb{R}_+^*$, such that*

$$\mathrm{TV}\left(u(t)\right) \le \mathrm{TV}(\bar{u}) , \quad \left\|u(t)\right\|_{\mathbf{L}^\infty(\mathbb{R};\mathbb{R})} \le \|\bar{u}\|_{\mathbf{L}^\infty(\mathbb{R};\mathbb{R})} \quad \textit{for all } t \in \mathbb{R}_+^* . \tag{5.4.1}$$

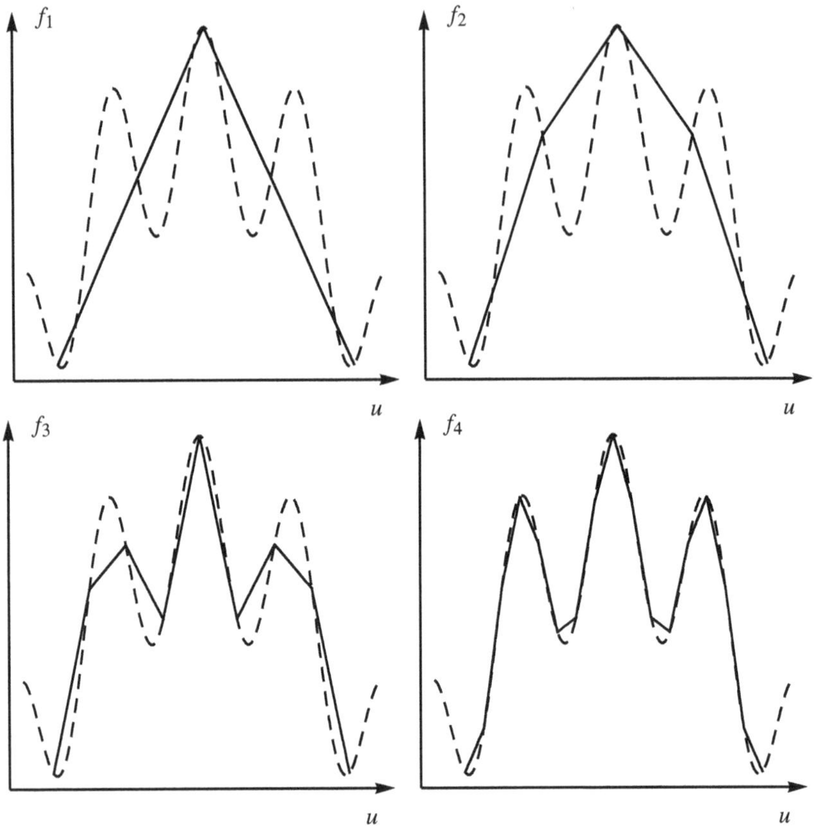

Fig. 5.4 The dashed lines represent a function f and the solid lines represent, respectively, its piecewise linear continuous approximations f_1, f_2, f_3, f_4 defined by (5.3.3).

Proof. Let $M = \|\overline{u}\|_{\mathbf{L}^\infty(\mathbb{R};\mathbb{R})}$. For any fixed $n \in \mathbb{N}$, let $\overline{u}_n$ be the piecewise constant approximation of $\overline{u}$ defined as in Sect. 5.3.1 and f_n be the piecewise linear continuous approximation of f defined as in Sect. 5.3.2. Then we are led to the Cauchy problem

$$\begin{cases} \partial_t u_n + \partial_x f_n(u_n) = 0 \\ u_n(t=0) = \overline{u}_n \ . \end{cases} \tag{5.4.2}$$

Let u_n be the piecewise constant entropy weak solution of (5.4.2) constructed as in Sect. 5.2. By Proposition 5.1, (5.3.2b) and (5.3.2c) it follows that for all $t \in \mathbb{R}_+$

$$\mathrm{TV}\left(u_n(t)\right) \leq \mathrm{TV}\left(\overline{u}\right) \qquad \text{and} \qquad \left\|u_n(t)\right\|_{\mathbf{L}^\infty(\mathbb{R};\mathbb{R})} \leq M \ . \tag{5.4.3}$$

By hypothesis, there exists $L \in \mathbb{R}_+$ such that

$$\left|f(v) - f(w)\right| \leq L\,|v-w| \qquad \text{for all } v,w \in [-M,M] \ .$$

By construction, L is a Lipschitz constant also for $f_n|_{[-M,M]}$. Therefore (5.2.1) implies that the propagation speed of all discontinuities of $u_n(t)$ is bounded by L. By using the bound (5.4.3) for the total variation, one obtains

$$\left\|u_n(s) - u_n(t)\right\|_{\mathbf{L}^1(\mathbb{R};\mathbb{R})} \leq L\,|s-t|\ \mathrm{TV}(\bar{u}) \qquad \text{for all } s,t \in \mathbb{R}_+^* \,. \tag{5.4.4}$$

We can thus apply Theorem 2.4 and deduce the existence of a subsequence u_m, which converges to a function u in $\mathbf{L}^1_{\mathbf{loc}}(\mathbb{R}_+^* \times \mathbb{R};\mathbb{R})$. Clearly (5.4.3) implies (5.4.1). Since $f_n \to f$ uniformly in $[-M,M]$ and recalling that each u_m is an entropy weak solution of (5.4.2), we obtain

$$\begin{aligned}
&\int_{\mathbb{R}_+}\int_{\mathbb{R}} \Big(|u-k|\ \partial_t\varphi + \mathrm{sgn}\,(u-k)\ \big(f(u)-f(k)\big)\ \partial_x\varphi\Big)\ \mathrm{d}x\ \mathrm{d}t \\
&= \lim_{m\to+\infty}\int_{\mathbb{R}_+}\int_{\mathbb{R}} \Big(|u_m-k|\ \partial_t\varphi + \mathrm{sgn}\,(u_m-k)\ \big(f_m(u_m)-f_m(k)\big)\ \partial_x\varphi\Big)\ \mathrm{d}x\ \mathrm{d}t \\
&\geq 0
\end{aligned}$$

for every $\varphi \in \mathbf{C}^\infty_\mathbf{c}(\mathbb{R}_+ \times \mathbb{R};\mathbb{R}_+^*)$. Finally, (5.4.4) and the property (5.3.2e) of the approximating sequence $\bar{u}_n$ imply that the initial condition (5.1.1b) is attained. This proves that u is an entropy weak solution of (5.1.1). □

Loosely speaking, the method of wave front tracking can be summarized in the following box:

(I) Give a Cauchy problem for a scalar one–dimensional conservation law

$$(*) \qquad \partial_t u + \partial_x f(u) = 0 \quad , \qquad u(t=0) = \bar{u}\,.$$

(II) Approximate the initial datum $\bar{u}$ by a piecewise constant function $\bar{u}_n$ and let $\mathscr{U}_n = \bar{u}(\mathbb{R})$.

(III) Approximate the flux f by a piecewise linear continuous function f_n with nodes in $\mathscr{U}_n$.

(IV) Compute the exact solution u_n to the (approximating) Cauchy problem

$$\partial_t u_n + \partial_x f_n(u_n) = 0 \quad , \qquad u_n(t=0) = \bar{u}_n\,.$$

(V) Let $n \to +\infty$ and obtain the solution of $(*)$ as the limit of u_n.

We have seen that the solution to a Riemann problem is always a monotone function. As a consequence, the solution computed by the wave front tracking method to a Cauchy problem for a scalar conservation law obeys a **maximum principle**; this means that the values attained by the solution belong to any connected interval containing the values attained by the initial datum.

5.5 Uniqueness

In this section we prove the classical theorem of Kružkov, providing an estimate of the $\mathbf{L}^1$ distance between any two bounded entropy weak solutions of the Cauchy problem (5.1.1). In particular, we will show the uniqueness of the solution to (5.1.1) within a class of $\mathbf{L}^\infty$ entropy weak solutions.

Theorem 5.2 (Kružkov [10]). *Let $f : \mathbb{R} \to \mathbb{R}$ be a locally Lipschitz continuous function. Let u, v be entropy weak solutions of the Cauchy problems*

$$\begin{array}{ll} \partial_t u + \partial_x f(u) = 0\,, & u(t=0) = \overline{u} \\ \partial_t v + \partial_x f(v) = 0\,, & v(t=0) = \overline{v}\,. \end{array} \tag{5.5.1}$$

Let $M, L \in \mathbb{R}_+$ be constants such that

$$\begin{array}{l} \left|u(t,x)\right| \leq M \quad , \quad \left|v(t,x)\right| \leq M \quad \textit{for all } (t,x) \in \mathbb{R}_+^* \times \mathbb{R} \\ \left|f(w) - f(z)\right| \leq L\,|w-z| \quad \textit{for all } w,z \in [-M,M]\,. \end{array} \tag{5.5.2}$$

Then, for every $R \in \mathbb{R}_+$ and $T > t_o \geq 0$, see Fig. 5.5, one has the estimate

$$\int_{|x|<R} \left|u(T,x) - v(T,x)\right|\, dx \leq \int_{|x|<R+(T-t_o)\,L} \left|u(t_o,x) - v(t_o,x)\right|\, dx\,. \tag{5.5.3}$$

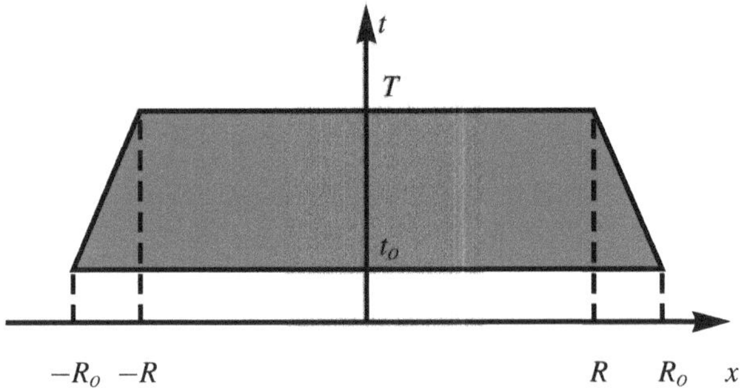

Fig. 5.5 The trapezoid T defined by (5.5.10) for $0 < t_o < T$, $R \in \mathbb{R}_+$ and $R_o = R + (T - t_o)\,L$

Proof. To prove the statement we use the **doubling of variables** technique. Introduce

$$\Xi(u,v) = \mathrm{sgn}(u-v)\ \big(f(u) - f(v)\big)\,. \tag{5.5.4}$$

By (5.1.2) the entropy weak solutions u and v satisfy the estimates

$$\int_{\mathbb{R}_+}\int_{\mathbb{R}} \Big[\left|u(t,x) - h\right|\, \partial_t \varphi(t,x,s,y) + \Xi\big(u(t,x),h\big)\ \partial_x \varphi(t,x,s,y)\Big]\, dx\ dt \geq 0 \tag{5.5.5}$$

$$\int_{\mathbb{R}_+}\int_{\mathbb{R}} \Big[\left|v(s,y) - k\right|\, \partial_s \varphi(t,x,s,y) + \Xi\big(v(s,y),k\big)\ \partial_y \varphi(t,x,s,y)\Big]\, dy\ ds \geq 0 \tag{5.5.6}$$

for any constants $h,k \in \mathbb{R}$ and $\varphi \in \mathbf{C_c^\infty}\left((\mathbb{R}_+ \times \mathbb{R})^2; \mathbb{R}_+^*\right)$. In (5.5.5), set $h = v(s,y)$ and then integrate with respect to (s,y) over $\mathbb{R}_+ \times \mathbb{R}$, while in (5.5.6) set $k = u(t,x)$ and then integrate with respect to (t,x) over $\mathbb{R}_+ \times \mathbb{R}$. The sum of the two obtained inequalities gives

$$\int\!\!\int\!\!\int\!\!\int \left[|u-v|\ (\partial_t\varphi + \partial_s\varphi) + \Xi(u,v)\ (\partial_x\varphi + \partial_y\varphi)\right] \mathrm{d}x\ \mathrm{d}t\ \mathrm{d}y\ \mathrm{d}s \geq 0\,. \quad (5.5.7)$$

Fig. 5.6 The solid line represents the mollifier δ defined by (5.5.8) and the dashed line represents δ_{ε_1} defined by (5.5.9a)

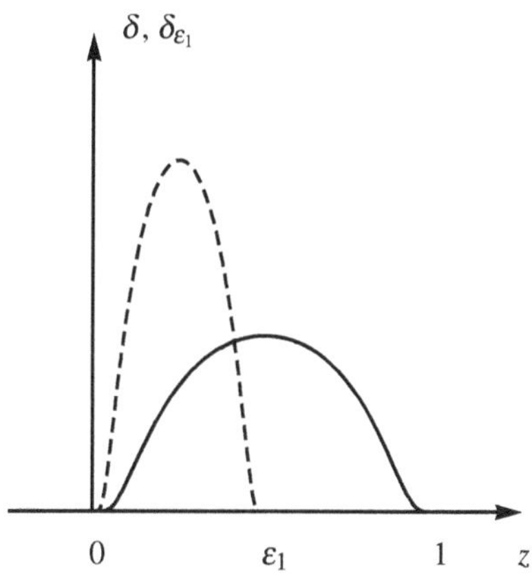

Introduce the function $\delta : \mathbb{R} \to \mathbb{R}$ defined by

$$\delta(z) = \begin{cases} \dfrac{\eta(z)}{\int_0^1 \eta(r)\ \mathrm{d}r} & \text{if } z \in [0,1] \\ 0 & \text{otherwise} \end{cases} \quad (5.5.8a)$$

$$\eta(z) = \exp\left(-\frac{1}{1-(2\,z-1)^2}\right)\,. \quad (5.5.8b)$$

Clearly δ is in $\mathbf{C_c^\infty}(\mathbb{R}; \mathbb{R}_+^*)$, $\operatorname{spt}\{\delta\} = [0,1]$ and $\int_{\mathbb{R}} \delta(z)\ \mathrm{d}z = 1$, see Fig. 5.6. For $\varepsilon_1, \varepsilon_2 \in \mathbb{R}_+$ sufficiently small, define

$$\delta_{\varepsilon_1}(z) = \frac{1}{\varepsilon_1}\,\delta\left(\frac{z}{\varepsilon_1}\right) \quad (5.5.9a)$$

$$\psi_{\varepsilon_2}(t,x) = \left(\int_{x-R-(T-t)\ L+\varepsilon_2}^{x+R+(T-t)\ L} \delta_{\varepsilon_2}(z)\ \mathrm{d}z\right)\left(\int_{t-T+\varepsilon_2}^{t-t_o} \delta_{\varepsilon_2}(z)\ \mathrm{d}z\right) \quad (5.5.9b)$$

$$\varphi(t,x,s,y) = \delta_{\varepsilon_1}(t-s)\ \delta_{\varepsilon_1}(x-y)\ \psi_{\varepsilon_2}(t,x)\,. \quad (5.5.9c)$$

We underline that for $\varepsilon \in \mathbb{R}_+$ sufficiently small $\delta_\varepsilon \in \mathbf{C}^\infty(\mathbb{R};[0,1/\varepsilon])$ approximates δ_0^D, the Dirac mass centered at the origin, $\psi_\varepsilon \in \mathbf{C}^\infty(\mathbb{R}^2;[0,1])$ approximates the characteristic function of the trapezoid

$$\mathsf{T} = \left\{(t,x) \in \mathbb{R}^2 \,:\, t_o < t < T,\ |x| < R + (T-t)\,L\right\}, \tag{5.5.10}$$

see Fig. 5.5, and that $\varphi \in \mathbf{C}^\infty_\mathbf{c}\left((\mathbb{R}_+ \times \mathbb{R})^2;\mathbb{R}_+^*\right)$ for $\varepsilon_1, \varepsilon_2 \in \mathbb{R}_+$ sufficiently small. Furthermore, a direct computation yields the equalities

$$(\partial_t\varphi + \partial_s\varphi)(t,x,s,y) = \delta_{\varepsilon_1}(t-s)\ \delta_{\varepsilon_1}(x-y)\ \partial_t\psi_{\varepsilon_2}(t,x) \tag{5.5.11a}$$
$$(\partial_x\varphi + \partial_y\varphi)(t,x,s,y) = \delta_{\varepsilon_1}(t-s)\ \delta_{\varepsilon_1}(x-y)\ \partial_x\psi_{\varepsilon_2}(t,x)\,. \tag{5.5.11b}$$

By choosing as a test function φ defined by (5.5.9), Eq. (5.5.7) writes

$$\int\!\!\int\!\!\int\!\!\int \left[|u-v|\ \partial_t\psi_{\varepsilon_2} + \Xi(u,v)\ \partial_x\psi_{\varepsilon_2}\right] \delta_{\varepsilon_1}(t-s)\ \delta_{\varepsilon_1}(x-y)\,\mathrm{d}x\ \mathrm{d}t\ \mathrm{d}y\ \mathrm{d}s \geq 0$$

and letting $\varepsilon_1 \searrow 0$, we obtain

$$\int_{\mathbb{R}_+}\int_{\mathbb{R}} \Big[\big|u(t,x)-v(t,x)\big|\ \partial_t\psi_{\varepsilon_2}(t,x) + \Xi\big(u(t,x),v(t,x)\big)\,\partial_x\psi_{\varepsilon_2}(t,x)\Big]\,\mathrm{d}x\ \mathrm{d}t \geq 0\,. \tag{5.5.12}$$

We can compute for $\partial_t\psi_{\varepsilon_2}(t,x)$ the following expression

$$\begin{aligned}
-L\Big(\delta_{\varepsilon_2}\big(x+R+(T-t)\,L\big) + \delta_{\varepsilon_2}\big(x-R-(T-t)\,L+\varepsilon_2\big)\Big)\left(\int_{t-T+\varepsilon_2}^{t-t_o}\delta_{\varepsilon_2}(z)\ \mathrm{d}z\right)&\\
+\left(\int_{x-R-(T-t)\,L+\varepsilon_2}^{x+R+(T-t)\,L}\delta_{\varepsilon_2}(z)\ \mathrm{d}z\right)\big(\delta_{\varepsilon_2}(t-t_o) - \delta_{\varepsilon_2}(t-T+\varepsilon_2)\big)&
\end{aligned}$$

and for $\partial_x\psi_{\varepsilon_2}(t,x)$ the following one

$$\big(\delta_{\varepsilon_2}(x+R+(T-t)\,L) - \delta_{\varepsilon_2}(x-R-(T-t)\,L+\varepsilon_2)\big)\left(\int_{t-T+\varepsilon_2}^{t-t_o}\delta_{\varepsilon_2}(z)\ \mathrm{d}z\right).$$

As a consequence, as $\varepsilon_2 \searrow 0$

$$\begin{aligned}
\partial_t\psi_{\varepsilon_2}(t,x) &\to -L\left(\delta^D_{-R-(T-t)\,L}(x) + \delta^D_{R+(T-t)\,L}(x)\right)\chi_{]t_o,T[}(t)\\
&\quad + \left(\delta^D_{t_o}(t) - \delta^D_T(t)\right)\chi_{]-R-(T-t)\,L,R+(T-t)\,L[}(x)\\
\partial_x\psi_{\varepsilon_2}(t,x) &\to \left(\delta^D_{-R-(T-t)\,L}(x) - \delta^D_{R+(T-t)\,L}(x)\right)\chi_{]t_o,T[}(t)
\end{aligned}$$

and therefore letting $\varepsilon_2 \searrow 0$ in (5.5.12) we obtain

$$\int_{-R-(T-t_o)\,L}^{R+(T-t_o)\,L} \left|u(t_o,x)-v(t_o,x)\right|\,\mathrm{d}x - \int_{-R}^{R} \left|u(T,x)-v(T,x)\right|\,\mathrm{d}x$$
$$+\mathscr{I}_-(R+(T-t)\,L)+\mathscr{I}_+(-R-(T-t)\,L)\geq 0\,,$$

where

$$\mathscr{I}_\pm(x)=\int_{t_o}^{T}\left(\pm\Xi\left(u(t,x),v(t,x)\right)-L\left|u(t,x)-v(t,x)\right|\right)\,\mathrm{d}t\,.$$

By (5.5.2) we have $\Xi(u,v)\leq L\,|u-v|$ and $\mathscr{I}_\pm(x)\leq 0$. Thus (5.5.3) is proved. □

The property (5.5.3) implies that the value at a point $(t,x)\in\mathbb{R}_+\times\mathbb{R}$ of an entropy weak solution u for the Cauchy problem (5.1.1) depends only on the restriction of $\overline{u}$ to the interval $[x-t\,L, x+t\,L]$. In other words, a sufficiently small perturbation of the initial datum with compact support disjoint from this interval does not modify the value $u(t,x)$. The **domain of dependence** of $(t,x)\in\mathbb{R}_+\times\mathbb{R}$ is the smallest set K such that, for every bounded function $\tilde{u}$ with compact support disjoint from K, the solution of the Cauchy problem with initial datum $\overline{u}+\varepsilon\,\tilde{u}$ coincides with $u(t,x)$ for $\varepsilon\in\mathbb{R}_+$ sufficiently small. The domain of dependence of (t,x) is thus included in $[x-t\,L, x+t\,L]$. These straightforward properties of entropy weak solutions are collected in the next proposition.

Proposition 5.2 ([10, 12]). *If u and v are the entropy weak solutions of* (5.5.1) *and*

$$L=\sup\left\{\left|a(w)\right|\,:\,w\in\left[\inf\{\overline{u},\overline{v}\},\sup\{\overline{u},\overline{v}\}\right]\right\}$$

is bounded, then:

(P1) For all $t\in\mathbb{R}_+$ and every $a,b\in\mathbb{R}$, with $a<b$, we have

$$\int_a^b \left|v(t,x)-u(t,x)\right|\,\mathrm{d}x\leq\int_{a-L\,t}^{b+L\,t}\left|\overline{v}(x)-\overline{u}(x)\right|\,\mathrm{d}x\,.$$

(P2) If u and v coincide on $\{x\in\mathbb{R}\,:\,|x-x_o|<d\}$, then u and v coincide on the whole triangle $\{(t,x)\in\mathbb{R}_+\times\mathbb{R}\,:\,|x-x_o|+L\,t<d\}$.
(P3) If $\overline{u}\geq\overline{v}$ a.e., then also $u(t)\geq v(t)$ a.e. for all $t\in\mathbb{R}_+$.
(P4) If $\overline{v}-\overline{u}\in\mathbf{L}^1(\mathbb{R};\mathbb{R})$, then for all $t\in\mathbb{R}_+$

$$\left\|v(t)-u(t)\right\|_{\mathbf{L}^1(\mathbb{R};\mathbb{R})}\leq\left\|\overline{v}-\overline{u}\right\|_{\mathbf{L}^1(\mathbb{R};\mathbb{R})}$$
$$\int_{\mathbb{R}}\left(v(t,x)-u(t,x)\right)\,\mathrm{d}x\leq\int_{\mathbb{R}}\left(\overline{v}(x)-\overline{u}(x)\right)\,\mathrm{d}x\,.$$

Corollary 5.1 (Uniqueness in $\mathbf{L}^\infty$). *Let $f\in\mathbf{C}^{0,1}(\mathbb{R};\mathbb{R})$ and u,v be bounded entropy weak solutions of* (5.5.1) *such that $(\overline{u}-\overline{v})\in\mathbf{L}^1(\mathbb{R};\mathbb{R})$. Then for every $t\in\mathbb{R}_+$ we have the estimate*

$$\int_{\mathbb{R}}\left|u(t,x)-v(t,x)\right|\,\mathrm{d}x\leq\int_{\mathbb{R}}\left|\overline{u}(x)-\overline{v}(x)\right|\,\mathrm{d}x\,. \tag{5.5.13}$$

In particular, for all initial data $\bar{u} \in \mathbf{L}^\infty(\mathbb{R};\mathbb{R})$, *the Cauchy problem* (5.1.1) *has at most one bounded entropy weak solution.*

Proof. By hypothesis, there exist constants $M, L \in \mathbb{R}_+$ for which (5.5.2) holds. For every $R, t \in \mathbb{R}_+$, by (5.5.3) one has

$$\int_{|x|\le R} \left| u(t,x) - v(t,x) \right| \, \mathrm{d}x \le \int_{|x|\le R+L\,t} \left| u(0,x) - v(0,x) \right| \, \mathrm{d}x \, .$$

Letting $R \to +\infty$ in the above estimate, we obtain (5.5.13), and hence the uniqueness of the solution. □

By Theorem 5.1, an entropy weak solution to the Cauchy problem (5.1.1) exists for all initial data in $\mathbf{BV} \cap \mathbf{L}^1(\mathbb{R};\mathbb{R})$. By Corollary 5.1, the $\mathbf{L}^1$ distance between such solutions does not increase in time. As a consequence, the solution operator can be extended by continuity to a much larger family of initial conditions. For instance, in [3, Theorem 6.3] is considered the case of an initial datum in $\mathbf{L}^1 \cap \mathbf{L}^\infty(\mathbb{R};\mathbb{R})$, while in [10, 11] in $\mathbf{L}^\infty(\mathbb{R};\mathbb{R})$.

References

1. Amadori, D.: Initial-boundary value problems for nonlinear systems of conservation laws. NoDEA Nonlinear Differential Equations Appl. 4(1), 1–42 (1997)
2. Amadori, D., Colombo, R.M.: Continuous dependence for 2×2 conservation laws with boundary. J. Differential Equations 138(2), 229–266 (1997)
3. Bressan, A.: Hyperbolic systems of conservation laws. Oxford Lecture Series in Mathematics and its Applications, vol. 20. Oxford University Press, Oxford (2000)
4. Chen, W., Wong, S.C., Shu, C.W., Zhang, P.: Front tracking algorithm for the Lighthill-Whitham-Richards traffic flow model with a piecewise quadratic, continuous, non-smooth and non-concave fundamental diagram. Int. J. Numer. Anal. Model. 6(4), 562–585 (2009)
5. Colombo, R.M.: Wave front tracking in systems of conservation laws. Appl. Math. 49(6), 501–537 (2004)
6. Colombo, R.M., Goatin, P., Rosini, M.D.: On the modelling and management of traffic. ESAIM: Mathematical Modelling and Numerical Analysis 45(05), 853–872 (2011)
7. Colombo, R.M., Rosini, M.D.: Pedestrian flows and non-classical shocks. Math. Methods Appl. Sci. 28(13), 1553–1567 (2005)
8. Dafermos, C.M.: Polygonal approximations of solutions of the initial value problem for a conservation law. J. Math. Anal. Appl. 38, 33–41 (1972)
9. Holden, H., Risebro, N.H.: Front tracking for hyperbolic conservation laws. Applied Mathematical Sciences, vol. 152. Springer, New York (2002)
10. Kružhkov, S.N.: First order quasilinear equations with several independent variables. Mat. Sb. (N.S.) 81(123), 228–255 (1970)
11. Morawetz, C.S.: Notes on time decay and scattering for some hyperbolic problems. In: Regional Conference Series in Applied Mathematics, vol. 19. SIAM, Providence (1975)
12. Temple, B.: No $\mathbf{L}^1$-contractive metrics for systems of conservation laws. Trans. Amer. Math. Soc. 288, 471–480 (1985)

Chapter 6
The Initial–Boundary Value Problem and the Constraint

Abstract. In this chapter we introduce the definitions of boundary data and of constraints. More precisely, we study the initial–boundary value problem in Sect. 6.2, the constrained Riemann problem in Sect. 6.3, the constrained Cauchy problem in Sect. 6.4 and finally the constrained initial–boundary value problem in the last section.

6.1 Introduction

Till now, we have considered the Cauchy problem, for which the physical domain is the whole space $\mathbb{R}$. However, to take into account real problems, such as the description of the vehicular traffic along a road with an entrance and an exit, we have rather to consider bounded domains of $\mathbb{R}$. Then it is not sufficient to add to the conservation law the initial condition, but also the boundary condition. The resulting problem is then called initial–boundary value problem. Aim of the first section is to understand under which boundary conditions the initial–boundary value problem is well–posed in a suitable sense [2]. Furthermore, if for instance the road under investigation has street lights and toll gates, then we are forced to introduce the concept of variable unilateral constraint [3]. This problem is solved by means of non–entropic shocks and also its well posedness is proved. The definition of the corresponding solution strictly rely on the properties of the flux, which is required to satisfy condition **(F)** given in Sect. 6.2. In particular, we study constrained Riemann problems in Sect. 6.3 and constrained Cauchy problems in Sect. 6.4. Finally, in the last section we apply the results achieved in the previous sections to constrained initial–boundary value problem [4].

M.D. Rosini: *Macroscopic Models for Vehicular Flows & Crowd Dynamics*, UCS, pp. 63–91.
DOI: 10.1007/978-3-319-00155-5_6

6.2 The Initial–Boundary Value Problem

We consider the problem of constructing a solution u (in a sense to be defined later, see Definition 6.1) to the one–dimensional **initial–boundary value problem (IBVP)** for a scalar conservation law in the "quarter plane" $\mathbb{R}^2_+$

$$\begin{aligned} &\partial_t u + \partial_x f(u) = 0 && (t,x) \in \mathbb{R}^2_+ && (6.2.1a)\\ &u(0,x) = u_o(x) && x \in \mathbb{R}_+ && (6.2.1b)\\ &f\left(u(t,0)\right) = q_b(t) && t \in \mathbb{R}_+ \,. && (6.2.1c)\end{aligned}$$

We assume for the flux that

(F) $f \in \mathbf{C^{0,1}}([0,R];[0,\overline{f}])$ satisfies $f(0) = f(R) = 0$ and there exists a $\overline{u} \in \,]0,R[$ such that $f'(u)\;(\overline{u} - u) > 0$ for almost all $u \in [0,R]$.

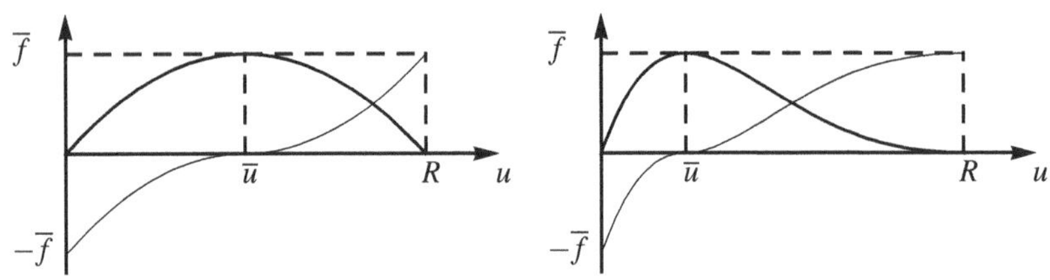

Fig. 6.1 In both diagrams, the thick solid line represents a flux f satisfying condition **(F)** and the thin solid line represents the corresponding function Ψ given by (6.2.7).

Above, see Fig. 6.1, $R \in \mathbb{R}_+$ denotes the maximal possible value taken by the unknown function $u\,:\; \mathbb{R}^2_+ \to [0,R]$ and $\overline{f} \in \mathbb{R}_+$ is the maximal flow. By **(F)**, $\overline{u} \in \,]0,R[$ stands for the unique value at which the flow $\overline{f}$ is attained, namely $\overline{f} = f\left(\overline{u}\right) = \max_{u \in [0,R]} f(u)$ and $f(u) < \overline{f}$ for all $u \in [0,R] \setminus \{\overline{u}\}$. As a consequence, the restriction f_* of f to $[0,\overline{u}]$, i.e. $f_* = f_{|[0,\overline{u}]}$, is invertible and its inverse f_*^{-1} satisfies

$$f_*^{-1}\,:\;[0,\overline{f}] \to [0,\overline{u}] \qquad \text{where} \qquad f_*^{-1}(q) = u \iff u \in [0,\overline{u}] \text{ and } f(u) = q\,.$$

The simplest case corresponds to have u_o and q_b given fixed constants. Then, the weak solution to the initial–boundary value problem (6.2.1) is given by the restriction to $\mathbb{R}^2_+$ of the weak solution to the Riemann problem

$$\begin{aligned} &\partial_t u + \partial_x f(u) = 0\\ &u(0,x) = \begin{cases} f_*^{-1}(q_b) & \text{if } x < 0\\ u_o & \text{if } x \geq 0\,. \end{cases} \end{aligned} \qquad (6.2.2)$$

As soon as the initial condition (6.2.1b) is given, the boundary condition (6.2.1c) can not be prescribed arbitrarily, as we shall see in the next example by considering the characteristics.

Example 6.1. Consider the initial–boundary value problem (6.2.1) associated to the flux

$$f(u) = \begin{cases} u & \text{if } u \in [0,1[\\ 2-u & \text{if } u \in [1,2] \,. \end{cases} \tag{6.2.3}$$

Let the data $u_o \in [0,2] \setminus \{1\}$ and $q_b \in [0,1[$ be two given fixed constants. Then, the weak solution to (6.2.1) is the restriction to $\mathbb{R}^2_+$ of the weak solution to the Riemann problem (6.2.2). By Proposition 3.1, the speed of the characteristics starting from

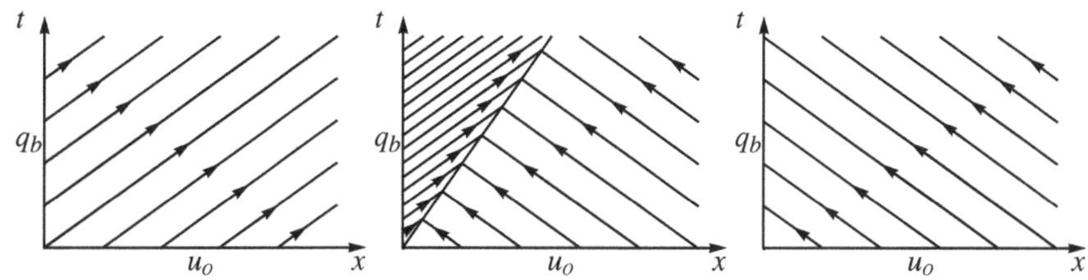

Fig. 6.2 The characteristic lines for the initial–boundary value problem (6.2.1), (6.2.3) in the case $u_o \in [0,1[$, left, in the case $u_o \in]1,2]$ and $q_b < f(u_o)$, center, and in the case $u_o \in]1,2]$ and $q_b \geq f(u_o)$, right.

the boundary $x = 0$ have speed equal to $\lambda = 1$, while the speed of the characteristics starting from $t = 0$ is equal to $\lambda = 1$ if $u_o \in [0,1[$ and equal to $\lambda = -1$ if $u_o \in]1,2]$.

As a consequence, if $u_o \in [0,1[$, see Fig. 6.2, left, or $u_o \in]1,2]$ and $q_b < f(u_o)$, see Fig. 6.2, center, then by Theorem 3.1 the weak solution to (6.2.1), (6.2.3) is

$$u(t,x) = \begin{cases} u_b & \text{if } x < \sigma\, t \\ u_o & \text{if } x \geq \sigma\, t \end{cases}, \qquad (t,x) \in \mathbb{R}^2_+ \,,$$

where $u_b = f_*^{-1}(q_b) \in [0,1[$ and $\sigma = \big(f(u_o) - q_b\big) / (u_o - u_b)$. Thus, one needs to prescribe the boundary condition (6.2.1c) on the boundary $x = 0$ to determine the solution.

On the other hand, if $u_o \in]1,2]$ and $q_b \geq f(u_o)$, then the characteristics starting from $t = 0$ impinge on the boundary $x = 0$, see Fig. 6.2, right. In this case, the solution is

$$u(t,x) = u_o \,, \qquad (t,x) \in \mathbb{R}^2_+ \,.$$

Therefore the information is carried from the initial datum (6.2.1b) and the solution does not depend on the boundary datum (6.2.1c). □

The above example points out that, in general, not only there may be a jump between $u_b(t)$ and $u(t,0)$ for a $t \in \mathbb{R}_+$, but this jump may well violate the Rankine–Hugoniot condition (3.4.4).

As pointed out by Bardos, LeRoux and Nedelec [2, Eq. (16.a)], the correct way to write the boundary condition (6.2.1c) is

$$\max_{k\in\mathscr{I}(u(x=0),u_b)} \left\{ \operatorname{sgn}\left(u\left(x=0\right)-u_b\right) \left(f\left(u\left(x=0\right)\right)-f(k)\right) \right\} = 0\,, \qquad (6.2.4)$$

where $u_b = f_*^{-1}(q_b)$ and $\mathscr{I}(u(x=0),u_b)$ is the closed interval between $u(x=0)$ and u_b. The interpretation of (6.2.4) can be easily explained by considering the propagation of the boundary datum, whose speed is given by the characteristic lines defined in Definition 3.3. Indeed, the maximum in (6.2.4) is ensured for $k = u(x=0)$ when the data are leaving the domain $\mathbb{R}^2_+$, and then the boundary condition is not necessarily satisfied. For entering data, we obtain $u(x=0) = u_b$, which obviously verifies (6.2.4).

Example 6.2. In the case considered in Example 6.1, the condition (6.2.4) writes

$$\max_{k\in\mathscr{I}(u(x=0),u_b)} \left\{ \left|u(x=0)-k\right| \right\} = \left|u(x=0)-u_b\right| = 0 \qquad \text{if } u(x=0)<1\,,$$

$$\begin{aligned} &\max\left\{ \max_{k\in[u_b,1]} \left\{f\left(u(x=0)\right)-k\right\}, \max_{k\in[1,u(x=0)]} \left\{k-u(x=0)\right\} \right\} \\ &= \max\left\{f\left(u(x=0)\right)-q_b, 0\right\} = 0 \qquad \text{if } u(x=0)>1\,. \end{aligned}$$

Therefore, condition (6.2.4) is ensured independently from the prescribed boundary condition u_b iff $u(x=0)>1$ and $f\left(u(x=0)\right) < q_b$, while, in all the other cases, this maximum is zero iff $f\left(u(x=0)\right)$ is equal to q_b. □

Definition 6.1. Let **(F)** hold, $u_o \in \mathbf{L}^\infty_{\mathbf{loc}}(\mathbb{R}_+;[0,R])$ and $q_b \in \mathbf{BV}(\mathbb{R}_+;[0,\overline{f}])$. A function $u \in \mathbf{C^0}\left(\mathbb{R}_+;\mathbf{L}^\infty_{\mathbf{loc}}(\mathbb{R}_+;[0,R])\right)$ is an **entropy weak solution** to the initial–boundary value problem (6.2.1) if for all $k \in [0,R]$ and all test function $\varphi \in \mathbf{C}^\infty_\mathbf{c}(\mathbb{R}^2;\mathbb{R}^*_+)$

$$\begin{aligned} 0 \le &\int_{\mathbb{R}_+}\int_{\mathbb{R}_+} \left(|u-k|\;\partial_t\varphi + \operatorname{sgn}(u-k)\;\left(f(u)-f(k)\right)\;\partial_x\varphi\right)\,\mathrm{d}x\,\mathrm{d}t \\ &+\int_{\mathbb{R}_+} \left|u_o(x)-k\right|\;\varphi(0,x)\,\mathrm{d}x \qquad (6.2.5)\\ &+\int_{\mathbb{R}_+} \operatorname{sgn}\left(f_*^{-1}\left(q_b(t)\right)-k\right)\;\left(f\left(u(t,0)\right)-f(k)\right)\;\varphi(t,0)\,\mathrm{d}t\,. \end{aligned}$$

Above and in the sequel, for any function v in $\mathbf{C^0}\left(\mathbb{R}_+;\mathbf{L}^\infty_{\mathbf{loc}}(\mathbb{R}_+;[0,R])\right)$, we denote by $v(t,0)$ the **measure theoretic trace** implicitly defined by

$$\lim_{\varepsilon\searrow 0}\frac{1}{\varepsilon}\int_{\mathbb{R}_+}\int_0^\varepsilon \left|v(t,x)-v(t,0)\right|\;\varphi(t,x)\,\mathrm{d}x\,\mathrm{d}t = 0$$

for all $\varphi \in \mathbf{C}^\infty_\mathbf{c}(\mathbb{R}^2;\mathbb{R})$. Remark that, by [1, Theorem 2.2], $v(t,0)$ exists and is finite.

Remark 6.1. Usually, in the more analytical literature, are considered boundary conditions of the form $u(t,0) = u_b(t)$, $t\in\mathbb{R}_+$, namely the boundary conditions amount to assign the value of the unknown along the boundary. On the contrary, in (6.2.1) we assign the *inflow*, which is better suited to traffic problems. This accounts for the appearance of f_*^{-1} in (6.2.5). It is immediate to prove that the boundary condition (6.2.1c) is equivalent to require that $u(t,0) = f_*^{-1}\left(q_b(t)\right)$, $t \in \mathbb{R}_+$. □

Remark 6.2. The first two lines in (6.2.5) originate from the classical Kružkov definition of entropy weak solution to the Cauchy problem, i.e. with no boundary, see Definition 3.6 and Theorem 3.2. The latter line is motivated by the boundary and is adapted from [2, Eq. (17)]. □

In view of considering problems with constraints, see Remark 6.4, we need to introduce the domain

$$\mathscr{D}^+ = \left\{ u \in \mathbf{L^1}\left(\mathbb{R}_+;[0,R]\right) \;:\; \Psi(u) \in \mathbf{BV}\left(\mathbb{R}_+;\left[-\overline{f},\overline{f}\right]\right)\right\}, \tag{6.2.6}$$

where $\Psi : [0,R] \to [-\overline{f},\overline{f}]$ is defined

$$\Psi(u) = \operatorname{sgn}(u-\overline{u})\left(\overline{f} - f(u)\right), \tag{6.2.7}$$

see Fig. 6.1. In fact, Ψ is one–to–one, but possibly singular at $u = \overline{u}$. Furthermore, if $u \in \mathbf{BV}(\mathbb{R};\mathbb{R})$, then $\mathrm{TV}(\Psi \circ u) \leq \left\| f' \right\|_{\mathbf{L}^\infty([0,R];\mathbb{R})} \mathrm{TV}(u)$, while $\mathrm{TV}(u)$ may well be infinite with $\Psi(u)$ finite, as in the case of $f(u) = u\,(u-1)$ and $u = \sum_{n=3}^{+\infty} \frac{1}{n}\, \chi_{\left[\frac{1}{2}+\frac{1}{n},\frac{1}{2}+\frac{2\,n+1}{2\,n\,(n+1)}\right[}$.

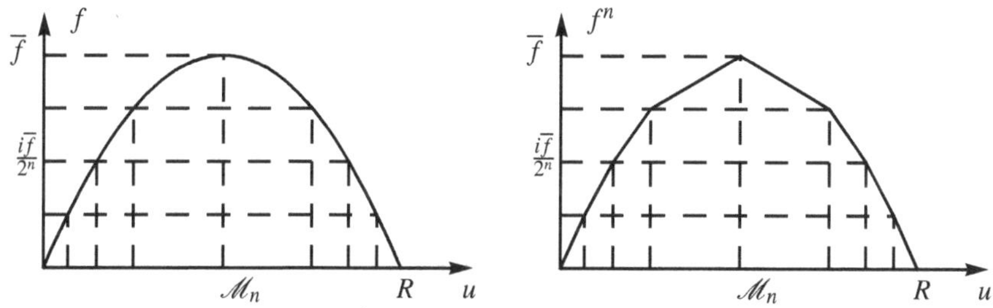

Fig. 6.3 Approximation of a strictly concave flux f satisfying **(F)** by a piecewise linear continuous function f^n

Theorem 6.1. *Let **(F)** hold and assume that*

(O) $u_o \in \mathscr{D}^+$;
(B) $q_b \in \mathbf{BV}(\mathbb{R}_+;[0,\overline{f}])$.

Then, there exists a unique entropy weak solution, u, to the initial–boundary value problem (6.2.1) *in the sense of Definition 6.1 and* $u(t) \in \mathscr{D}^+$ *for all* $t \in \mathbb{R}_+$. *Furthermore, if* u' *is the solution to* (6.2.1) *associated with the pair* $(u_o', q_b') \in \mathscr{D}^+ \times \mathbf{BV}(\mathbb{R}_+;[0,\overline{f}])$, *then the following Lipschitz estimate holds for all* $t \in \mathbb{R}_+$

$$\left\|u(t)-u'(t)\right\|_{\mathbf{L^1}(\mathbb{R}_+,\mathbb{R})} \leq \left\|u_o-u'_o\right\|_{\mathbf{L^1}(\mathbb{R}_+,\mathbb{R})} + \left\|q_b-q'_b\right\|_{\mathbf{L^1}([0,t],\mathbb{R})} . \tag{6.2.8}$$

Proof. For any fixed $n \in \mathbb{N}$, introduce in $[0,R]$ the meshes $\mathcal{M}_n = f^{-1}(2^{-n}\mathbb{N}^*\overline{f})$. Let $f^n \in \mathbf{PLC}([0,R];[0,\overline{f}])$ coincide with f on $\mathcal{M}_n$ and be differentiable in $]0,R[\setminus \mathcal{M}_n$, see Fig. 6.3. In order to simplify the study, we assume also that f is strictly concave, namely $f''(u)<0$. Let $q_b^n \in \mathbf{PC}\left(\mathbb{R}_+;f(\mathcal{M}_n)\right)$ coincide with q_b on $f(\mathcal{M}_n)$, in the sense that $q_b(t) = q_b^n(t)$ whenever $q_b(t) \in f(\mathcal{M}_n)$. In the same way, let u_o^n be in $\mathbf{PC}(\mathbb{R}_+;\mathcal{M}_n)$ and such that $u_o^n(x) = u_o(x)$ whenever $u_o(x) \in \mathcal{M}_n$. Clearly, f^n satisfies **(F)**, $q_b^n \in \mathbf{BV}(\mathbb{R}_+;[0,\overline{f}])$ and $u_o^n \in \mathcal{D}^+$ because so do, respectively, f, q_b and u_o. The functions u_o^n and q_b^n can be represented as

$$u_o^n = \sum_{\alpha\in\mathbb{N}} u_{o\alpha}^n \, \chi_{]x_{o\alpha-1}^n,x_{o\alpha}^n]} \qquad \text{with} \qquad u_{o\alpha}^n \in \mathcal{M}_n \tag{6.2.9a}$$

$$q_b^n = \sum_{\beta\in\mathbb{N}} q_{b\beta}^n \, \chi_{]t_{b\beta-1}^n,t_{b\beta}^n]} \qquad \text{with} \qquad q_{b\beta}^n \in f(\mathcal{M}_n) \tag{6.2.9b}$$

and we agree that $x_{o0}^n = t_{b0}^n = 0$. All the approximations above are meant in the strong $\mathbf{L^1}$–topology, that is

$$\lim_{n\to+\infty} \left(\left\|u_o^n - u_o\right\|_{\mathbf{L^1}(\mathbb{R}_+;\mathbb{R})} + \left\|q_b^n - q_b\right\|_{\mathbf{L^1}(\mathbb{R}_+;\mathbb{R})}\right) = 0 .$$

We now follow the classical **wave front tracking** technique. In particular, an approximate solution u^n to (6.2.1) is computed as the exact entropy weak solution to the initial–boundary value problem

$$\begin{cases} \partial_t u^n + \partial_x f^n(u^n) = 0 & (t,x) \in \mathbb{R}_+^2 \\ u^n(0,x) = u_o^n(x) & x \in \mathbb{R}_+ \\ f^n\left(u^n(t,0)\right) = q_b^n(t) & t \in \mathbb{R}_+ . \end{cases} \tag{6.2.10}$$

In other words, u^n is obtained gluing the solutions to the Riemann problems at the points $(0,x_{o\alpha}^n)$ of jump of u_o^n, at the points $(t_{b\beta}^n,0)$ where the boundary datum changes and at all interaction points where two or more waves intersect, or one or more waves hit the boundary. For the construction of the solution to standard Riemann problems with a piecewise linear and continuous flow we refer to Sect. 5.2. For the solution to the Riemann problem at the boundary, we refer to (6.2.2).

Iteratively solving Riemann problems at the boundary and at the interactions, we define u^n, which is an entropy weak solution to (6.2.10) and an approximate solution to (6.2.1). We prove that $u^n(t) \in \mathcal{D}_n^+$, where

$$\mathcal{D}_n^+ = \left\{u \in \mathbf{PC}(\mathbb{R}_+;\mathcal{M}_n) \, : \, \Psi(u) \in \mathbf{BV}(\mathbb{R}_+;\mathbb{R})\right\} . \tag{6.2.11}$$

To this aim, introduce the set

$$\overline{\mathcal{D}_n^+} = \mathcal{D}_n^+ \times \mathbf{PC}\left(\mathbb{R}_+;f\left(\mathcal{M}_n\right)\right) \tag{6.2.12}$$

and a semigroup $\overline{Sb}^n : \mathbb{R}_+ \times \overline{\mathscr{D}}_n^+ \to \overline{\mathscr{D}}_n^+$ by setting $\overline{Sb}_t^n(u_o^n, q_b^n) = \left(u^n(t), \mathscr{T}_t q_b^n\right)$, where $\mathscr{T}$ is the translator operator defined by (2.1.5).

For any $(u^n, q_b^n) \in \overline{\mathscr{D}}_n^+$, written as in (6.2.9), define the Glimm type functional

$$\Upsilon_b\left(u^n, q_b^n\right) = \sum_{\alpha\in\mathbb{N}} \left|\Psi(u_{\alpha+1}^n) - \Psi(u_\alpha^n)\right| + 2\sum_{\beta\in\mathbb{N}} \left|q_{b\beta+1}^n - q_{b\beta}^n\right| + \gamma_b\left(u^n, q_b^n\right) \quad (6.2.13)$$

where Ψ is as in (6.2.7) and γ_b is defined

$$\gamma_b\left(u^n, q_b^n\right) = \begin{cases} 0 & \text{if } u^n(0+) = f_*^{-1}\left(q_b^n(0)\right) \\ 2\left(\overline{f} - q_b^n(0)\right) & \text{otherwise}. \end{cases} \quad (6.2.14)$$

Long but elementary computations show that at any interaction, the functional Υ_b either decreases by at least $2^{-n}\overline{f}$, or remains constant while the total number of waves in the approximate solution does not increase (this may happen in interactions away from the boundary). In particular, recalling the study of the interactions away from the boundary done in Sect. 5.2, we immediately see that if two waves interact in a point $(\bar{t}, \bar{x})$ of the interior of the domain, a single shock is created and

$$\begin{aligned} \Delta\Upsilon_b(\bar{t}) &= \Upsilon_b\left(u^n(\bar{t}+), q_b^n(\bar{t}+)\right) - \Upsilon_b\left(u^n(\bar{t}-), q_b^n(\bar{t}-)\right) \\ &= \left|\Psi(u_{\alpha-1}^n) - \Psi(u_{\alpha+1}^n)\right| - \left|\Psi(u_{\alpha-1}^n) - \Psi(u_\alpha^n)\right| - \left|\Psi(u_\alpha^n) - \Psi(u_{\alpha+1}^n)\right| \le 0. \end{aligned}$$

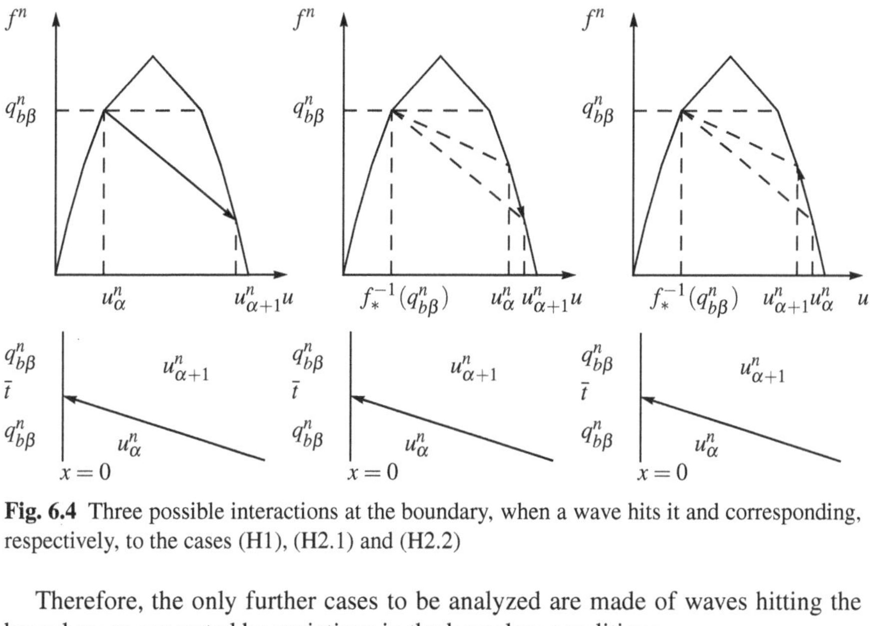

Fig. 6.4 Three possible interactions at the boundary, when a wave hits it and corresponding, respectively, to the cases (H1), (H2.1) and (H2.2)

Therefore, the only further cases to be analyzed are made of waves hitting the boundary or generated by variations in the boundary conditions.

(H) Assume that a wave between u^n_α and $u^n_{\alpha+1}$ hits the boundary at time $\bar{t} \in]t^n_{b\beta-1}, t^n_{b\beta}]$.

(H1) If $u^n_\alpha = f_*^{-1}(q^n_{b\beta})$, then must be $u^n_{\alpha+1} > \bar{u}$ and $f(u^n_{\alpha+1}) < q^n_{b\beta}$. In this case the wave hits the boundary and disappears, see Fig. 6.4, left. Therefore

$$\Delta \Upsilon_b(\bar{t}) = 2\left(\bar{f} - q^n_{b\beta}\right) - \left|\Psi(u^n_{\alpha+1}) - \Psi(u^n_\alpha)\right| = f(u^n_{\alpha+1}) - q^n_{b\beta} \leq -2^{-n}.$$

(H2) If $u^n_\alpha \neq f_*^{-1}(q^n_{b\beta})$, then must be $u^n_\alpha > \bar{u}$ and $f(u^n_\alpha) \leq q^n_{b\beta}$.

(H2.1) If $u^n_{\alpha+1} > u^n_\alpha$, then the wave disappears after hitting the boundary, see Fig. 6.4, center. In this case

$$\Delta \Upsilon_b(\bar{t}) = -\left|\Psi(u^n_{\alpha+1}) - \Psi(u^n_\alpha)\right| = f(u^n_{\alpha+1}) - f(u^n_\alpha) \leq -2^{-n}.$$

(H2.2) If $u^n_{\alpha+1} < u^n_\alpha$ and $f(u^n_{\alpha+1}) \leq q^n_{b\beta}$, then the hitting wave disappears, see Fig. 6.4, right, and

$$\Delta \Upsilon_b(\bar{t}) = -\left|\Psi(u^n_{\alpha+1}) - \Psi(u^n_\alpha)\right| = f(u^n_\alpha) - f(u^n_{\alpha+1}) \leq -2^{-n}.$$

(H2.3) If $u^n_{\alpha+1} < u^n_\alpha$ and $f(u^n_{\alpha+1}) > q^n_{b\beta}$, then must be $f(u^n_\alpha) = q^n_{b\beta}$, from the boundary starts a wave, see Fig. 6.5, left, and

$$\begin{aligned}\Delta \Upsilon_b(\bar{t}) = &\left|\Psi\left(f_*^{-1}(q^n_{b\beta})\right) - \Psi(u^n_{\alpha+1})\right| - \left|\Psi(u^n_{\alpha+1}) - \Psi(u^n_\alpha)\right| \\ &-2\left(\bar{f} - q^n_{b\beta}\right) = 2\left(q^n_{b\beta} - f(u^n_{\alpha+1})\right) \leq -2^{-n}.\end{aligned}$$

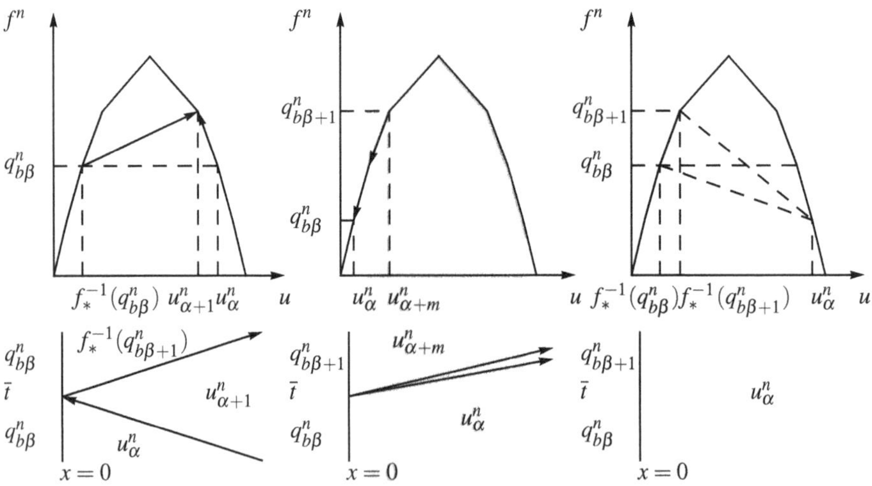

Fig. 6.5 Left: The interaction considered in (H2.3). Center and right: Two possible interactions at the boundary, when the boundary datum increases and corresponding, respectively, to the cases (B1.1) and (B1.2).

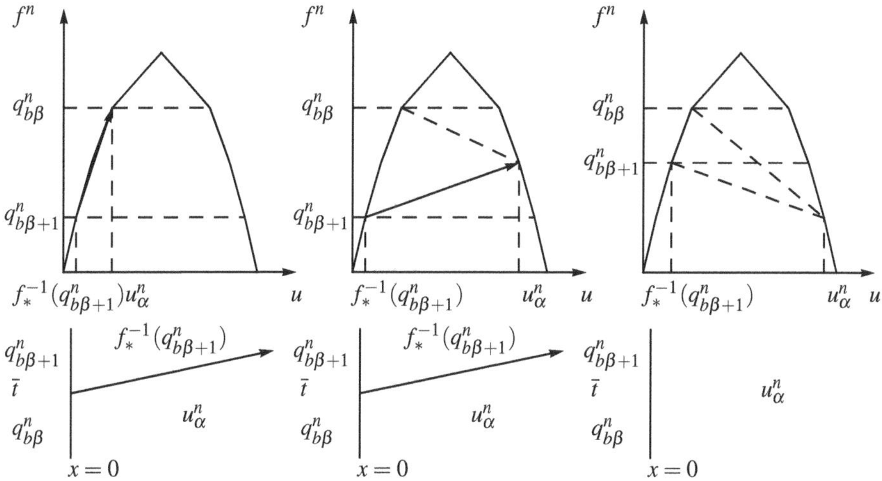

Fig. 6.6 Three possible interactions at the boundary, when the boundary datum decreases and corresponding, respectively, to the cases (B2.1), (B2.2) and (B2.3)

(B) At time $\bar{t} = t^n_{b\beta}$ the boundary condition changes. Let $u^n_\alpha = u(t^n_{b\beta}-, 0+)$.

(B1) Assume that $q^n_{b\beta} < q^n_{b\beta+1}$.

(B1.1) If $u^n_\alpha = f_*^{-1}(q^n_{b\beta})$, then at time $\bar{t}$ start from the boundary $(m-1)$ waves between $u^n_{\alpha+m-i}$ and $u^n_{\alpha+m-1-i}$, $i = 0, \ldots, m-1$, being $u^n_{\alpha+m} = f_*^{-1}(q^n_{b\beta+1})$, see Fig. 6.5, center. In this case

$$\begin{aligned}\Delta\Upsilon_b(\bar{t}) &= \sum_{i=0}^{m-1} \left|\Psi(u^n_{\alpha+i}) - \Psi(u^n_{\alpha+i+1})\right| - 2\left|q^n_{b\beta+1} - q^n_{b\beta}\right| \\ &= \sum_{i=0}^{m-1} \left(f(u^n_{\alpha+i+1}) - f(u^n_{\alpha+i})\right) - 2\left(q^n_{b\beta+1} - q^n_{b\beta}\right) \\ &= -\left(q^n_{b\beta+1} - q^n_{b\beta}\right) \le -2^{-n}.\end{aligned}$$

(B1.2) If $u^n_\alpha \neq f_*^{-1}(q^n_{b\beta})$, then must be $u^n_\alpha > \bar{u}$ and $f(u^n_\alpha) \le q^n_{b\beta}$, see Fig. 6.5, right. In this case no wave appears and

$$\begin{aligned}\Delta\Upsilon_b(\bar{t}) &= 2\left(\bar{f} - q^n_{b\beta+1}\right) - 2\left|q^n_{b\beta+1} - q^n_{b\beta}\right| - 2\left(\bar{f} - q^n_{b\beta}\right) \\ &= -4\left(q^n_{b\beta+1} - q^n_{b\beta}\right) \le -2^{-n}.\end{aligned}$$

(B2) Assume that $q^n_{b\beta} > q^n_{b\beta+1}$.

(B2.1) If $u^n_\alpha = f_*^{-1}(q^n_{b\beta})$, then an increasing wave between $f_*^{-1}(q^n_{b\beta+1})$ and u^n_α starts from the boundary at time $\bar{t}$, see Fig. 6.6, left. The functional Υ_b changes as follows

$$\begin{aligned}\Delta\Upsilon_b(\bar t) &= \left|\Psi\left(f_*^{-1}(q^n_{b\beta+1})\right) - \Psi(u^n_\alpha)\right| - 2\left|q^n_{b\beta+1} - q^n_{b\beta}\right| \\ &= q^n_{b\beta+1} - q^n_{b\beta} \le -2^{-n}\,.\end{aligned}$$

(B2.2) If $u^n_\alpha \neq f_*^{-1}(q^n_{b\beta})$ and $f(u^n_\alpha) > q^n_{b\beta+1}$, then must be $u^n_\alpha > \bar u$, $f(u^n_\alpha) \le q^n_{b\beta}$ and an increasing wave between $f_*^{-1}(q^n_{b\beta+1})$ and u^n_α starts from the boundary at time $\bar t$, see Fig. 6.6, center. Furthermore

$$\begin{aligned}\Delta\Upsilon_b(\bar t) = &\left|\Psi\left(f_*^{-1}(q^n_{b\beta+1})\right) - \Psi(u^n_\alpha)\right| - 2\left|q^n_{b\beta+1} - q^n_{b\beta}\right| \\ &-2\left(f(\bar u) - q^n_{b\beta}\right) = q^n_{b\beta+1} - f(u^n_\alpha) \le -2^{-n}\,.\end{aligned}$$

(B2.3) If $u^n_\alpha \neq f_*^{-1}(q^n_{b\beta})$ and $f(u^n_\alpha) \le q^n_{b\beta+1}$, then must be $u^n_\alpha > \bar u$, $f(u^n_\alpha) \le q^n_{b\beta}$, no wave starts from the boundary, see Fig. 6.6, right, and

$$\Delta\Upsilon_b(\bar t) = 2\left(\bar f - q^n_{b\beta+1}\right) - 2\left|q^n_{b\beta+1} - q^n_{b\beta}\right| - 2\left(\bar f - q^n_{b\beta}\right) = 0\,.$$

Hence, the map $[t \mapsto \Upsilon_b\left(\overline{Sb}^n_t\left(u^n_o, q^n_b\right)\right)]$ is non–increasing. Clearly, $\mathrm{TV}\left(\Psi(u^n)\right) + \mathrm{TV}(q^n_b) \le \Upsilon_b(u^n, q^n_b)$, so that Theorem 2.4 can be applied proving that $\Psi(u^n)$ converges a.e. to a limit $\psi \in \mathbf{C}^{0,1}\left(\mathbb{R}_+; \mathbf{BV}\left(\mathbb{R}_+; [-\bar f, \bar f]\right)\right)$. Since Ψ is invertible, also u^n converges a.e. to $u = \Psi^{-1}(\psi)$ and $u \in \mathbf{C}^{0,1}\left(\mathbb{R}_+; \mathbf{L}^\infty\left(\mathbb{R}_+; [0,R]\right)\right)$. To prove that u solves (6.2.1) in the sense of Definition 6.1, we have to check the validity of (6.2.5). Due to its local nature, we consider the different lines in (6.2.5) separately.

1. If $\mathrm{spt}\{\varphi\}$ is separated from $x = 0$, we need to deal only with the first two lines, essentially with Kružkov definition of solution, see Definition 3.6 and Theorem 3.2. They are non–negative for any n because, by construction, u^n is an entropy weak solution to (6.2.10). Therefore, they are non–negative also at the limit.
2. If $\mathrm{spt}\{\varphi\}$ intersect the line $x = 0$, then we deal only with the boundary, in particular with the third line in (6.2.5). Then, u^n exactly satisfies the boundary condition, at least for all but countably many times. Hence, also at the limit u satisfies the boundary condition.

The uniqueness of the solution, as well as the Lipschitz estimate (6.2.8), is proved by means of the **doubling of variables** technique, introduced in [5], and used in [2] in the case with boundary. Let $u(t,x) = \lim_{n\to+\infty} \overline{Sb}^n_t\left(u_o, q_b\right)(x)$ and $u'(s,y) = \lim_{n\to+\infty} \overline{Sb}^n_s\left(u'_o, q'_b\right)(y)$. Take $T \in \mathbb{R}_+$, $h,k \in [0,R]$ and $\varphi \in \mathbf{C}^\infty_\mathbf{c}(\mathbb{R}^4_+; \mathbb{R}^*_+)$. By (6.2.5), we have

$$\int_{\mathbb{R}_+}\int_{\mathbb{R}_+} \left[\left|u(t,x) - h\right|\,\partial_t\varphi(t,x,s,y) + \Xi\left(u(t,x), h\right)\,\partial_x\varphi(t,x,s,y)\right] dx\; dt \ge 0 \tag{6.2.15}$$

$$\int_{\mathbb{R}_+}\int_{\mathbb{R}_+}\left[\left|u'(s,y)-k\right|\,\partial_s\varphi(t,x,s,y)+\Xi\left(u'(s,y),k\right)\,\partial_y\varphi(t,x,s,y)\right]\mathrm{d}y\ \mathrm{d}s\geq 0\,, \tag{6.2.16}$$

where $\Xi\,:\,\mathbb{R}^2\to\mathbb{R}$ is defined by (5.5.4). Set now $h=u'(s,y)$ in (6.2.15) and integrate with respect to (s,y). Analogously, set $k=u(t,x)$ in (6.2.16) and integrate with respect to (t,x). Summing the obtained inequalities, we get

$$\iiiint\left[\left|u-u'\right|\,(\partial_t\varphi+\partial_s\varphi)+\Xi(u,u')\,(\partial_x\varphi+\partial_y\varphi)\right]\mathrm{d}x\ \mathrm{d}t\ \mathrm{d}y\ \mathrm{d}s\geq 0\,. \tag{6.2.17}$$

For $\varepsilon_1,\varepsilon_2\in\mathbb{R}_+$ sufficiently small, define

$$\psi_{\varepsilon_2}(t,x)=\left(\int_{x-\varepsilon_2^{-1}+\varepsilon_2}^{x}\delta_{\varepsilon_2}(z)\ \mathrm{d}z\right)\left(\int_{t-T+\varepsilon_2}^{t}\delta_{\varepsilon_2}(z)\ \mathrm{d}z\right) \tag{6.2.18a}$$

$$\varphi(t,x,s,y)=\delta_{\varepsilon_1}(t-s)\ \delta_{\varepsilon_1}(x-y)\ \psi_{\varepsilon_2}(t,x)\,, \tag{6.2.18b}$$

where δ_ε is defined by (5.5.9a). We underline that for $\varepsilon\in\mathbb{R}_+$ sufficiently small, $\psi_\varepsilon\in\mathbf{C}^\infty(\mathbb{R}^2;[0,1])$ approximates the characteristic function of $\{(t,x)\in\mathbb{R}^2\ :\ 0<t<T,\ x\in\mathbb{R}_+\}$. By using (6.2.18) and (5.5.11) in (6.2.17) and letting $\varepsilon_1\searrow 0$ we obtain

$$\int_{\mathbb{R}_+}\int_{\mathbb{R}_+}\left[\left|u(t,x)-v(t,x)\right|\,\partial_t\psi_{\varepsilon_2}(t,x)+\Xi\left(u(t,x),v(t,x)\right)\partial_x\psi_{\varepsilon_2}(t,x)\right]\mathrm{d}x\ \mathrm{d}t\geq 0\,.$$

Since as $\varepsilon_2\searrow 0$

$$\begin{aligned}
\partial_t\psi_{\varepsilon_2}(t,x)&=\left(\delta_{\varepsilon_2}(t)-\delta_{\varepsilon_2}(t-T+\varepsilon_2)\right)\left(\int_{x-\varepsilon_2^{-1}+\varepsilon_2}^{x}\delta_{\varepsilon_2}(z)\ \mathrm{d}z\right)\\
&\to\left(\delta_0^D(t)-\delta_T^D(t)\right)\ \chi_{\mathbb{R}_+}(x)\\
\partial_x\psi_{\varepsilon_2}(t,x)&=\left(\delta_{\varepsilon_2}(x)-\delta_{\varepsilon_2}(x-\varepsilon_2^{-1}+\varepsilon_2)\right)\left(\int_{t-T+\varepsilon_2}^{t}\delta_{\varepsilon_2}(z)\ \mathrm{d}z\right)\\
&\to\left(\delta_{0+}^D(x)-\delta_{+\infty}^D(x)\right)\ \chi_{]0,T[}(t)\,,
\end{aligned}$$

by letting $\varepsilon_2\searrow 0$ in the last estimate we obtain

$$\int_{\mathbb{R}_+}\left(\left|u_o(x)-u_o'(x)\right|-\left|u(T,x)-u'(T,x)\right|\right)\mathrm{d}x+\int_0^T\Xi\left(u(t,0+),u'(t,0+)\right)\mathrm{d}t\geq 0\,.$$

To complete the proof of (6.2.8), it remains to show that $\Xi\left(u(t,0+),u'(t,0+)\right)\leq\left|q_b(t)-q_b'(t)\right|$. Due to the symmetries of Ξ, it is not limitative to assume that $u(t,0+)<u'(t,0+)$, namely $\Xi\left(u(t,0+),u'(t,0+)\right)=f\left(u'(t,0+)\right)-f\left(u(t,0+)\right)$. If $u(t,0+)>\bar{u}$, then $\Xi\left(u(t,0+),u'(t,0+)\right)<0$. If $u(t,0+)\leq\bar{u}<u'(t,0+)$, then $u(t,0+)=f_*^{-1}\left(q_b(t)\right)$, $f\left(u'(t,0+)\right)\leq q_b'(t)$ and therefore $\Xi\left(u(t,0+),u'(t,0+)\right)\leq$

$q'_b(t) - q_b(t)$. If $u'(t,0+) \leq \bar{u}$, then $u(t,0+) = f_*^{-1}(q_b(t))$, $u'(t,0+) = f_*^{-1}(q'_b(t))$ and therefore $\Xi\left(u(t,0+),u'(t,0+)\right) = q'_b(t) - q_b(t)$. □

An alternative proof of Theorem 6.1 can be found in [2], where the existence of the solution is established by means of the vanishing viscosity method.

6.3 The Constrained Riemann Problem

In this section we introduce the concept of constraint by considering constrained Riemann problems of the form

$$\partial_t u + \partial_x f(u) = 0 \qquad (t,x) \in \mathbb{R}_+ \times \mathbb{R} \tag{6.3.1a}$$

$$u(0,x) = \begin{cases} u^l & \text{if } x < 0 \\ u^r & \text{if } x \geq 0 \end{cases} \qquad x \in \mathbb{R} \tag{6.3.1b}$$

$$f\left(u(t,0)\right) \leq q_c \qquad t \in \mathbb{R}_+\,, \tag{6.3.1c}$$

where the flux f satisfies the condition **(F)** given in Sect. 6.2, while the data $u^l, u^r \in [0,R]$ and the value of the constraint $q_c \in [0,\bar{f}]$ are constant. Let $\mathscr{R}_e$ be the entropic Riemann solver for (6.3.1a), (6.3.1b) given in Definition 4.5. Then,

$$\mathscr{R}_c[u^l,u^r](x) = \begin{cases} \mathscr{R}_e[u^l,u^r](x) & \text{if } \; f\left(\mathscr{R}_e[u^l,u^r](0)\right) \leq q_c \\ \begin{cases} \mathscr{R}_e[u^l,\hat{u}](x) & \text{if } x < 0 \\ \mathscr{R}_e[\check{u},u^r](x) & \text{if } x \geq 0 \end{cases} & \text{otherwise} \end{cases} \tag{6.3.2}$$

is a Riemann solver associated to the constrained Riemann problem (6.3.1). Above,

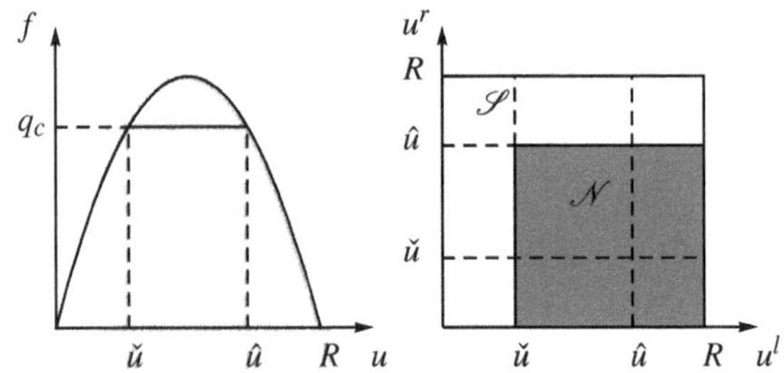

Fig. 6.7 Left: The constraint $[t \mapsto q_c(t)]$ and the functions $[t \mapsto \hat{u}(t)]$ and $[t \mapsto \check{u}(t)]$. Right: Representation of the standard and non–standard regions, respectively denoted $\mathscr{S}$ and $\mathscr{N}$, defined in (6.3.3).

$\check{u} = f_*^{-1}(q_c)$ and $\hat{u}$, with $\hat{u} \geq \bar{u}$, are the solutions to $f(\rho) = q_c$, see Fig. 6.7, left. Observe that $\check{u}$ and $\hat{u}$ coalesce iff $q_c = \bar{f}$. On the other hand, as soon as $q_c < \bar{f}$ and the constraint is effective, a stationary decreasing non–entropic shock between $\check{u}$ and $\hat{u}$ arises at $x = 0$ and the weak solution $[(t,x) \mapsto \mathscr{R}_c[u^l,u^r](x/t)]$ is not entropic.

According to (6.3.2), the constraint (6.3.1c) may well cause sharp increases in the total variation of the solution as, for instance, in the next example:

Example 6.3. If $u_o(x) = \bar{u}$ and $q_c(t) = f(\bar{u})/2$, then the solution to (6.3.1) performs two entropic shocks interspersed with a non–entropic shock and its total variation jumps from 0 to $2(\hat{u} - \check{u})$. □

Let $\mathscr{B} = \left(\{\check{u}\} \times [0,\hat{u}]\right) \cup \left([\check{u},R] \times \{\hat{u}\}\right)$ and introduce the **standard region** $\mathscr{S}$ and the **non–standard region** $\mathscr{N}$, see Fig. 6.7, right, defined by

$$\mathscr{N} =]\check{u},R] \times [0,\hat{u}[\ , \qquad \mathscr{S} = [0,R]^2 \setminus (\mathscr{N} \cup \mathscr{B})\ . \tag{6.3.3}$$

Proposition 6.1. *The Riemann solver $\mathscr{R}_c$ defined by* (6.3.2) *satisfies the following properties for all $u^l, u^r \in [0,R]$:*

(RS1) $\mathscr{R}_c[u^l,u^r] \neq \mathscr{R}_e[u^l,u^r]$ iff $(u^l,u^r) \in \mathscr{N}$.
(RS2) $[(t,x) \mapsto \mathscr{R}_c[u^l,u^r](x/t)]$ is a self similar weak solution to (6.3.1).
(RS3) $\mathscr{R}_c[u^l,u^r] \in \mathbf{BV}(\mathbb{R};[0,R])$.
(RS4) $\mathscr{R}_c[u^l,u^r]$ satisfies the constraint (6.3.1c) *in the sense that*

$$\lim_{x \searrow 0} f\left(\mathscr{R}_c[u^l,u^r](x)\right) \leq q_c \quad \textit{and} \quad \lim_{x \nearrow 0} f\left(\mathscr{R}_c[u^l,u^r](x)\right) \leq q_c\ . \tag{6.3.4}$$

(RS5) $\mathscr{R}_c$ is consistent in the sense of Definition 4.6.
(RS6) $\mathscr{R}_c : [0,R]^2 \to \mathbf{L^1_{loc}}(\mathbb{R};\mathbb{R})$ is uniformly continuous.

Proof. For notational convenience, consider the characteristic speed $\lambda(\rho) = f'(\rho)$ and the speed $\Lambda(u^l,u^r) = \left(f(u^l) - f(u^r)\right)/(u^l - u^r)$ of a (possible non–entropic) shock between u^l and u^r.

(RS1) If $(u^l,u^r) \in \mathscr{N}$, then, differently from the standard solution, the corresponding solution performs a stationary non–entropic shock from $\hat{u}$ to $\check{u}$, as shown for the case $f'' < 0$ in Fig. 6.9, left and center, Fig. 6.10, right, Fig. 6.11, center and right. On the other hand, if $(u^l,u^r) \in \mathscr{S} \cup \mathscr{B}$, then $\mathscr{R}_c[u^l,u^r] = \mathscr{R}_e[u^l,u^r]$, as shown for the case $f'' < 0$ in Fig. 6.8, Fig. 6.9, right, Fig. 6.10, left and center, Fig. 6.11, left.

(RS2) Self similarity is obvious. Off from $x = 0$, $\mathscr{R}_c$ yields weak solution because so does $\mathscr{R}_e$. Along $x = 0$, the Rankine–Hugoniot condition (3.4.4) is satisfied, since the jump at $x = 0$ is a (possibly non–entropic) stationary shock.

(RS3) This property is immediate, since $\mathscr{R}_e$ is in **BV** and $\mathscr{R}_c$ amounts to juxtapose standard solutions.

(RS4) Note first that both limits exist and are finite since $\mathscr{R}_c[u^l,u^r] \in \mathbf{BV}(\mathbb{R};\mathbb{R})$. For simplicity, we assume that $f'' < 0$ and consider only the left limit, since the right one is essentially analogous.

(a) If $\mathscr{R}_c[u^l,u^r] = \mathscr{R}_e[u^l,u^r]$, then by definition (6.3.2), condition (6.3.4) holds, see Fig. 6.8, Fig. 6.9, right, Fig. 6.10, left and center, Fig. 6.11, left.

(b) If $\mathscr{R}_c[u^l,u^r] \neq \mathscr{R}_e[u^l,u^r]$, then $\mathscr{R}_c[u^l,\hat{u}]$ is either an entropic shock with negative speed, see Fig. 6.9, left and center, Fig. 6.11, right, or a rarefaction with negative speed, see Fig. 6.10, right, Fig. 6.11, center, and therefore $\lim_{x \nearrow 0} \mathscr{R}_c[u^l,u^r](x) = \hat{u}$.

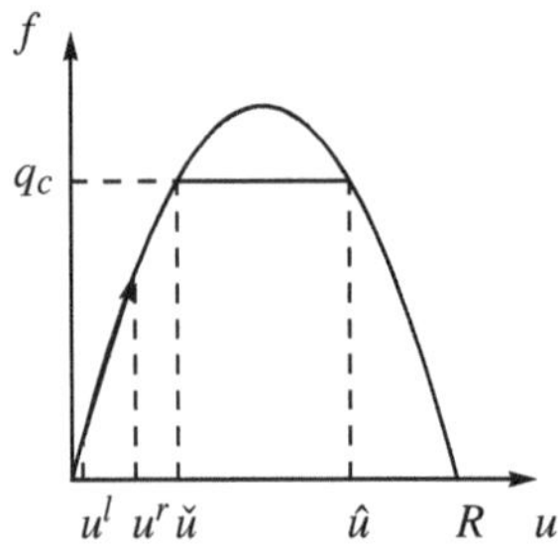

Fig. 6.8 If $u^l < u^r$ and $u^l \leq \check{u}$, then $\mathscr{R}_c[u^l, u^r]$ coincides with $\mathscr{R}_e[u^l, u^r]$ and performs an entropic shock

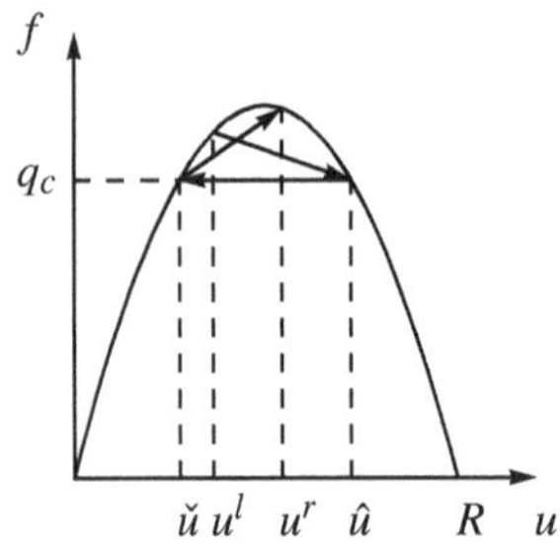

Fig. 6.9 Left and center: If $\check{u} < u^l, u^r < \hat{u}$, then $\mathscr{R}_c[u^l, u^r]$ performs an entropic shock from u^l to $\hat{u}$, followed by a non–entropic shock from $\hat{u}$ to $\check{u}$, followed by an entropic shock from $\check{u}$ to u^r. Right: If $\hat{u} \leq u^l < u^r$, then $\mathscr{R}_c[u^l, u^r]$ coincides with $\mathscr{R}_e[u^l, u^r]$ and performs an entropic shock

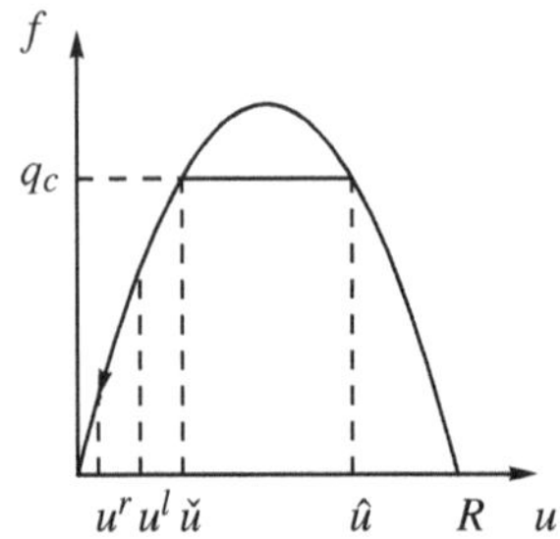

Fig. 6.10 Left and center: If $u^r < u^l$ and $u^l \leq \check{u}$ or $\hat{u} \leq u^r$, then $\mathscr{R}_c[u^l, u^r]$ coincides with $\mathscr{R}_e[u^l, u^r]$ and performs a rarefaction. Right: If $u^r < \check{u}$ and $\hat{u} < u^l$, then $\mathscr{R}_c[u^l, u^r]$ performs a rarefaction from u^l to $\hat{u}$, followed by a non–entropic shock from $\hat{u}$ to $\check{u}$, followed by a rarefaction from $\check{u}$ to u^r.

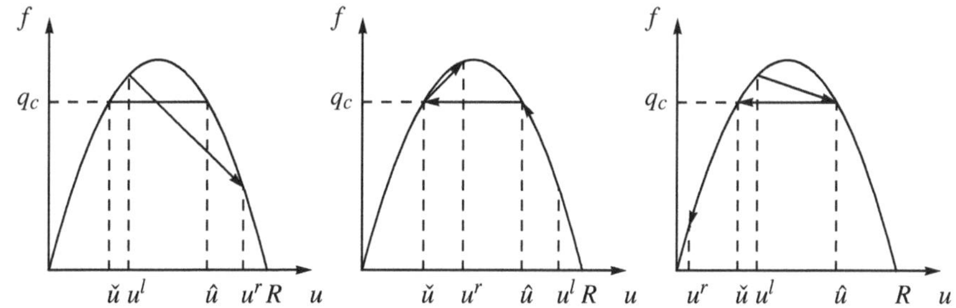

Fig. 6.11 Left: If $\check{u} \le u^l \le \hat{u} \le u^r$, then $\mathscr{R}_c[u^l,u^r]$ coincides with $\mathscr{R}_e[u^l,u^r]$ and performs an entropic shock. Center: If $\check{u} \le u^r < \hat{u} \le u^l$, then $\mathscr{R}_c[u^l,u^r]$ performs a possible null rarefaction from u^l to $\hat{u}$, followed by a non–entropic shock from $\hat{u}$ to $\check{u}$, followed by a possible null entropic shock from $\check{u}$ to u^r. Right: If $u^r < \check{u} < u^l < \hat{u}$, then $\mathscr{R}_c[u^l,u^r]$ performs an entropic shock from u^l to $\hat{u}$, followed by a non–entropic shock from $\hat{u}$ to $\check{u}$, followed by rarefaction from $\check{u}$ to u^r.

(RS5) Consistency directly follows from the analogous property of the entropic Riemann solver.

(RS6) For any fixed $\varepsilon \in \mathbb{R}_+$ and $a,b \in \mathbb{R}$ with $a < b$, we have to prove that there exists $\delta \in \mathbb{R}_+$ such that $\left\|\mathscr{R}_c[u_2^l,u_2^r] - \mathscr{R}_c[u_1^l,u_1^r]\right\|_{\mathbf{L^1}([a,b];\mathbb{R})} \le \varepsilon$ for all $\left|u_2^{lr} - u_1^{lr}\right| < \delta$. We consider separately different cases.

(s.1) If $(u_1^l,u_1^r) \in \{\check{u}\} \times [0,\check{u}[$ and $(u_2^l,u_2^r) \in \mathscr{N}$ are sufficiently close, see Fig. 6.12, left, then

$$\int_a^b \left|\mathscr{R}_c[u_2^l,u_2^r] - \mathscr{R}_c[\check{u},u_1^r]\right| \,\mathrm{d}x \le \left(u_2^l - \check{u} + \left|u_2^r - u_1^r\right|\right)\,(b-a) \\ -(\hat{u}-\check{u})\;\Lambda(u_2^l,\hat{u})\,,$$

which is arbitrarily small as (u_2^l,u_2^r) approaches (u_1^l,u_1^r).

(s.2) If $(u_1^l,u_1^r) \in \{\check{u}\}\times]\check{u},\hat{u}[$ and $(u_2^l,u_2^r) \in \mathscr{N}$ are sufficiently close, see Fig. 6.12, right, then

$$\int_a^b \left|\mathscr{R}_c[\check{u},u_1^r] - \mathscr{R}_c[u_2^l,u_2^r]\right| \,\mathrm{d}x \le \left(u_2^l - \check{u} + \left|u_2^r - u_1^r\right|\right)\,(b-a) \\ -(\hat{u}-\check{u})\;\Lambda(u_2^l,\hat{u}) + \left(u_2^r + u_1^r - 2\check{u}\right)\,\left|\Lambda(\check{u},u_2^r) - \Lambda(\check{u},u_1^r)\right|\,,$$

which is arbitrarily small as (u_2^l,u_2^r) approaches (u_1^l,u_1^r).

(s.3) If $(u_1^l,u_1^r) \in]\check{u},\hat{u}[\times\{\hat{u}\}$ and $(u_2^l,u_2^r) \in \mathscr{N}$ are sufficiently close, see Fig. 6.13, left, then

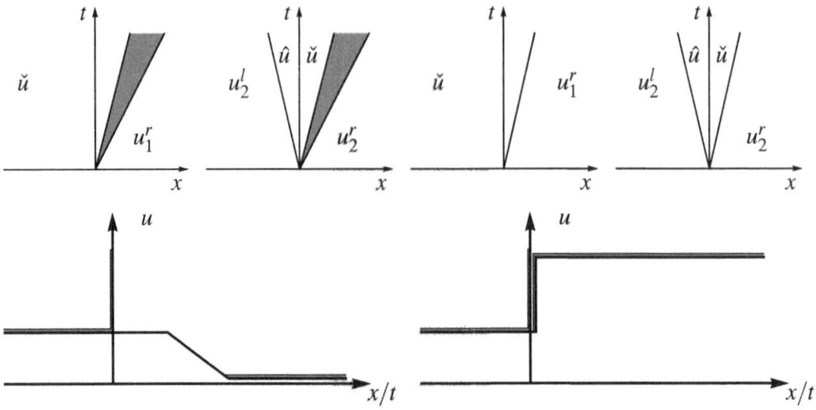

Fig. 6.12 Representation in the (x,t)–plane, above, and in the $(x/t,u)$–plane, below, of the solutions $[(t,x) \mapsto \mathscr{R}_c[u_1^l, u_1^r](x/t)]$ and $[(t,x) \mapsto \mathscr{R}_c[u_2^l, u_2^r](x/t)]$ when $u_1^r, u_2^r < u_1^l = \check{u} < u_2^l < \hat{u}$, left, and when $u_1^l = \check{u} < u_2^l, u_1^r, u_2^r < \hat{u}$, right.

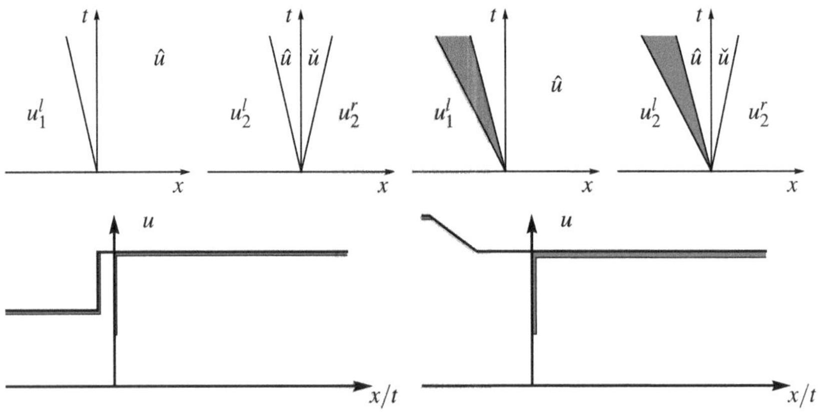

Fig. 6.13 Representation in the (x,t)–plane, above, and in the $(x/t,u)$–plane, below, of the solutions $[(t,x) \mapsto \mathscr{R}_c[u_1^l, u_1^r](x/t)]$ and $[(t,x) \mapsto \mathscr{R}_c[u_2^l, u_2^r](x/t)]$ when $\check{u} < u_1^l, u_2^l, u_2^r < \hat{u} = u_1^r$, left, and when $\check{u} < u_2^r < u_1^r = \hat{u} < u_1^l, u_2^l$, right.

$$\int_a^b \left| \mathscr{R}_c[\check{u}, u_1^r] - \mathscr{R}_c[u_2^l, u_2^r] \right| \mathrm{d}x \leq \left(\hat{u} - u_2^r + \left| u_2^l - u_1^l \right| \right) (b-a)$$
$$+ \left(2\hat{u} - u_2^l - u_1^l \right) \left| \Lambda(\hat{u}, u_2^l) - \Lambda(\hat{u}, u_1^l) \right| + (\hat{u} - \check{u})\, \Lambda(\check{u}, u_2^r) \,,$$

which is arbitrarily small as (u_2^l, u_2^r) approaches (u_1^l, u_1^r).

(s.4) If $(u_1^l, u_1^r) \in]\hat{u}, R] \times \{\hat{u}\}$ and $(u_2^l, u_2^r) \in \mathscr{N}$ are sufficiently close, see Fig. 6.13, right, then

$$\int_a^b \left|\mathscr{R}_c[\check{u}, u_1^r] - \mathscr{R}_c[u_2^l, u_2^r]\right| \, \mathrm{d}x \le \left(\hat{u} - u_2^r + \left|u_2^l - u_1^l\right|\right) (b-a) \\ +(\hat{u} - \check{u})\, \Lambda(\check{u}, u_2^r) \,,$$

which is arbitrarily small as (u_2^l, u_2^r) approaches (u_1^l, u_1^r).

Finally, the remaining cases follow from the properties of the entropic Riemann solver. □

6.4 The Constrained Cauchy Problem

Consider now the **constrained Cauchy problem**, or **constrained initial value problem**,

$$\partial_t u + \partial_x f(u) = 0 \qquad (t,x) \in \mathbb{R}_+ \times \mathbb{R} \qquad (6.4.1a)$$

$$u(0,x) = u_o(x) \qquad x \in \mathbb{R} \qquad (6.4.1b)$$

$$f\left(u(t,x_c)\right) \le q_c(t) \qquad t \in \mathbb{R}_+ \,, \qquad (6.4.1c)$$

where $x = x_c \in \mathbb{R}$ is the location where the constraint is placed. It is not limitative to assume that $x_c = 0$.

Definition 6.2. A map $u \in \mathbf{C^0}\left(\mathbb{R}_+; \mathbf{L^\infty_{loc}}(\mathbb{R}; [0,R])\right)$ is a **weak solution** to the constrained Cauchy problem (6.4.1) if for every test function $\varphi \in \mathbf{C_c^\infty}(\mathbb{R}^2; \mathbb{R}_+^*)$ and for every $k \in [0,R]$

$$\begin{aligned} 0 \le & \int_{\mathbb{R}_+} \int_{\mathbb{R}} \left(|u-k|\, \partial_t \varphi + \mathrm{sgn}(u-k)\, \left(f(u) - f(k)\right)\, \partial_x \varphi \right) \, \mathrm{d}x \; \mathrm{d}t \\ & + \int_{\mathbb{R}} \left|u_o(x) - k\right| \, \varphi(0,x)\, \mathrm{d}x \\ & + 2 \int_{\mathbb{R}_+} \left(1 - \frac{q_c(t)}{f(\bar{u})}\right) f(k)\, \varphi(t,0)\, \mathrm{d}t \end{aligned} \qquad (6.4.2)$$

and $f\left(u(t,0-)\right) = f\left(u(t,0+)\right) \le q_c(t)$ for almost all $t \in \mathbb{R}_+$.

Above we denote by $u(t,0\pm)$ the **measure theoretic trace** implicitly defined by

$$\lim_{\varepsilon \searrow 0} \frac{1}{\varepsilon} \int_{\mathbb{R}_+} \int_0^\varepsilon \left|u(t,x) - u(t,0+)\right| \, \varphi(t,x)\, \mathrm{d}x\, \mathrm{d}t = 0$$

$$\lim_{\varepsilon \searrow 0} \frac{1}{\varepsilon} \int_{\mathbb{R}_+} \int_{-\varepsilon}^0 \left|u(t,x) - u(t,0-)\right| \, \varphi(t,x)\, \mathrm{d}x\, \mathrm{d}t = 0$$

for all $\varphi \in \mathbf{C_c^\infty}(\mathbb{R}^2; \mathbb{R})$. Remark that both traces at $x = 0$ exist and are finite by [1, Theorem 2.2].

Definition 6.2 selects the solution that allows the maximal flow through the constraint, for a non–entropy stationary shock at $x = 0$ separating states $\check{u}$ and $\hat{u}$ with $\check{u} \le \bar{u} \le \hat{u}$, $f(\check{u}) = f(\hat{u}) = q_c$ turns out to be non–entropic.

Remark 6.3. We underline that the first two lines in (6.4.2) are motivated by the Cauchy problem (6.4.1a), (6.4.1b), see Theorem 3.2. The latter line accounts for the constraint (6.4.1c), as in [3, definitons 3.1 and 3.2]. Finally, the last condition selects, for instance, the solution performed in the Example 6.3 rather than the constant weak solution $u(t,x) \equiv \bar{u}$, that also satisfies (6.4.2). For other equivalent formulations in the case of unilateral constraints, we refer to [1, Proposition 2.6]. □

Remark 6.4. As the Example 6.3 suggests, a time dependent constraint (6.4.1c) and the consequent definition (6.3.2) may well lead to solutions with unbounded total variation even when the data are in **BV**. We introduce the domain

$$\mathscr{D} = \left\{ u \in \mathbf{L^1}\left(\mathbb{R};[0,R]\right) \ : \ \Psi(u) \in \mathbf{BV}\left(\mathbb{R};\left[-\bar{f},\bar{f}\right]\right)\right\} \tag{6.4.3}$$

to overcome this difficulty. Above Ψ is defined by (6.2.7). □

Theorem 6.2. *Let* **(F)** *hold. Assume moreover that*

(O)' $u_o \in \mathscr{D}$;
(C) $q_c \in \mathbf{BV}(\mathbb{R}_+;[0,\bar{f}])$.

Then there exists a unique weak solution, u, to the constrained Cauchy problem (6.4.1) *in the sense of Definition 6.2 and $u(t) \in \mathscr{D}$ for all $t \in \mathbb{R}_+$. Furthermore, if u' is the solution to* (6.4.1) *associated with the pair $(u'_o, q'_c) \in \mathscr{D} \times \mathbf{BV}(\mathbb{R}_+;[0,\bar{f}])$, then the following Lipschitz estimate holds for all $t \in \mathbb{R}_+$*

$$\left\|u(t) - u'(t)\right\|_{\mathbf{L^1}(\mathbb{R},\mathbb{R})} \leq \left\|u_o - u'_o\right\|_{\mathbf{L^1}(\mathbb{R},\mathbb{R})} + 2\left\|q_c - q'_c\right\|_{\mathbf{L^1}([0,t],\mathbb{R})}. \tag{6.4.4}$$

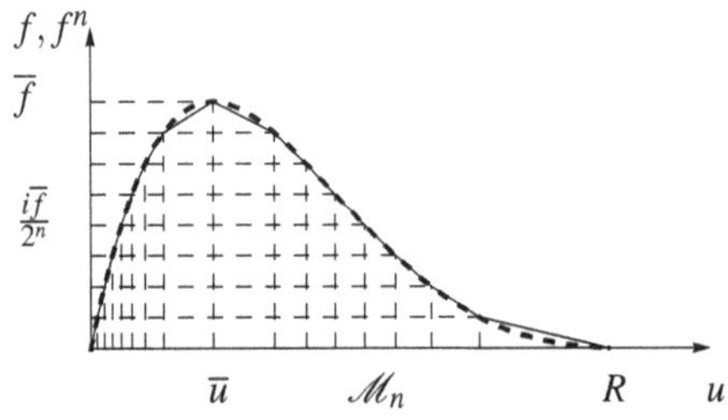

Fig. 6.14 Notation used for the proof of Theorem 6.2. The thick dashed line represents a flux f satisfying **(F)** and the solid line represents its approximation f^n.

Proof. Fix $n \in \mathbb{N}$. Consider the mesh $\mathscr{M}_n$ and the approximating flux f^n defined as in the proof of Theorem 6.1. Let $q_c^n \in \mathbf{PC}\left(\mathbb{R}_+; f(\mathscr{M}_n)\right)$ coincide with q_c on $f(\mathscr{M}_n) = 2^{-n}\mathbb{N}^*\bar{f} \cap [0,\bar{f}]$, in the sense that $q_c(t) = q_c^n(t)$ whenever $q_c(t) \in f(\mathscr{M}_n)$. Note that q_c^n satisfies **(C)**, because so does q_c. Let $u_o^n \in \mathscr{D}$ be in $\mathbf{PC}(\mathbb{R};\mathscr{M}_n)$ and such that $u_o^n(x) = u_o(x)$ whenever $u_o(x) \in \mathscr{M}_n$. The functions q_c^n and u_o^n can be represented as

$$u_o^n = \sum_{\alpha\in\mathbb{Z}} u_{o\alpha}^n \, \chi_{]x_{o\alpha-1}^n, x_{o\alpha}^n]} \qquad \text{with} \qquad u_{o\alpha}^n \in \mathscr{M}_n \tag{6.4.5a}$$

$$q_c^n = \sum_{\beta\in\mathbb{N}} q_{c\beta}^n \, \chi_{]t_{c\beta-1}^n, t_{c\beta}^n]} \qquad \text{with} \qquad q_{c\beta}^n \in f(\mathscr{M}_n) \tag{6.4.5b}$$

and we agree that $x_{o0}^n = t_{c0}^n = 0$. All the approximations above are meant in the strong $\mathbf{L^1}$–topology, that is

$$\lim_{n\to+\infty} \left(\left\| u_o^n - u_o \right\|_{\mathbf{L}^1(\mathbb{R};\mathbb{R})} + \left\| q_c^n - q_c \right\|_{\mathbf{L}^1(\mathbb{R}_+;\mathbb{R})} \right) = 0 \,.$$

We now follow the classical **wave front tracking** technique. In particular, an approximate solution u^n to (6.4.1) is computed as the exact entropy weak solution to the constrained Cauchy problem

$$\begin{aligned} &\partial_t u^n + \partial_x f^n(u^n) = 0 && (t,x) \in \mathbb{R}_+ \times \mathbb{R} && (6.4.6a)\\ &u^n(0,x) = u_o^n(x) && x \in \mathbb{R} && (6.4.6b)\\ &f^n\left(u^n(t,0)\right) \leq q_c^n(t) && t \in \mathbb{R}_+ \,. && (6.4.6c) \end{aligned}$$

In other words, u^n is obtained gluing the solutions to the Riemann problems at the points $(0, x_{o\alpha}^n)$ of jump of u_o^n, at the points $(t_{c\beta}^n, 0)$ where the constraint changes and at all interaction points where two or more waves intersect, or one or more waves hit the constraint. For the construction of the solution to standard Riemann problems with a piecewise linear and continuous flow we refer to Sect. 5.2. The constrained Riemann problem is solved in detail in Sect. 6.3.

Iteratively solving Riemann problems at interactions and at the constraint, we define a u^n which is an entropy weak solution to (6.4.6) and an approximate solution to (6.4.1). We prove that $u^n(t) \in \mathscr{D}_n = \left\{ u \in \mathbf{PC}(\mathbb{R}; \mathscr{M}_n) \, : \, \Psi(u) \in \mathbf{BV}(\mathbb{R};\mathbb{R}) \right\}$. To this aim, introduce the semigroup $\overline{Sc}^n \, : \, \mathbb{R}_+ \times \overline{\mathscr{D}}_n \to \overline{\mathscr{D}}_n$ defined by $\overline{Sc}_t^n(u_o^n, q_c^n) = \left(u^n(t), \mathscr{T}_t q_c^n\right)$, where $\overline{\mathscr{D}}_n = \mathscr{D}_n \times \mathbf{PC}\left(\mathbb{R}_+; f\left(\mathscr{M}_n\right)\right)$ and $\mathscr{T}$ is the translation operator. For any $(u^n, q_c^n) \in \overline{\mathscr{D}}_n$, written as in (6.4.5), define the Glimm type functional

$$\Upsilon_c\left(u^n, q_c^n\right) = \sum_{\alpha\in\mathbb{Z}} \left| \Psi(u_{\alpha+1}^n) - \Psi(u_\alpha^n) \right| + 5 \sum_{\beta\in\mathbb{N}} \left| q_{c\beta+1}^n - q_{c\beta}^n \right| + \gamma_c\left(u^n, q_c^n\right) \tag{6.4.7}$$

where Ψ is as in (6.2.7) and γ_c is defined

$$\gamma_c\left(u^n, q_c^n\right) = \begin{cases} 0 & \text{if } u^n(0-) > \bar{u} > u^n(0+) \text{ and} \\ & f\left(u^n(0+)\right) = f\left(u^n(0-)\right) = q_c^n(0) \\ 4\left(\bar{f} - q_c^n(0)\right) & \text{otherwise}\,. \end{cases} \tag{6.4.8}$$

Long but elementary computations show that at any interaction, the functional Υ_c either decreases by at least $2^{-n}\bar{f}$, or remains constant and the total number of waves in the approximate solution does not increase. The proof is obtained considering different cases separately. We will consider interaction points $(\bar{x}, \bar{t})$ with $\bar{x} \in \mathbb{R}_-^*$, the case $\bar{x} \in \mathbb{R}_+^*$ being symmetric. Up to refine the mesh $\mathscr{M}_n$, it is not restrictive to assume that at any interaction time either two waves intersect, see case I.1, or

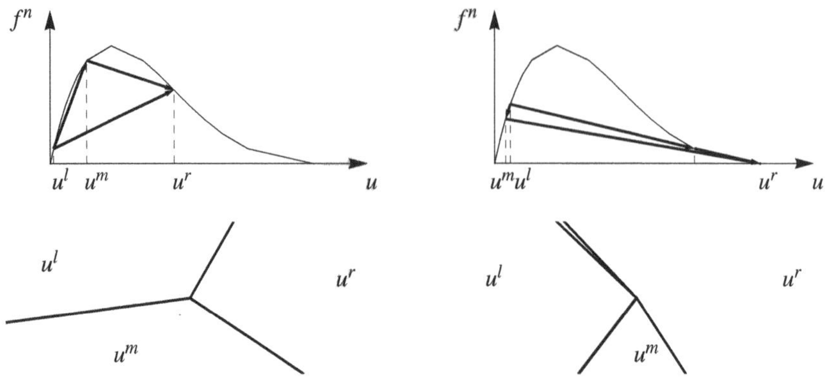

Fig. 6.15 Left: Interaction with $\bar{x} \neq 0$ of two waves with jumps of the same sign. Right: Interaction with $\bar{x} \neq 0$ of two waves with jumps of opposite sign.

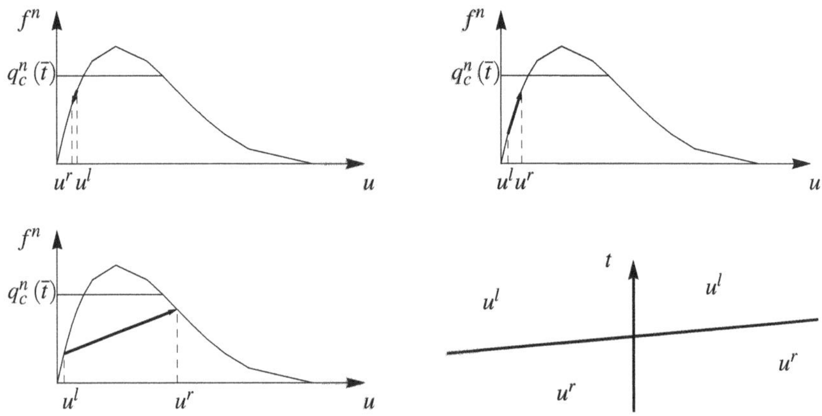

Fig. 6.16 Interaction with $\bar{x} = 0$, $u^m = u^r$ and $f(u^l) \leq q_c^n(\bar{t})$

a single wave hits $x = 0$, see case I.2, or the value of the constraint changes, see case I.3.

I.1 If $\bar{x} < 0$, as already observed in the proof of Proposition 5.1, we have only two possibilities:

I.1.1 If the jumps of the two incoming waves have the same sign, then the number of waves diminishes and the total variation of the solution does not change after the interaction, see Fig. 6.15, left.

I.1.2 If the jumps of the two incoming waves have opposite sign, then the number of waves may increase, however the total variation must decrease after the interaction owing to a cancelation effect, see Fig. 6.15, right.

I.2 If a wave between u^l and u^m hits the constraint coming from the left, i.e. $\bar{x} = 0$, and $u(\bar{t}-, 0+) = u^r$, then it is possible to distinguish the following cases:

I.2.1 If $u^m = u^r$, then necessarily $f^n(u^r) \leq q_c^n(\bar{t})$.

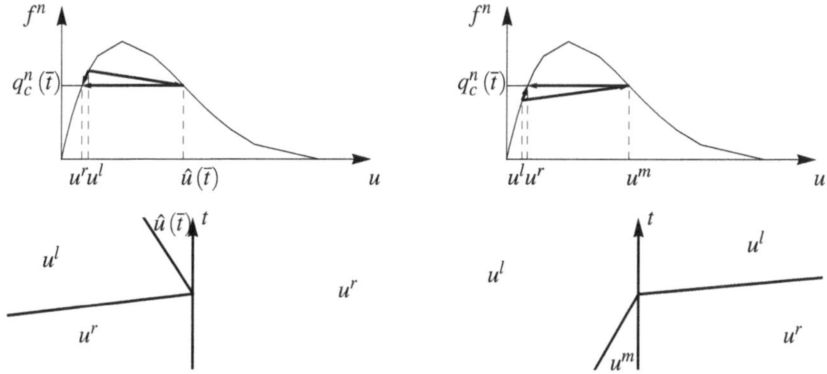

Fig. 6.17 Left: Interaction with $\bar{x} = 0$, $u^m = u^r$ and $f(u^l) > q_c^n(\bar{t})$. Right: Interaction with $\bar{x} = 0$, $u^m > u^r$ and $f(u^l) < q_c^n(\bar{t})$.

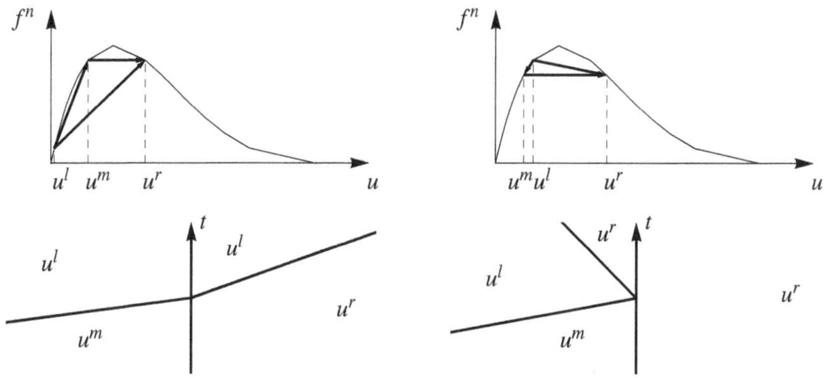

Fig. 6.18 Left: Interaction with $\bar{x} = 0$, $f(u^m) = f(u^r) \leq q_c^n(\bar{t})$ and $u^l < u^r < u^m$. Right: Interaction with $\bar{x} = 0$, $f(u^m) = f(u^r) \leq q_c^n(\bar{t})$ and $u^r < u^l < u^m$.

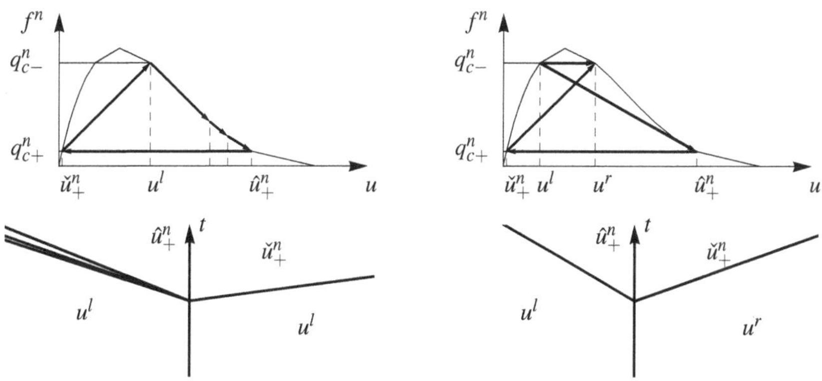

Fig. 6.19 Left: Interaction with $\bar{x} = 0$, $u^l = u^r$ and $q_{c-}^n \geq f^n(u^l) > q_{c+}^n$. Right: Interaction with $\bar{x} = 0$, $u^l < u^r$ and $q_{c-}^n \geq f^n(u^l) = f^n(u^r) > q_{c+}^n$.

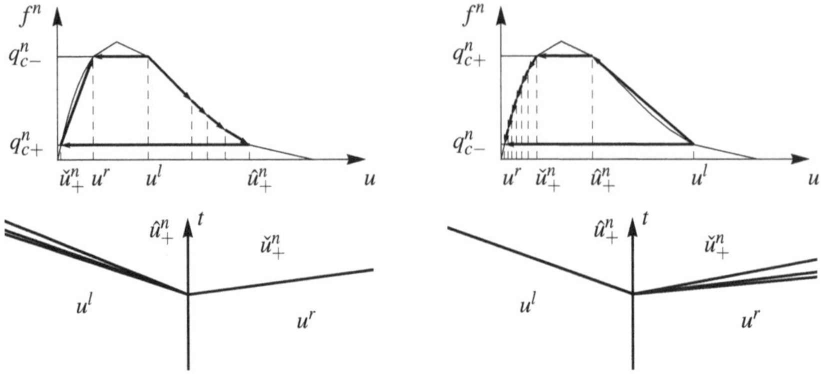

Fig. 6.20 Left: Interaction with $\bar{x}=0$, $u^r < u^l$ and $q^n_{c-} = f^n(u^l) > q^n_{c+}$. Right: Interaction with $\bar{x}=0$, $u^r < u^l$ and $q^n_{c-} = f^n(u^l) < q^n_{c+}$.

I.2.1.1 If $f^n(u^l) \leq q^n_c(\bar{t})$, then the wave crosses the constraint, no wave is created and each of the three terms in (6.4.7) does not change after the interaction, see Fig. 6.16.

I.2.1.2 If $f^n(u^l) > q^n_c(\bar{t})$, see Fig. 6.17, left, then it must be $u^r = f_*^{-1}\left(q^n_c(\bar{t})\right)$, the wave is reflected by the constraint, no wave is created and

$$\begin{aligned}\Delta\Upsilon_c(\bar{t}) &= \Upsilon_c\left(u^n(\bar{t}+), q^n_c(\bar{t}+)\right) - \Upsilon_c\left(u^n(\bar{t}-), q^n_c(\bar{t}-)\right)\\ &= \left|\Psi(u^l) - \Psi\left(\hat{u}^n(\bar{t})\right)\right| + \left|\Psi\left(\hat{u}^n(\bar{t})\right) - \Psi(u^r)\right| - \left|\Psi(u^l) - \Psi(u^r)\right|\\ &\quad - 4\left(\bar{f} - q^n_c(\bar{t})\right) = 2\left(q^n_c(\bar{t}) - f^n(u^l)\right) = -2^{1-n}\bar{f},\end{aligned}$$

where $\hat{u}^n(\bar{t}) \geq \bar{u}$ is defined by $f\left(\hat{u}^n(\bar{t})\right) = q^n_c(\bar{t})$.

I.2.2 If $u^m > u^r$, see Fig. 6.17, right, then necessarily $f^n(u^m) = f^n(u^r) = q^n_c(\bar{t})$, $u^l < u^r$ and therefore

$$\begin{aligned}\Delta\Upsilon_c(\bar{t}) = &\left|\Psi(u^l) - \Psi(u^r)\right| - \left|\Psi(u^l) - \Psi(u^m)\right| - \left|\Psi(u^m) - \Psi(u^r)\right|\\ &+4\left(\bar{f} - q^n_c(\bar{t})\right) = 0\,.\end{aligned}$$

I.2.3 If $u^r > u^m$, then necessarily $f^n(u^m) = f^n(u^r) \leq q^n_c(\bar{t})$ and it is possible to distinguish two cases:

I.2.3.1 If $u^l < u^m$, see Fig. 6.18, left, then the wave crosses the constraint, no wave is created and

$$\Delta\Upsilon_c(\bar{t}) = \left|\Psi(u^l) - \Psi(u^r)\right| - \left|\Psi(u^l) - \Psi(u^m)\right| - \left|\Psi(u^m) - \Psi(u^r)\right| = 0\,.$$

I.2.3.2 If $u^l > u^m$, see Fig. 6.18, right, then the wave is reflected by the constraint, no wave is created and

$$\begin{aligned}\Delta\Upsilon_c(\bar t) &= \left|\Psi(u^l)-\Psi(u^r)\right| - \left|\Psi(u^l)-\Psi(u^m)\right| - \left|\Psi(u^m)-\Psi(u^r)\right| \\ &= 2\left(f^n(u^m)-f^n(u^l)\right) = -2^{1-n}\,\bar f\,.\end{aligned}$$

I.3 If the value of the constraint changes, namely $q_c^n(\bar t-) \neq q_c^n(\bar t+)$, then it is possible to distinguish the following cases. For notational convenience, let us introduce $u^{lr} = u^n(\bar t-, 0\mp)$, $q_{c\pm}^n = q_c^n(\bar t\pm)$, $\check u_\pm^n = f_*^{-1}(q_{c\pm}^n)$ and $\hat u_\pm^n \geq \bar u$ defined by $f^n(\hat u_\pm^n) = q_{c\pm}^n$.

I.3.1 If $u^l = u^r$, then necessarily $f^n(u^r) \leq q_{c-}^n$. If $q_{c+}^n \geq f^n(u^r)$, then no wave is created and each of the three terms in (6.4.7) does not change after the interaction. If $q_{c+}^n < f^n(u^r)$, see Fig. 6.19, left, then waves exit the point $(\bar t, 0)$ on both sides of the constraint, and

$$\begin{aligned}\Delta\Upsilon_c(\bar t) = &\left|\Psi(\hat u_+^n)-\Psi(u^l)\right| + \left|\Psi(\check u_+^n)-\Psi(\hat u_+^n)\right| + \left|\Psi(u^l)-\Psi(\check u_+^n)\right| \\ &-5\left|q_{c+}^n - q_{c-}^n\right| - 4\left(\bar f - q_{c-}^n\right) = q_{c+}^n - q_{c-}^n \leq -2^{-n}\,\bar f\,.\end{aligned}$$

I.3.2 If $u^l < u^r$, then necessarily $f^n(u^l) = f^n(u^r) \leq q_{c-}^n$. If $q_{c+}^n \geq f^n(u^r)$, then no wave is created and each of the three terms in (6.4.7) does not change after the interaction. If $q_{c+}^n < f^n(u^r)$, see Fig. 6.19, right, then waves exit the point $(\bar t, 0)$ on both sides of the constraint, and

$$\begin{aligned}\Delta\Upsilon_c(\bar t) = &\left|\Psi(\hat u_+^n)-\Psi(u^l)\right| + \left|\Psi(\check u_+^n)-\Psi(\hat u_+^n)\right| + \left|\Psi(u^r)-\Psi(\check u_+^n)\right| \\ -&\left|\Psi(u^r)-\Psi(u^l)\right| - 5\left|q_{c+}^n - q_{c-}^n\right| - 4\left(\bar f - q_{c-}^n\right) = q_{c+}^n - q_{c-}^n \leq -2^{-n}\,\bar f\,.\end{aligned}$$

I.3.3 If $u^l > u^r$, then necessarily $f^n(u^l) = f^n(u^r) = q_{c-}^n$ and there are two possible cases:

I.3.3.1 If $q_{c-}^n > q_{c+}^n$, see Fig. 6.20, left, then possibly more than two waves exit the point $(\bar t, 0)$ and

$$\begin{aligned}\Delta\Upsilon_c(\bar t) = &\left|\Psi(\hat u_+^n)-\Psi(u^l)\right| + \left|\Psi(\check u_+^n)-\Psi(\hat u_+^n)\right| + \left|\Psi(u^r)-\Psi(\check u_+^n)\right| \\ &-\left|\Psi(u^r)-\Psi(u^l)\right| - 5\left|q_{c+}^n - q_{c-}^n\right| = q_{c+}^n - q_{c-}^n \leq -2^{-n}\,\bar f\,.\end{aligned}$$

I.3.3.2 If $q_{c-}^n < q_{c+}^n$, see Fig. 6.20, right, then possibly more than two waves exit the point $(\bar t, 0)$ and

$$\begin{aligned}\Delta\Upsilon_c(\bar t) = &\left|\Psi(\hat u_+^n)-\Psi(u^l)\right| + \left|\Psi(\check u_+^n)-\Psi(\hat u_+^n)\right| + \left|\Psi(u^r)-\Psi(\check u_+^n)\right| \\ &-\left|\Psi(u^r)-\Psi(u^l)\right| - 5\left|q_{c+}^n - q_{c-}^n\right| = 5\left(q_{c-}^n - q_{c+}^n\right) \leq -2^{-n}\,\bar f\,.\end{aligned}$$

Hence, the map $[t \mapsto \Upsilon_c\left(\overline{Sc}_t^n(u^n, q_c^n)\right)]$ is non–increasing and therefore, proceeding as in the proof of Theorem 6.1, it can be proved that u^n converges a.e. to

$u \in \mathbf{C}^{0,1}\left(\mathbb{R}_+; \mathbf{L}^\infty\left(\mathbb{R}; [0,R]\right)\right)$. To prove that u solves (6.4.1) in the sense of Definition 6.2, we first check the validity of (6.4.2). Due to its local nature, we consider the different lines in (6.4.2) separately:

1. If $\mathrm{spt}\{\varphi\}$ is separated from $x = 0$, we need to deal only with the first two lines, essentially with Kružkov definition of solution, see Definition 3.6 and Theorem 3.2. They are non–negative for any n because, by construction, u^n is an entropy weak solution to (6.4.6).
2. If $\mathrm{spt}\{\varphi\}$ intersects the line $x = 0$, then we are left to verify that u satisfies the constraint. Let $B_\pm = \mathrm{spt}\{\varphi\} \cap \{\pm x > 0\}$. Since all the other possible discontinuities are entropic in the classical sense, it is sufficient to consider the case $u(t,x) = \hat{u}(t)$ for all $(t,x) \in B_-$ and $u(t,x) = \check{u}(t)$ for all $(t,x) \in B_+$, where $\check{u}(t) = f_*^{-1}\left(q_c(t)\right)$ and $\hat{u}(t) \geq \bar{u}$ is defined by $f\left(\hat{u}(t)\right) = q_c(t)$. Integrating by parts the left hand side of (6.4.2) one gets

$$
\begin{aligned}
0 \leq & \int_{\mathbb{R}_+} \left(\mathrm{sgn}\left(\hat{u}(t) - k\right) - \mathrm{sgn}\left(\check{u}(t) - k\right)\right) \left(q_c(t) - f(k)\right) \varphi(t,0)\, \mathrm{d}t \\
& + 2 \int_{\mathbb{R}_+} \left(1 - \frac{q_c(t)}{f(\bar{u})}\right) f(k)\, \varphi(t,0)\, \mathrm{d}t\ .
\end{aligned}
$$

Since $\varphi(t,0) \in \mathbb{R}_+^*$, it is sufficient to prove that

$$
\left(\mathrm{sgn}\left(\hat{u}(t) - k\right) - \mathrm{sgn}\left(\check{u}(t) - k\right)\right) \left(q_c(t) - f(k)\right) + 2\left(1 - \frac{q_c(t)}{f(\bar{u})}\right) f(k) \geq 0
$$

for almost all $(t,0) \in \mathrm{spt}\{\varphi\}$ and for all $k \in [0,R]$. It is easy to check that

$$
\begin{aligned}
&\left(1 - \frac{q_c(t)}{f(\bar{u})}\right) f(k) \geq 0 && \text{if } f(k) \leq q_c(t) \\
&q_c(t)\left(1 - \frac{f(k)}{f(\bar{u})}\right) f(k) \geq 0 && \text{if } f(k) > q_c(t)\ .
\end{aligned}
$$

Finally, note first that, by construction, $f\left(u^n(t,0-)\right) = f\left(u^n(t,0+)\right) \leq q_c^n(t)$ holds for all $n \in \mathbb{N}$ and all but countably many $t \in \mathbb{R}_+$. Then, the existence of the traces at the limit follows from [1, Theorem 2.2].

The uniqueness of the solution, as well as the Lipschitz estimate (6.4.4), is proved by means of the **doubling of variables** technique, introduced in [5], used in the framework of constrained scalar conservation laws in [1, 3]. Let $T \in \mathbb{R}_+$ be a fixed constant. By construction $u(t,x) = \lim_{n\to+\infty} \overline{Sc_t^n}(u_o, q_c)(x)$ and $u'(s,y) = \lim_{n\to+\infty} \overline{Sc_s^n}(u_o', q_c')(y)$ satisfy (6.4.2), and therefore, for any $k,l \in [0,R]$ we have

$$
\int_{\mathbb{R}_+}\int_{\mathbb{R}} \left(\left|u(t,x) - k\right| \partial_t \varphi(t,x,s,y) + \Xi\left(u(t,x),k\right) \partial_x \varphi(t,x,s,y)\right) \mathrm{d}x\ \mathrm{d}t \geq 0 \tag{6.4.9}
$$

$$\int_{\mathbb{R}_+}\int_{\mathbb{R}}\Big(\left|u'(s,y)-l\right|\,\partial_s\varphi(t,x,s,y)+\Xi(u'(s,y),l)\,\partial_y\varphi(t,x,s,y)\Big)\,\mathrm{d}y\;\mathrm{d}s\geq 0\,, \tag{6.4.10}$$

where $\Xi\in\mathbf{Lip}(\mathbb{R}^2;\mathbb{R})$ is given by (5.5.4) and $\varphi\in\mathbf{C}^\infty_\mathbf{c}(\mathbb{R}^4;\mathbb{R})$ is given by

$$\psi_{\varepsilon_2}(t,x)=\left(\int_{-x-\varepsilon_2^{-1}+\varepsilon_2}^{-x}\delta_{\varepsilon_2}(z)\,\mathrm{d}z+\int_{x-\varepsilon_2^{-1}+\varepsilon_2}^{x}\delta_{\varepsilon_2}(z)\,\mathrm{d}z\right)\left(\int_{t-T+\varepsilon_2}^{t}\delta_{\varepsilon_2}(z)\,\mathrm{d}z\right) \tag{6.4.11a}$$

$$\varphi(t,x,s,y)=\delta_{\varepsilon_1}(t-s)\;\delta_{\varepsilon_1}(x-y)\;\psi_{\varepsilon_2}(t,x) \tag{6.4.11b}$$

where $\varepsilon_1,\varepsilon_2\in\mathbb{R}_+$ are sufficiently small fixed constants and δ_ε is defined by (5.5.9a). We underline that for $\varepsilon\in\mathbb{R}_+$ sufficiently small, $\psi_\varepsilon\in\mathbf{C}^\infty(\mathbb{R}^2;[0,1])$ approximates the characteristic function of $\left\{(t,x)\in\mathbb{R}^2\,:\,0<t<T,\ x\in\mathbb{R}\setminus\{0\}\right\}$. Set now $k=u'(s,y)$ in (6.4.9) and integrate with respect to (s,y). Analogously, set $l=u(t,x)$ in (6.4.10) and integrate with respect to (t,x). By summing the obtained inequalities, using (5.5.11) and letting $\varepsilon_1\searrow 0$ we get

$$\int_{\mathbb{R}_+}\int_{\mathbb{R}}\Big[\left|u(t,x)-v(t,x)\right|\,\partial_t\psi_{\varepsilon_2}(t,x)+\Xi\left(u(t,x),v(t,x)\right)\partial_x\psi_{\varepsilon_2}(t,x)\Big]\,\mathrm{d}x\;\mathrm{d}t\geq 0\,.$$

and letting also ε_2 goes to zero

$$\int_{\mathbb{R}}\left|u_o(x)-u'_o(x)\right|\,\mathrm{d}x-\int_{\mathbb{R}}\left|u(T,x)-u'(T,x)\right|\,\mathrm{d}x+\mathscr{I}\geq 0\,,$$

where

$$\mathscr{I}=\int_0^T\Big(\Xi\left(u,u'\right)(t,0+)-\Xi\left(u,u'\right)(t,0-)\Big)\,\mathrm{d}t\leq 2\int_0^T\left|q_c-q'_c\right|(t)\,\mathrm{d}t$$

and this ends the proof. □

6.5 The Constrained Initial–Boundary Value Problem

In this section we study the **constrained initial–boundary value problem** [4] for a scalar conservation law

$$\begin{aligned}
&\partial_t u+\partial_x f(u)=0 && (t,x)\in\mathbb{R}^2_+ && (6.5.1a)\\
&u(0,x)=u_o(x) && x\in\mathbb{R}_+ && (6.5.1b)\\
&f\left(u(t,0)\right)=q_b(t) && t\in\mathbb{R}_+ && (6.5.1c)\\
&f\left(u(t,x_c)\right)\leq q_c(t) && t\in\mathbb{R}_+\,, && (6.5.1d)
\end{aligned}$$

where $x = x_c \in \mathbb{R}_+$ denotes the position of the constraint. The next definition of solution to (6.5.1) is obtained by suitably merging the Definition 6.1 of solution to an initial–boundary value problem, with the Definition 6.2 of solution to a constrained Cauchy problem.

Definition 6.3. A map $u \in \mathbf{C^0}\left(\mathbb{R}_+;\mathbf{L^\infty_{loc}}(\mathbb{R}_+;[0,R])\right)$ is a **weak solution** to (6.5.1) if for every test function $\varphi \in \mathbf{C^\infty_c}(\mathbb{R}^2;\mathbb{R}^*_+)$ and for every $k \in [0,R]$

$$\begin{aligned} 0 \leq & \int_{\mathbb{R}_+}\int_{\mathbb{R}_+} \Big(\left|u-k\right| \partial_t\varphi + \operatorname{sgn}(u-k)\ \big(f(u)-f(k)\big)\,\partial_x\varphi\Big)\ \mathrm{d}x\ \mathrm{d}t \\ & + \int_{\mathbb{R}_+} \left|u_o(x)-k\right|\ \varphi(0,x)\ \mathrm{d}x \\ & + \int_{\mathbb{R}_+} \operatorname{sgn}\left(f_*^{-1}\left(q_b(t)\right)-k\right)\left(f\left(u(t,0+)\right)-f(k)\right)\varphi(t,0)\ \mathrm{d}t \\ & + 2\int_{\mathbb{R}_+}\left(1-\frac{q_c(t)}{f\left(\bar{u}\right)}\right) f(k)\,\varphi(t,x_c)\ \mathrm{d}t \end{aligned} \tag{6.5.2}$$

and $f\left(u(t,x_c-)\right) = f\left(u(t,x_c+)\right) \leq q_c(t)$ for almost all $t \in \mathbb{R}_+$.

Above we denote by $u(t,x_c\pm)$ the measure theoretic traces.

Remark 6.5. The first three lines in (6.5.2) are motivated by the initial–boundary value problem (6.5.1a), (6.5.1b), (6.5.1c), see Definition 6.1. The latter line accounts for the constraint (6.5.1d), see Definition 6.2. □

It is immediate to prove that any solution to (6.2.1) in the sense of Definition 6.1 that satisfies $f\left(u(t,x_c-)\right) = f\left(u(t,x_c+)\right) \leq q_c(t)$, also satisfies (6.5.1) in the sense of Definition 6.3. Similarly, by the local nature of Definition 6.3, any solution to (6.5.1) in the sense of Definition 6.3 also solves (6.2.1) in the sense of Definition 6.1, in a neighborhood of any $(t,x) \in \mathbb{R}^2_+$, with $x \neq x_c$.

Theorem 6.3. *Let **(F)**, **(O)**, **(B)**, **(C)** hold. Then, there exists a unique solution* $u = u(t,x;u_o,q_b,q_c)$ *to* (6.5.1) *in the sense of Definition* 6.3 *and* $u(t;u_o,q_b,q_c) \in \mathscr{D}^+$ *for all* $t \in \mathbb{R}_+$, *where* $\mathscr{D}^+$ *is defined by* (6.2.6). *Moreover, if* u_o, u'_o *both satisfy **(O)**,* q_b, q'_b *satisfy **(B)** and* q_c, q'_c *satisfy **(C)**, the corresponding solutions* u, u' *satisfy for every* $t \in \mathbb{R}_+$ *the Lipschitz estimate*

$$\begin{aligned} \left\|u(t)-u'(t)\right\|_{\mathbf{L^1}(\mathbb{R}_+;\mathbb{R})} \leq & \left\|u_o-u'_o\right\|_{\mathbf{L^1}(\mathbb{R}_+;\mathbb{R})} + \left\|q_b-q'_b\right\|_{\mathbf{L^1}([0,t];\mathbb{R})} \\ & + 2\left\|q_c-q'_c\right\|_{\mathbf{L^1}([0,t];\mathbb{R})}\,. \end{aligned} \tag{6.5.3}$$

Proof. By adopting the same notations used in the proofs of Theorem 6.1 and Theorem 6.2, we approximate the constrained initial–boundary value problem (6.5.1) with

$$\begin{cases} \partial_t u^n + \partial_x f^n(u^n) = 0 & (t,x) \in \mathbb{R}^2_+ \\ u^n(0,x) = u^n_o(x) & x \in \mathbb{R}_+ \\ f^n\left(u^n(t,0)\right) = q^n_b(t) & t \in \mathbb{R}_+ \\ f^n\left(u^n(t,x_c)\right) \leq q^n_c(t) & t \in \mathbb{R}_+\,. \end{cases} \tag{6.5.4}$$

An approximate solution u^n to (6.5.1) is then computed via the classical **wave front tracking** technique as the exact weak solution to the above constrained initial–boundary value problem. In other words, u^n is obtained gluing the solutions to the Riemann problems at the points $(0, x^n_{o\alpha})$ of jump of u^n_o, at the points $(t^n_{b\beta}, 0)$ where the boundary datum changes, at the points $(t^n_{c\beta}, x_c)$ where the constraint changes and at all interaction points where two or more waves intersect, or one or more waves hit the boundary or the constraint. For the construction of the solution to standard Riemann problems with a piecewise linear and continuous flow we refer to Sect. 5.2. The constrained Riemann problem is solved in detail in Sect. 6.3. For the solution to the Riemann problem at the boundary, we refer to (6.2.2).

Iteratively solving Riemann problems at the boundary, at interactions and at the constraint, we construct an entropy weak solution u^n to (6.5.4). We prove that $u^n(t) \in \mathscr{D}^+_n$, where $\mathscr{D}^+_n$ is defined by (6.2.11). To this aim, introduce the set

$$\overline{\mathscr{D}}^+_n = \mathscr{D}^+_n \times \mathbf{PC}\left(\mathbb{R}_+; f(\mathscr{M}_n)\right) \times \mathbf{PC}\left(\mathbb{R}_+; f(\mathscr{M}_n)\right)$$

and a semigroup $\overline{S}^n : \mathbb{R}_+ \times \overline{\mathscr{D}}^+_n \to \overline{\mathscr{D}}^+_n$ setting $\overline{S}^n_t(u^n_o, q^n_b, q^n_c) = \left(u^n(t), \mathscr{T}_t q^n_b, \mathscr{T}_t q^n_c\right)$, where $\mathscr{T}$ is the translation operator.

On any $(u^n, q^n_b, q^n_c) \in \overline{\mathscr{D}}^+_n$, written as in (6.2.9), (6.4.5b), define the Glimm type functional

$$\begin{aligned} \Upsilon\left(u^n, q^n_b, q^n_c\right) = & \sum_{\alpha\in\mathbb{N}} \left|\Psi(u^n_{\alpha+1}) - \Psi(u^n_\alpha)\right| + 2\sum_{\beta\in\mathbb{N}} \left|q^n_{b\beta+1} - q^n_{b\beta}\right| \\ & + 5\sum_{\beta\in\mathbb{N}} \left|q^n_{c\beta+1} - q^n_{c\beta}\right| + \gamma_b\left(u^n, q^n_b\right) + \gamma_c\left(u^n, q^n_c\right), \end{aligned} \tag{6.5.5}$$

where Ψ, γ_b and γ_c are, respectively, as in (6.2.7), (6.2.14) and (6.4.8). By the study of the interactions done in the proofs of Theorem 6.1 and Theorem 6.2, the functional Υ either decreases by at least $2^{-n}\overline{f}$, or remains constant while the total number of waves in the approximate solution does not increase (this may happen in interactions away both from the boundary and from the constraint). Hence, the map $\left[t \mapsto \Upsilon\left(\overline{S}^n_t(u^n, q^n_o, q^n_c)\right)\right]$ is non–increasing. Clearly, $\mathrm{TV}\left(\Psi(u^n)\right) + \mathrm{TV}(q^n_b) + \mathrm{TV}(q^n_c) \le \Upsilon(u^n, q^n_b, q^n_c)$, so that Theorem 2.4 can be applied proving that $\Psi(u^n)$ converges a.e. to a limit $\psi \in \mathbf{C}^{0,1}\left(\mathbb{R}_+; \mathbf{BV}(\mathbb{R}_+; [-\overline{f}, \overline{f}])\right)$. Since Ψ is invertible, also u^n converges a.e. to $u = \Psi^{-1}(\psi)$ and $u \in \mathbf{C}^{0,1}\left(\mathbb{R}_+; \mathbf{L}^\infty(\mathbb{R}_+; [0, R])\right)$. To prove that u solves (6.5.1) in the sense of Definition 6.3 is enough to proceed as for the proofs of Theorem 6.1 and Theorem 6.2.

The uniqueness of the solution, as well as the Lipschitz estimate (6.5.3), is established by means of the doubling of variables technique, already used in the proofs of Theorem 6.1 and Theorem 6.2. □

The next corollary ensures that the through flow at a point $\overline{x}$ is a Lipschitz function in $\mathbf{L}^1$ of the inflow q_b and of the constraint q_c.

Corollary 6.1. *Let f satisfy **(F)**, q_b, q_b' satisfy **(B)**, q_c, q_c' satisfy **(C)** and $u_o(x) \equiv 0$. Assume that there exists τ_o such that $\mathrm{spt}\{q_b\} \cup \mathrm{spt}\{q_b'\} \subseteq [0, \tau_o]$ and that there exists a positive τ_c such that*

$$\int_0^{\tau_c} f\left(u(t, x_c+)\right) \, \mathrm{d}t = \int_0^{\tau_o} q_b(t) \, \mathrm{d}t \quad \textit{and} \quad \int_0^{\tau_c} f\left(u'(t, x_c+)\right) \, \mathrm{d}t = \int_0^{\tau_o} q_b'(t) \, \mathrm{d}t \,, \tag{6.5.6}$$

where u and u' are the weak solutions to (6.5.1) corresponding, respectively, to q_b, q_c and to q_b', q_c'. Then, for any $\bar{x} > x_c$ and $t \in \mathbb{R}_+^*$, the following Lipschitz estimate holds

$$\begin{aligned} \left\| f\left(u(\cdot, \bar{x}-)\right) - f\left(u'(\cdot, \bar{x}-)\right) \right\|_{\mathbf{L}^1([0,t];\mathbb{R})} \le & \left\| q_b - q_b' \right\|_{\mathbf{L}^1([0,\min\{t,\tau_o\}];\mathbb{R})} \\ & + 2\left\| q_c - q_c' \right\|_{\mathbf{L}^1([0,\min\{t,\tau_c\}];\mathbb{R})} \,. \end{aligned} \tag{6.5.7}$$

Proof. Fix $t \in \mathbb{R}_+$ and $\bar{x} > x_c$. Starting from the entropy condition (6.5.2) in Definition 6.3, the doubling of variables method applied to the domain $\mathbb{R} \times \mathbb{R}_+ \setminus \{x = x_c\}$, yields the inequality

$$\int_{\mathbb{R}_+} \int_{\mathbb{R}_+} \left(\left| u - u' \right| \partial_t \varphi + \Xi(u, u') \, \partial_x \varphi \right) \, \mathrm{d}y \, \mathrm{d}s \ge 0$$

for all $\varphi \in \mathbf{C}_\mathbf{c}^\infty \left((\mathbb{R}_+)^2 \setminus \left(\mathbb{R}_+ \times \{x = x_c\}\right); \mathbb{R}_+ \right)$, where Ξ is defined by (5.5.4). Replacing φ by a sequence of approximations φ_ε of the characteristic function of the set $]0,t[\times \left(]0,\bar{x}[\setminus \{x_c\}\right)$, and passing to the limit as $\varepsilon \to 0$, we get

$$\int_0^t \Xi(u,u')(s,\bar{x}-) \, \mathrm{d}s + \int_0^x \left| u - u' \right|(t-, y) \, \mathrm{d}y \le \int_0^t \Xi(u,u')(s, 0+) \, \mathrm{d}s + II_{x_c} \,, \tag{6.5.8a}$$

where

$$II_{x_c} = \int_0^t \left(\Xi(u,u')(s, x_c+) - \Xi(u,u')(s, x_c-) \right) \mathrm{d}s \le 2 \int_0^t \left| q_c - q_c' \right|(s) \, \mathrm{d}s \,. \tag{6.5.8b}$$

We now observe that, since the flows are exiting through $x = \bar{x}$, i.e. $\frac{df}{dr}\left(u(t, \bar{x}-)\right) \ge 0$ and $\frac{df}{dr}\left(u'(t, \bar{x}-)\right) \ge 0$, we can deduce the identity

$$\int_0^t \Xi(u,u')(s, \bar{x}-) \, \mathrm{d}s = \int_0^t \left| f(u) - f(u') \right|(s, \bar{x}-) \, \mathrm{d}s \,. \tag{6.5.8c}$$

On the other side, at $x = 0$, we have

$$\int_0^t \Xi(u,u')(s, 0+) \, \mathrm{d}s \le \int_0^t \left| q_b - q_b' \right|(s) \, \mathrm{d}s \,. \tag{6.5.8d}$$

Finally, combining the equations in (6.5.8), we get (6.5.7). □

References

1. Andreianov, B., Goatin, P., Seguin, N.: Finite volume schemes for locally constrained conservation laws. Numerische Mathematik 115, 609–645 (2010)
2. Bardos, C., le Roux, A.Y., Nédélec, J.C.: First order quasilinear equations with boundary conditions. Comm. Partial Differential Equations 4(9), 1017–1034 (1979)
3. Colombo, R.M., Goatin, P.: A well posed conservation law with a variable unilateral constraint. J. Differential Equations 234(2), 654–675 (2007)
4. Colombo, R.M., Goatin, P., Rosini, M.D.: Conservation laws with unilateral constraints in traffic modeling. In: Mussone, L., Crisalli, U. (eds.) Transport Management and Land-Use Effects in Presence of Unusual Demand, Atti del Convegno SIDT 2009 (June 2009)
5. Kružhkov, S.N.: First order quasilinear equations with several independent variables. Mat. Sb. (N.S.) 81(123), 228–255 (1970)

Chapter 7
One–Dimensional Systems of Conservation Laws

Abstract. In this chapter we introduce the definitions of hyperbolicity and strict hyperbolicity and generalize the definitions given in the previous chapters to the case of general systems of conservation laws in one space dimension. In the last two sections we also show how to construct a weak solution to the Cauchy problem for a strictly hyperbolic system of conservation laws with constant coefficient and to the Riemann problem for a strictly hyperbolic system of conservation laws with genuinely nonlinear or linearly degenerate characteristic fields. We prove also the local existence of a weak solution to the Riemann problem for a general system, when the initial states are sufficiently close.

7.1 Introduction

This section concerns with **systems of** N **conservation laws** in one space dimension of the form

$$\begin{aligned} \partial_t u_1 + \partial_x f_1(u_1,\ldots,u_N) &= 0 \\ &\vdots \\ \partial_t u_N + \partial_x f_N(u_1,\ldots,u_N) &= 0 \end{aligned} \tag{7.1.1}$$

with $f_j : \mathbb{R}^N \to \mathbb{R}$, $j = 1,\ldots,N$, that are assumed to be $\mathbf{C}^2$. Introducing the **conserved quantity** $u = (u_1,\ldots,u_N)^T : \mathbb{R}_+ \times \mathbb{R} \to \mathbb{R}^N$ and the **flux function** $f = (f_1,\ldots,f_N)^T : \mathbb{R}^N \to \mathbb{R}^N$, system (7.1.1) can be written in the **conservative form**

$$\partial_t u + \partial_x f(u) = 0_N \tag{7.1.2}$$

or in the **quasilinear form**

$$\partial_t u + a(u)\, \partial_x u = 0_N \,, \tag{7.1.3}$$

M.D. Rosini: *Macroscopic Models for Vehicular Flows & Crowd Dynamics*, UCS, pp. 93–110.
DOI: 10.1007/978-3-319-00155-5_7

where $a(u) = Df(u) = \left(\partial_{u_j} f_i(u)\right)_{1 \le i,j \le N}$ is the Jacobian matrix of the map f at the point u.

Definition 7.1. The system (7.1.2) is said **hyperbolic** if $a(u)$ has N real eigenvalues $\lambda_1(u) \le \lambda_2(u) \le \ldots \le \lambda_N(u)$, together with a basis of right eigenvectors $\{r_1(u), r_2(u), \ldots, r_N(u)\}$. The system (7.1.1) is said **strictly hyperbolic** if it is hyperbolic and if, for every $u \in \mathbb{R}^N$, the eigenvalues of the matrix $a(u)$ are all distinct, say $\lambda_1(u) < \lambda_2(u) < \ldots < \lambda_N(u)$. A point where two or more eigenvalues collide and the algebraic multiplicity exceeds the geometric one is called **umbilical point**.

The eigenvalues λ_j, $j = 1, \ldots, N$, associated to an hyperbolic system of conservation laws (7.1.1) are called **wave speeds** or **characteristic speeds** , while the maps $[u \mapsto (\lambda_j(u), r_j(u))]$, $j = 1, \ldots, N$, are referred to as **characteristic fields**. By the assumption on the regularity of f, the characteristic fields are $\mathbf{C}^1$ functions.

Remark 7.1. In the scalar case $N = 1$, there is a single eigenvalue $\lambda_1(u) = f'(u) = a(u)$ and we set $r_1(u) = 1$. In fact, every scalar conservation law is hyperbolic. □

By definition, for any $j, k = 1, \ldots, N$ with $j \neq k$ and for all $u \in \mathbb{R}^N$

$$a(u)\, r_j(u) = \lambda_j(u)\, r_j(u), \quad l_j(u)^T\, a(u) = \lambda_j(u)\, l_j(u)^T, \quad l_j(u) \cdot r_k(u) = 0, \tag{7.1.4}$$

where $l_j(u) \in \mathbb{R}^N$ is the left eigenvector of $a(u)$ associated to $\lambda_j(u)$, i.e., $l_j(u)$ is a right eigenvector of $a(u)^T$.

Definition 7.2. We say that the jth characteristic field of (7.1.1) is **genuinely nonlinear** if

$$\nabla\lambda_j(u) \cdot r_j(u) \neq 0 \qquad \text{for all } u \in \mathbb{R}^N$$

and **linearly degenerate** if

$$\nabla\lambda_j(u) \cdot r_j(u) = 0 \qquad \text{for all } u \in \mathbb{R}^N .$$

The eigenvectors are defined up to a multiplicative constant; therefore, if the jth characteristic field is genuinely nonlinear, there is a canonical choice to normalize the right and left eigenvectors $r_j(u)$ and $l_j(u)$ in such a way that for all $u \in \mathbb{R}^N$

$$\nabla\lambda_j(u) \cdot r_j(u) = 1 \qquad \text{and} \qquad l_j(u) \cdot r_j(u) = 1 . \tag{7.1.5}$$

Proposition 7.1 (Boillat [1]). *If λ is a characteristic speed of the system* (7.1.1), *whose multiplicity is a constant greater than or equal to* 2, *then the corresponding characteristic fields are linearly degenerate.*

Proof. By hypotheses we can choose two linearly independent eigenvectors r and s associated to the eigenvalue λ. Differentiating the relation $a\, r = \lambda\, r$ in the direction s, we have for any $i = 1, \ldots, N$

$$\sum_{j,k=1}^{N} \partial_{u_k} a_{ij}(u)\; s_k(u)\; r_j(u) + \sum_{j,k=1}^{N} a_{ij}(u)\; \partial_{u_k} r_j(u)\; s_k(u)$$
$$= \sum_{k=1}^{N} \partial_{u_k} \lambda(u)\; s_k(u)\; r_i(u) + \sum_{k=1}^{N} \lambda(u)\; \partial_{u_k} r_i(u)\; s_k(u)\,.$$

Analogously, differentiating the relation $a\,s = \lambda\, s$ in the direction r, we have for any $i = 1,\ldots,N$

$$\sum_{j,k=1}^{N} \partial_{u_k} a_{ij}(u)\; r_k(u)\; s_j(u) + \sum_{j,k=1}^{N} a_{ij}(u)\; \partial_{u_k} s_j(u)\; r_k(u)$$
$$= \sum_{k=1}^{N} \partial_{u_k} \lambda(u)\; r_k(u)\; s_i(u) + \sum_{k=1}^{N} \lambda(u)\; \partial_{u_k} s_i(u)\; r_k(u)\,.$$

Since $\nabla a(u)$ is a symmetric bilinear form, subtracting these equalities we can eliminate the term corresponding to $\nabla a(u)$ and obtain

$$\sum_{j,k=1}^{N} \left(a_{ij}(u) - \lambda(u)\; \delta_{ij}^{K}\right)\; \left(\partial_{u_k} r_j(u)\; s_k(u) - \partial_{u_k} s_j(u)\; r_k(u)\right)$$
$$= \left(\sum_{k=1}^{N} \partial_{u_k} \lambda(u)\; s_k(u)\right)\; r_i(u) - \left(\sum_{k=1}^{N} \partial_{u_k} \lambda(u)\; r_k(u)\right)\; s_i(u)\,,$$

or equivalently

$$\left(a(u) - \lambda(u)\,\mathrm{Id}_N\right)\; \{r,s\} = \left(\nabla\lambda(u)\cdot s(u)\right)\; r(u) - \left(\nabla\lambda(u)\cdot r(u)\right)\; s(u)\,,$$

where $\{r,s\}$ denotes the **Poisson bracket** of the vector fields r and s. The right–hand side of the above equality is also an eigenvector associated to λ, with the result that $\left(a(u) - \lambda(u)\,\mathrm{Id}_N\right)^2\; \{r,s\} = 0_N$. The equality of the algebraic and geometric multiplicity of λ ensures that also $\left(a(u) - \lambda(u)\,\mathrm{Id}_N\right)\; \{r,s\} = 0_N$. Therefore we find the relation $\left(\nabla\lambda(u)\cdot s(u)\right)\; r(u) - \left(\nabla\lambda(u)\cdot r(u)\right)\; s(u) = 0_N$ and the linear independence of r and s implies the nullity of the coefficients. □

Remark 7.2. Observe that if $[s \mapsto v(s)]$ is an **integral curve** of a genuinely nonlinear (respectively linearly degenerate) vector field r_j, namely $v'(\xi) = r_j\left(v(\xi)\right)$, then $[\xi \mapsto \lambda_j\left(v(\xi)\right)]$ is strictly monotone (respectively constant). □

Definition 7.3. A $\mathbf{C}^1$ function $w\;:\;\mathbb{R}^N \to \mathbb{R}$ is a ***j*–Riemann invariant** if for all $u \in \mathbb{R}^N$ it satisfies

$$\nabla w(u)\cdot r_j(u) = 0 \qquad \text{for all } u \in \mathbb{R}^N\,.$$

Proposition 7.2. *Any j–Riemann invariant is constant along the integral curves of the vector field r_j.*

Proof. In fact, directly by definition we have that $\frac{\mathrm{d}}{\mathrm{d}\xi} w\left(v(\xi)\right) = \nabla w\left(v(\xi)\right) \cdot v'(\xi) = \nabla w\left(v(\xi)\right) \cdot r_j\left(v(\xi)\right) = 0$. □

Definition 7.4. Let $\overline{u} \in \mathbf{L}^\infty_{\mathbf{loc}}(\mathbb{R};\mathbb{R}^N)$. A function $u \in \mathbf{C^0}\left(\mathbb{R}^*_+;\mathbf{L}^\infty_{\mathbf{loc}}(\mathbb{R};\mathbb{R}^N)\right)$ is a **weak solution** to the Cauchy problem

$$\partial_t u + \partial_x f(u) = 0_N \tag{7.1.6a}$$
$$u(0,x) = \overline{u}(x) \tag{7.1.6b}$$

if for all function $\varphi \in \mathbf{C}^\infty_\mathbf{c}(\mathbb{R}^*_+ \times \mathbb{R};\mathbb{R}^N)$

$$\int_{\mathbb{R}_+}\int_{\mathbb{R}} \left(u \cdot \partial_t \varphi + f(u) \cdot \partial_x \varphi\right)\, \mathrm{d}x\, \mathrm{d}t + \int_{\mathbb{R}} \overline{u}(x) \cdot \varphi(0,x)\, \mathrm{d}x = 0\,. \tag{7.1.7}$$

Remark 7.3. The hyperbolicity and the notions of genuinely nonlinear or linearly degenerate characteristic fields are invariant under the change of unknown $[u \mapsto v = \varphi(u)]$ for any diffeomorphism $\varphi\,:\, \mathbb{R}^N \to \mathbb{R}^N$. On the other hand, the definition of weak solution is not preserved by the diffeomorphisms, even in the one–dimensional case (see Example 3.3). □

Analogously to the scalar case, a piecewise $\mathbf{C^1}$ function u is a weak solution of (7.1.6) iff is a solution of (7.1.6) in the usual sense in the domains where it is smooth and, furthermore, there exists a finite number of smooth curves $\Gamma\,:\, x = x(t)$ of discontinuity for u along which the **Rankine–Hugoniot (jump) condition**

$$(u_+ - u_-)\,\sigma = f(u_+) - f(u_-) \tag{7.1.8}$$

holds. As usual, above we use the standard notation $u_\pm(t,x) = \lim_{\varepsilon \searrow 0} u(t, x \pm \varepsilon)$ and $\sigma(t) = x'(t)$.

Remark 7.4. The vector equations in (7.1.8) form a set of N scalar equations relating the right and the left states $u_+, u_- \in \mathbb{R}^N$ with the speed of propagation $\sigma \in \mathbb{R}$ of the discontinuity. □

Remark 7.5. Introduce the **averaged matrix** $a(u,v) = \int_0^1 a\left(\gamma\, u + (1-\gamma)\, v\right)\, \mathrm{d}\gamma$. The Rankine–Hugoniot conditions (7.1.8) can be written in the following form

$$\begin{aligned}(u_+ - u_-)\,\sigma = f(u_+) - f(u_-) &= \int_0^1 \frac{\mathrm{d}}{\mathrm{d}\gamma} f\left(\gamma\, u_+ + (1-\gamma)\, u_-\right)\, \mathrm{d}\gamma \\ &= \int_0^1 a\left(\gamma\, u_+ + (1-\gamma)\, u_-\right)\, (u_+ - u_-)\, \mathrm{d}\gamma = a(u_+,u_-)\, (u_+ - u_-)\,.\end{aligned}$$

In other words, the Rankine–Hugoniot conditions (7.1.8) are equivalent to require that $(u_+ - u_-)$ is an eigenvector of the matrix $a(u_-,u_+)$ and that the speed σ coincides with the corresponding eigenvalue. In particular, if f is Lipschitz continuous, then $|\sigma| \leq \mathscr{L}ip(f)$. □

As in the scalar case, the class of weak solutions is not appropriate because the weak solution of the Cauchy problem (7.1.6) is not in general unique. For this reason, we must impose an entropy criterion in order to select a (hopefully) unique solution.

Definition 7.5. A continuously differentiable convex function $U : \mathbb{R}^N \to \mathbb{R}$ is called an **entropy** for the system of conservation laws (7.1.1) if there exists a function $F : \mathbb{R}^N \to \mathbb{R}$, called **entropy flux**, such that for all $u \in \mathbb{R}^N$

$$\nabla U(u)^T \, a(u) = \nabla F(u)^T \, . \tag{7.1.9}$$

In this case, (U,F) is called an **entropy pair** for the system (7.1.1).

Remark 7.6. Observe that (7.1.9) can be regarded as a first–order system of N equations for the two scalar variables U and F. For $N \geq 3$, this system is overdeterminate. Thus, in general one should expect that do not exist any entropy pair if $N \geq 3$. However, in all practical examples derived from mechanics or physics, it is possible to find an entropy pair that has a physical meaning. A classification of hyperbolic systems with respect to their entropies is given in [10]. □

Remark 7.7. Equation (7.1.9) implies that any smooth solution u of (7.1.1) satisfies the additional conservation law

$$\partial_t U(u) + \partial_x F(u) = 0 \, . \tag{7.1.10}$$

Indeed, (7.1.9) together with (7.1.1) yield

$$\begin{aligned} \partial_t U(u) + \partial_x F(u) &= \nabla U(u) \cdot \partial_t u + \nabla F(u) \cdot \partial_x u \\ &= \nabla U(u) \cdot \big(-a(u)\partial_x u\big) + \nabla F(u) \cdot \partial_x u = 0 \, . \end{aligned}$$

On the other hand, as already observed in Remark 3.6 for the scalar case, when u is a weak solution of (7.1.1), in general it does not provide a weak solution to (7.1.10). It can be proved, see for instance [4, Theorem 3.3], that any weak solution of (7.1.1) obtained via the vanishing viscosity method satisfies in the sense of distribution the entropy condition

$$\partial_t U(u) + \partial_x F(u) \leq 0$$

for any entropy pair (U,F). This leads us to introduce the following definition of entropy weak solution. □

Definition 7.6. A weak solution u of the Cauchy problem (7.1.6), $\overline{u} \in \mathbf{L}^\infty_{\mathbf{loc}}(\mathbb{R};\mathbb{R}^N)$, is called an **entropy weak solution** if satisfies for any entropy pair (U,F) and for all test functions $\varphi \in \mathbf{C}^\infty_\mathbf{c}(\mathbb{R}^*_+ \times \mathbb{R};\mathbb{R}^*_+)$ the **entropy inequality**

$$\int_{\mathbb{R}_+} \int_{\mathbb{R}} \big(U(u) \; \partial_t \varphi + F(u) \; \partial_x \varphi\big) \; \mathrm{d}x \; \mathrm{d}t + \int_{\mathbb{R}} U\left(\overline{u}(x)\right) \; \varphi(0,x) \; \mathrm{d}x \geq 0 \, . \tag{7.1.11}$$

By arguing as in the scalar case, see Theorem 3.3, it is a simple matter to check that a piecewise $\mathbf{C}^1$ function u is an entropy weak solution of (7.1.6) iff

(i) u is a smooth solution of (7.1.6) in the domains where u is $\mathbf{C}^1$.
(ii) u satisfies the Rankine–Hugoniot condition (7.1.8) along its curves of discontinuity.
(iii) If $\Gamma : x = x(t)$ is a curve of discontinuity for u and $\sigma(t) = x'(t)$, then for any entropy pair (U,F), u satisfies along Γ the **entropy jump condition**

$$\big(U(u_+) - U(u_-)\big)\ \sigma \geq F(u_+) - F(u_-)\ . \tag{7.1.12}$$

For later use, let us introduce the **characteristic curves** $C_j = \{(t,x) \in \mathbb{R}^*_+ \times \mathbb{R}^N : x = x(t)\}$ of the jth characteristic field associated to a solution u of (7.1.2) as the integral curves of the ordinary differential system

$$x'(t) = \lambda_j\big(u(t,x)\big)\ .$$

We conclude the section underlining that the method of characteristics as well as the method of wave front tracking to construct weak solutions remain effective also for sufficiently general one–dimensional systems of conservation laws, see a proof in [2, Chap. 7], [3] or [5, Chap. 6]. However, in the next two sections we consider only the case of a Cauchy problem for a strictly hyperbolic system of conservation laws with constant coefficient and of a Riemann problem for a strictly hyperbolic system of conservation laws with genuinely nonlinear or linearly degenerate characteristic fields.

7.2 Strictly Hyperbolic Linear Systems with Constant Coefficients

Consider the Cauchy problem for a strictly hyperbolic linear system of conservation laws with constant coefficients

$$\partial_t u + a\ \partial_x u = 0_N\ , \qquad u(t=0) = \overline{u}\ , \tag{7.2.1}$$

where $u = (u_1, \ldots, u_N)^T$ is a column vector of $\mathbb{R}^N$ and a is an $N \times N$ constant real matrix having N distinct real eigenvalues

$$\lambda_1 < \lambda_2 < \ldots < \lambda_N\ .$$

Let $l_j, r_j \in \mathbb{R}^N$ be, respectively, the left and right eigenvectors corresponding to the eigenvalue λ_j. Since the eigenvalues are strictly distinct, $\{r_1, r_2, \ldots, r_N\}$ and $\{l_1, l_2, \ldots, l_N\}$ are two bases in $\mathbb{R}^N$. Moreover, we can normalize the left eigenvectors in such a way that

$$l_j \cdot r_k = \delta^K_{jk}\ . \tag{7.2.2}$$

By multiplying (7.2.1) on the left by l_j, the system (7.2.1) decouples into the following N independent scalar Cauchy problems with $\alpha_j = l_j \cdot u$ as the unknown and $\overline{\alpha}_j = l_j \cdot \overline{u}$ as the initial datum

$$\partial_t \alpha_j + \lambda_j \, \partial_x \alpha_j = 0 \quad , \qquad \alpha_j(t=0) = \overline{\alpha}_j \quad , \qquad j = 1, \ldots, N \, .$$

As proved in the Example 3.1, the solutions to the above Cauchy problem is the traveling wave $\alpha_j(t,x) = \overline{\alpha}_j(x - \lambda_j \, t)$. By using the characteristic decomposition (2.4.5) for $u = \sum_{j=1}^{N} \alpha_j \, r_j$, it results that

$$u(t,x) = \sum_{j=1}^{N} l_j \cdot \overline{u}(x - \lambda_j \, t) \, r_j \tag{7.2.3}$$

provides a weak solution to (7.2.1). We underline that in general (7.2.3) is not an entropy weak solution.

Consider in particular the Cauchy problem (7.2.1) with a Riemann data such as

$$\overline{u}(x) = \begin{cases} u^l & \text{if } x < 0 \\ u^r & \text{if } x \geq 0 \, . \end{cases} \tag{7.2.4}$$

Then the solution to (7.2.1), (7.2.4) is given by

$$u(t,x) = \begin{cases} w_0 = u^l & \text{if } x < \lambda_1 \, t \\ w_1 & \text{if } \lambda_1 \, t \leq x < \lambda_2 \, t \\ \vdots & \\ w_{N-1} & \text{if } \lambda_{N-1} \, t \leq x < \lambda_N \, t \\ w_N = u^r & \text{if } x \geq \lambda_N \, t \, , \end{cases} \tag{7.2.5a}$$

where

$$w_m = u^l + \sum_{j=1}^{m} \beta_i \, r_j \, , \qquad m \in \{1, \ldots, N-1\} \tag{7.2.5b}$$

with $\beta_j = l_j \cdot (u^r - u^l)$. This points out that, in general, the initial discontinuity breaks up into N discontinuities, which propagate with the characteristic speeds λ_j, $j = 1, \ldots, N$, see Fig. 7.1. Note that the Rankine–Hugoniot relation (7.1.8) is satisfied across the line of discontinuity $x = \lambda_m \, t$ since

$$(w_m - w_{m-1}) \, \lambda_m = \beta_m \, \lambda_m \, r_m = \beta_m \, a \, r_m = a \, (w_m - w_{m-1}) \, .$$

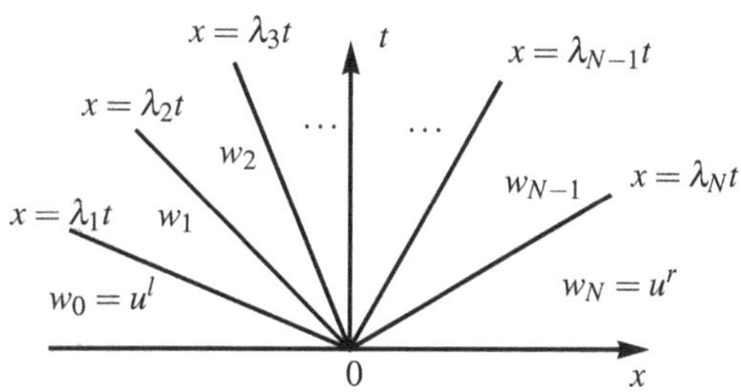

Fig. 7.1 Representation in the (x,t)–plane of the solution (7.2.5) to the Riemann problem for the strictly hyperbolic linear system of conservation laws with constant coefficient (7.2.1), (7.2.4).

7.3 Riemann Problems

In this section we are looking for piecewise smooth solutions $u : \mathbb{R}_+ \times \mathbb{R} \to \mathbb{R}^N$ to Riemann problems for nonlinear strictly hyperbolic systems of the form

$$\partial_t u + \partial_x f(u) = 0_N \quad , \qquad u(t=0) = \begin{cases} u^l & \text{if } x < 0 \\ u^r & \text{if } x \geq 0\,. \end{cases} \tag{7.3.1}$$

We assume that $u^l \neq u^r$, because otherwise the solution to (7.3.1) is trivially the constant function $u(t,x) \equiv u^l = u^r$. By assumption the eigenvalues of the Jacobian matrix $a(u)$ of $f(u)$ are such that $\lambda_1(u) < \lambda_2(u) < \ldots < \lambda_N(u)$ and the corresponding eigenvectors $r_1(u), r_2(u), \ldots, r_N(u)$ are independent.

7.3.1 *Rarefaction Waves*

At first, we restrict the attention to **self similar solutions** of (7.3.1), i.e., solutions of the form

$$u(t,x) = \phi\left(x/t\right)\,. \tag{7.3.2}$$

If ϕ is $\mathbf{C^1}$, then these solutions satisfy the ordinary differential equation

$$-\frac{x}{t^2}\,\phi'\left(x/t\right) + \frac{1}{t}\,a\Big(\phi\left(x/t\right)\Big)\;\phi'\left(x/t\right) = 0_N\,,$$

so that, multiplying by t and letting $\xi = x/t$, we obtain the equation

$$\Big[a\Big(\phi\left(\xi\right)\Big) - \xi\;\mathrm{Id}_N\Big]\;\phi'\left(\xi\right) = 0_N\,.$$

Since $\phi(-\infty) = u^l \neq u^r = \phi(+\infty)$, we have $\phi'(\xi) \not\equiv 0$ and the above equation implies that there exists an index $j \in \{1,\ldots,N\}$ and a smooth function $\alpha : \mathbb{R} \to \mathbb{R}$ such that

$$\phi'(\xi) = \alpha(\xi)\; r_j\left(\phi(\xi)\right) \tag{7.3.3a}$$

$$\xi = \lambda_j\left(\phi(\xi)\right)\,. \tag{7.3.3b}$$

If $\phi'(\xi)$ is non–zero on an interval, since the eigenvalues are distinct, the index j does not depend on ξ in that interval. If we differentiate (7.3.3b) with respect to ξ and then use (7.3.3a), we get

$$1 = \nabla\lambda_j\left(\phi(\xi)\right)\cdot\phi'(\xi) = \alpha(\xi)\;\nabla\lambda_j\left(\phi(\xi)\right)\cdot r_j\left(\phi(\xi)\right)\,. \tag{7.3.4}$$

Equation (7.3.4) can not be solved if the jth characteristic field is linearly degenerate. On the other hand, if the jth characteristic field is genuinely nonlinear, by the normalization (7.1.5) and Eq. (7.3.4) we get $\alpha(\xi) \equiv 1$ and (7.3.3) becomes

$$\phi'(\xi) = r_j(\phi(\xi)) \tag{7.3.5a}$$
$$\lambda_j(\phi(\xi)) = \xi \,. \tag{7.3.5b}$$

In particular, (7.3.5a) implies that ϕ is an integral curve of the vector field r_j and therefore $[\xi \mapsto \lambda_j(\phi(\xi))]$ is strictly monotone, see Remark 7.2. This together with (7.3.5) imply that $\phi\left(\lambda_j(u^l)\right) = u^l$ and $\phi\left(\lambda_j(u^r)\right) = u^r$. In conclusion, the above analysis proves that the function

$$u(t,x) = \begin{cases} u^l & \text{if } x < \lambda_j(u^l)\,t \\ \phi\left(x/t\right) & \text{if } \lambda_j(u^l)\,t \le x < \lambda_j(u^r)\,t \\ u^r & \text{if } x \ge \lambda_j(u^r)\,t \end{cases} \tag{7.3.6}$$

is a continuous self similar weak solution of (7.3.1). More precisely, u is piecewise $\mathbf{C}^1$ and, as a consequence, it is an entropy weak solution of (7.3.1).

Definition 7.7. A self similar weak solution (7.3.6) of the Riemann problem (7.3.1) is called ***j*–centered rarefaction wave** connecting the states u^l and u^r, and the point at which the discontinuity occurs is called **center of the wave**.

Concerning the existence of j–rarefaction waves, we have the following local result.

Theorem 7.1. *Assume that the jth characteristic field is genuinely nonlinear with the normalization* (7.1.5). *Given a state u^l, there exists a curve $\mathscr{R}_j(u^l)$ of states defined by* (7.3.8) *that can be connected to u^l on the right by a j–rarefaction wave. Moreover, $\mathscr{R}_j(u^l)$ has a parametrization $[\varepsilon \mapsto \Phi_j(\varepsilon)]$ defined for $\varepsilon \in \mathbb{R}_+^*$ small enough, such that*

$$\Phi_j(\varepsilon) = u^l + \varepsilon\, r_j(u^l) + \frac{\varepsilon^2}{2}\,\nabla r_j(u^l)\, r_j(u^l) + O(\varepsilon^3)\,. \tag{7.3.7}$$

Finally, for any j–Riemann invariant w, we have $w\left(\Phi_j(\varepsilon)\right) = w(u^l)$.

Proof. The ordinary differential system

$$\phi'(\xi) = r_j(\phi(\xi))\,, \qquad \xi > \lambda_j(u^l) \tag{7.3.8a}$$
$$\phi\left(\lambda_j(u^l)\right) = u^l \tag{7.3.8b}$$

has a $\mathbf{C}^1$ solution for $\lambda_j(u^l) \le \xi \le \lambda_j(u^l) + \varepsilon_o$, with $\varepsilon_o \in \mathbb{R}_+$ small enough. By using (7.3.8a) and the normalization (7.1.5), we have

$$\frac{d}{d\xi}\lambda_j(\phi(\xi)) = \nabla\lambda_j(\phi(\xi)) \cdot \phi'(\xi) = \nabla\lambda_j(\phi(\xi)) \cdot r_j(\phi(\xi)) = 1\,.$$

If we integrate the above equation with respect to ξ and use (7.3.8b), then we get

$$\xi - \lambda_j(u^l) = \int_{\lambda_j(u^l)}^{\xi} \mathrm{d}\xi = \lambda_j\left(\phi(\xi)\right) - \lambda_j\left(\phi\left(\lambda_j(u^l)\right)\right) = \lambda_j\left(\phi(\xi)\right) - \lambda_j(u^l)$$

so that $\lambda_j\left(\phi(\xi)\right) = \xi$. Hence, the function ϕ is indeed the solution of (7.3.5) for which (7.3.8b) holds.

Next, we observe that $\mathscr{R}_j(u^l) = \{\phi(\xi) \,:\, \lambda_j(u^l) \leq \xi \leq \lambda(u^l) + \varepsilon_o\}$. Hence, it is sufficient to set $\Phi_j(\varepsilon) = \phi(\lambda_j(u^l) + \varepsilon)$, $0 \leq \varepsilon \leq \varepsilon_o$. By (7.3.8) we have

$$\begin{aligned}
\Phi_j(0) &= \phi\left(\lambda_j(u^l)\right) = u^l \\
\Phi_j'(0) &= \phi'\left(\lambda_j(u^l)\right) = r_j\left(v\left(\lambda_j(u^l)\right)\right) = r_j(u^l) \\
\Phi_j''(0) &= \nabla r_j\left(\phi\left(\lambda_j(u^l)\right)\right)\, \phi'\left(\lambda_j(u^l)\right) = \nabla r_j(u^l)\, r_j(u^l)\,.
\end{aligned}$$

This proves the expansion (7.3.7). In particular, $\mathscr{R}_j(u^l)$ is an integral curve of r_j, which is tangent to $r_j(u^l)$ at the point u^l and therefore, by Proposition 7.2, any j–Riemann invariant is constant along it. □

The curve $\mathscr{R}_j(u^l)$ is called a *j*–**rarefaction curve**, see Fig. 7.2.

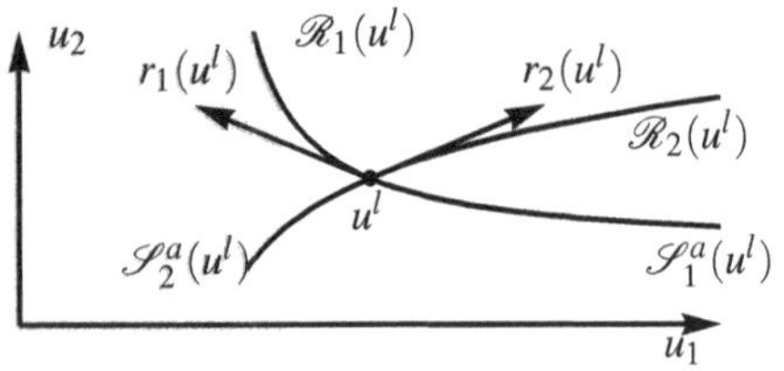

Fig. 7.2 Representation in the (u_1, u_2)–phase plane of the curves $\mathscr{S}_j^a(u^l)$, $\mathscr{R}_j(u^l)$, $j = 1, 2$, in the genuinely nonlinear case with $N = 2$

7.3.2 Shock Waves and Contact Discontinuities

Let us recall that, given two distinct states u^l and u^r of $\mathbb{R}^N$, a piecewise constant discontinuous function of the following form

$$u(t,x) = \begin{cases} u^l & \text{if } x < \sigma\, t \\ u^r & \text{if } x \geq \sigma\, t \end{cases} \tag{7.3.9}$$

is a weak solution of (7.3.1) if the speed of propagation σ satisfies the Rankine–Hugoniot condition (7.1.8).

Definition 7.8. The **Rankine–Hugoniot set** of u^l is the set of all states $u \in \mathbb{R}^N$ such that there exists $\sigma = \sigma(u^l, u) \in \mathbb{R}$ for which (7.3.9) is a weak solution to the Riemann problem (7.3.1).

The structure of the Rankine–Hugoniot set of u^l is given by the following theorem.

Theorem 7.2. *[Lax [7]] Let u^l be in $\mathbb{R}^N$. The Rankine–Hugoniot set of u^l is locally made of N smooth curves $\mathscr{S}_j(u^l)$, $j = 1,\dots,N$. Moreover, each $\mathscr{S}_j(u^l)$ has a parametrization $[\varepsilon \mapsto \Psi_j(\varepsilon)]$ defined for $|\varepsilon|$ small enough, such that*

$$\Psi_j(\varepsilon) = u^l + \varepsilon\, r_j(u^l) + \frac{\varepsilon^2}{2}\, \nabla r_j(u^l)\, r_j(u^l) + O(\varepsilon^3) \tag{7.3.10a}$$

and

$$\sigma\left(u^l, \Psi_j(\varepsilon)\right) = \lambda_j(u^l) + \frac{\varepsilon}{2}\, \nabla\lambda_j(u^l)\cdot r_j(u^l) + O(\varepsilon^2)\,. \tag{7.3.10b}$$

Proof. As proved in Remark 7.5, the Rankine–Hugoniot relation (7.1.8) is equivalent to require that $(u^r - u^l)$ is an eigenvector to the averaged matrix $a(u^l, u^r)$ having the speed of propagation $\sigma = \sigma(u^l, u^r)$ as the corresponding eigenvalue. Note that the matrix $a(u^l, u^l) = a(u^l)$ has N real distinct eigenvalues $\lambda_1(u^l) < \lambda_2(u^l) < \ldots < \lambda_N(u^l)$, and the function $[u \mapsto a(u^l, u)]$ is continuous. Thus, by using a continuity argument, there exists a neighborhood $\mathscr{N}$ of u^l and N real functions $[u \mapsto \lambda_j(u^l, u)]$, $j = 1,\dots,N$, defined in $\mathscr{N}$ such that $\lambda_j(u^l, u)$, $j = 1,\dots,N$, are the N distinct real eigenvalues of $a(u^l, u)$ with $\lambda_j(u^l, u^l) = \lambda_j(u^l)$. Therefore, a state $u \in \mathscr{N}$ belongs to the Rankine–Hugoniot set of u^l iff there exists an index $j \in \{1,\dots,N\}$ such that $\sigma(u^l, u) = \lambda_j(u^l, u)$ and $(u - u^l)$ is a corresponding right eigenvector of $a(u^l, u)$. If $k \neq j$ and $l_k(u^l, u)$ is the left eigenvector corresponding to $\lambda_k(u^l, u)$, i.e., $l_k(u^l, u)^T\, a(u^l, u) = \lambda_k(u^l, u)\, l_k(u^l, u)^T$, then

$$l_k(u^l, u)\cdot(u - u^l) = 0\,. \tag{7.3.11}$$

The jth eigenvector of $a(u^l, u)$ is precisely the non–trivial solution of (7.3.11) for all $k \neq j$. This gives a system of $N - 1$ scalar nonlinear algebraic equations in the N unknowns $u = (u_1, \dots, u_N)^T$, which can be written in the form

$$M_j(u) = L_j(u)\,(u - u^l) = 0\,,$$

where

$$L_j(u) = \Big(l_1(u^l, u)\ \ldots\ l_{j-1}(u^l, u)\ l_{j+1}(u^l, u)\ \ldots\ l_N(u^l, u)\Big)^T\,.$$

It results

$$M_j(u^l) = 0\,, \qquad \nabla M_j(u^l) = L_j(u^l)\,.$$

Moreover, since the vectors $l_j(u^l, u^l) = l_j(u^l)$, $j = 1,\dots,N$, are linearly independent, the $(N-1)\times N$ matrix $L_j(u)$ has rank $N - 1$. Hence, by the implicit function theorem (Theorem 2.2) there exists a one–parameter family $\mathscr{S}_j(u^l) : u = \Psi_j(\theta)$ of solutions to (7.3.11), $|\theta|$ small enough, such that

$$l_k(u^l, \Psi_j(\theta))\cdot(\Psi_j(\theta) - u^l) = 0 \tag{7.3.12}$$

and

$$\Psi_j(0) = u^l \qquad \text{and} \qquad \sigma\left(u^l, \Psi_j(0)\right) = \lambda_j(u^l)\,. \tag{7.3.13a}$$

Therefore, we have proved that the Rankine–Hugoniot set of u^l is locally made of the N curves $\mathscr{S}_j(u^l)$. It remains to check the expansions (7.3.10). From (7.3.12) and (7.3.13a) it follows that for all $k \neq j$

$$0 = \lim_{\theta \to 0} l_k \left(u^l, \Psi_k(\theta)\right)^T \left(\frac{\Psi_j(\theta) - \Psi_j(0)}{\theta}\right) = l_k(u^l)^T \, \Psi_j'(0) \, ,$$

so that $\Psi_j'(0)$ is collinear to $r_j(u^l)$. Hence, it is not limitative to assume that

$$\Psi_j'(0) = r_j(u^l) \, . \tag{7.3.13b}$$

By introducing the notation

$$a_j(\theta) = a\left(\Psi_j(\theta)\right) \, , \qquad \sigma_j(\theta) = \lambda_j\left(u^l, \Psi_j(\theta)\right) \, ,$$

the first condition in (7.1.4) write

$$a_j \, r_j(\Psi_j) = \lambda_j(\Psi_j) \, r_j(\Psi_j) \, , \qquad \sigma_j \, (\Psi_j - u^l) = f(\Psi_j) - f(u^l) \, .$$

By differentiating one time the first equation above and two times the second equation above, we get

$$\begin{aligned} a_j' \, r_j(\Psi_j) + a_j \, \nabla r_j(\Psi_j) \, \Psi_j' &= \nabla\lambda_j(\Psi_j) \cdot \Psi_j' \, r_j(\Psi_j) + \lambda_j(\Psi_j) \, \nabla r_j(\Psi_j) \, \Psi_j' \\ \sigma_j'' \, (\Psi_j - u^l) + 2 \, \sigma_j' \, \Psi_j' + \sigma_j \, \Psi_j'' &= a_j' \, \Psi_j' + a_j \, \Psi_j'' \end{aligned}$$

and computing at $\theta = 0$, from (7.3.13b) it follows that

$$\begin{aligned} \left[a(u^l) - \lambda_j(u^l) \, \mathrm{Id}_N\right] \, \nabla r_j(u^l) \, r_j(u^l) + a_j'(0) \, r_j(u^l) - \nabla\lambda_j(u^l) \cdot r_j(u^l) \, r_j(u^l) &= 0_N \\ \left[a(u^l) - \lambda_j(u^l) \, \mathrm{Id}_N\right] \, \Psi_j''(0) + a_j'(0) \, r_j(u^l) - 2 \, \sigma_j'(0) \, r_j(u^l) &= 0_N \, . \end{aligned}$$

Hence, by subtracting the above equalities, we get

$$\begin{aligned} &\left[a(u^l) - \lambda_j(u^l) \, \mathrm{Id}_N\right] \, \left(\Psi_j''(0) - \nabla r_j(u^l) \, r_j(u^l)\right) + \nabla\lambda_j(u^l) \cdot r_j(u^l) \, r_j(u^l) \\ &= 2 \, \sigma_j'(0) \, r_j(u^l) \, . \end{aligned}$$

Now, multiplying the above equality by $l_j(u^l)^T$ on the left and using the normalization (7.2.2) gives

$$\sigma_j'(0) = \frac{1}{2} \, \nabla\lambda_j(u^l) \cdot r_j(u^l) \tag{7.3.13c}$$

and therefore

$$\left[a(u^l) - \lambda_j(u^l) \, \mathrm{Id}_N\right] \, \left(\Psi_j''(0) - \nabla r_j(u^l) \, r_j(u^l)\right) = 0_N \, .$$

As a consequence, there exists $\beta \in \mathbb{R}$ such that

$$\Psi_j''(0) - \nabla r_j(u^l)\, r_j(u^l) = \beta\, r_j(u^l)\,. \tag{7.3.13d}$$

Next, by using (7.3.13) and making the change of parametrization $\theta = \varepsilon - \frac{1}{2}\,\beta\,\varepsilon^2$ we obtain

$$\begin{aligned}
\Psi_j(\varepsilon) &= u^l + \theta\, r_j(u^l) + \frac{\theta^2}{2}\,\Big(\nabla r_j(u^l)\, r_j(u^l) + \beta\, r_j(u^l)\Big) + O(\theta^3) \\
&= u^l + \varepsilon\, r_j(u^l) + \frac{\varepsilon^2}{2}\,\nabla r_j(u^l)\, r_j(u^l) + O(\varepsilon^3) \\
\sigma\Big(u^l, \Psi_j(\varepsilon)\Big) &= \lambda_j(u^l) + \frac{\theta}{2}\,\nabla\lambda_j(u^l)\cdot r_j(u^l) + O(\theta^2) \\
&= \lambda_j(u^l) + \frac{\varepsilon}{2}\,\nabla\lambda_j(u^l)\cdot r_j(u^l) + O(\varepsilon^2)
\end{aligned}$$

i.e., the expansions in (7.3.10). □

Consider the case where the jth characteristic field is genuinely nonlinear. The curve $\mathscr{S}_j(u^l)$ is then called a j–**shock curve** and, by using the normalization (7.1.5), (7.3.10b) can be written

$$\sigma\Big(u^l, \Psi_j(\varepsilon)\Big) = \lambda_j(u^l) + \frac{\varepsilon}{2} + O(\varepsilon^2)\,.$$

If u^r belongs to the j–shock curve $\mathscr{S}_j(u^l)$, or equivalently if u^l belongs to the j–shock curve $\mathscr{S}_j(u^r)$, a weak solution to (7.3.1) of the form (7.3.9) is called j–**shock wave**.

Remark 7.8. We underline that if the jth characteristic field is genuinely nonlinear, then to all the states u of the j–shock curve $\mathscr{S}_j(u^l)$ correspond a weak solution which, however, is not in general entropic, see Theorem 7.4. □

Let us next turn to the linearly degenerate case.

Theorem 7.3. *If the jth characteristic field is linearly degenerate, the curve $\mathscr{S}_j(u^l)$ given by Theorem 7.2 coincides with the integral curve of r_j and*

$$\sigma\Big(u^l, \Psi_j(\varepsilon)\Big) = \lambda_j\left(\Psi_j(\varepsilon)\right) = \lambda_j(u^l)\,. \tag{7.3.14}$$

Finally, for any j–Riemann invariant w, we have

$$w\left(\Psi_j(\varepsilon)\right) = w(u^l)\,. \tag{7.3.15}$$

Proof. Let v be the integral curve of r_j passing through u^l, i.e., $[\xi \mapsto v(\xi)]$ is the solution of

$$v'(\xi) = r_j\left(v(\xi)\right) \quad , \qquad v(0) = u^l\,.$$

Then

$$\left(f\left(v(\xi)\right) - f(u^l) - \lambda_j\left(v(\xi)\right)\left(v(\xi) - u^l\right)\right)\Big|_{\xi=0} = 0_N\,,$$

and by the linear degeneracy of the jth characteristic field

$$\begin{aligned}
&\frac{\mathrm{d}}{\mathrm{d}\xi}\left(f\left(v(\xi)\right) - f(u^l) - \lambda_j\left(v(\xi)\right)\left(v(\xi) - u^l\right)\right)\\
&= \left(a\left(v(\xi)\right) - \lambda_j\left(v(\xi)\right)\mathrm{Id}_N\right) v'(\xi) - \nabla\lambda_j\left(v(\xi)\right)\cdot v'(\xi)\,\left(v(\xi) - u^l\right)\\
&= \left(a\left(v(\xi)\right) - \lambda_j\left(v(\xi)\right)\mathrm{Id}_N\right) r_j\left(v(\xi)\right) - \nabla\lambda_j\left(v(\xi)\right)\cdot r_j\left(v(\xi)\right)\left(v(\xi) - u^l\right) = 0_N,
\end{aligned}$$

therefore

$$f\left(v(\xi)\right) - f(u^l) = \lambda_j\left(v(\xi)\right)\left(v(\xi) - u^l\right)\,.$$

Hence, the Rankine–Hugoniot condition (7.1.8) holds along v with constant speed $\sigma\left(u^l, v(\xi)\right) = \lambda_j\left(v(\xi)\right)$. As a consequence $[\xi \mapsto v(\xi)]$ coincides with $[\xi \mapsto \Psi_j(\xi)]$. Furthermore, by Proposition 7.2, any j–Riemann invariant is constant along it. □

If the jth characteristic field is linearly degenerate and $u^r \in \mathscr{S}_j(u^l)$ or, equivalently, $u^l \in \mathscr{S}_j(u^r)$, then a weak solution to (7.3.1) of the form (7.3.9) with $\sigma = \lambda_j(u^l) = \lambda_j(u^r) = \overline{\lambda}_j$, i.e.

$$u(t,x) = \begin{cases} u^l & \text{if } x < \overline{\lambda}_j\, t \\ u^r & \text{if } x \geq \overline{\lambda}_j\, t\,, \end{cases}$$

is called ***j*–contact discontinuity**.

The following definition introduces an admissibility condition for the discontinuities, which generalize the Definition 3.7 given in the scalar case.

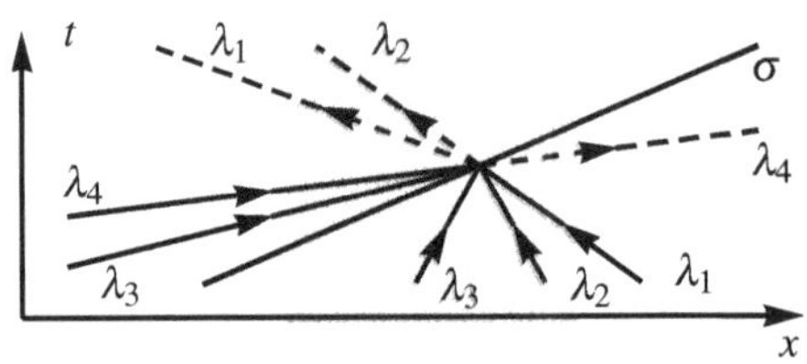

Fig. 7.3 Representation in the (x,t)–plane of a 3–shock satisfying the Lax condition (7.3.16a) in the case $N = 4$ and a genuinely nonlinear 3rd characteristic field. The solid line is the discontinuity, the solid vectors the incoming characteristics and the dashed vectors the outgoing characteristics.

Definition 7.9. The discontinuity (7.3.9) satisfies the **Lax condition** if there exists an index $j \in \{1, 2, \ldots, N\}$ such that either

$$\lambda_j(u^r) < \sigma < \lambda_j(u^l) \quad , \quad \lambda_{j-1}(u^l) < \sigma < \lambda_{j+1}(u^r) \tag{7.3.16a}$$

if the jth characteristic field is genuinely nonlinear; or

$$\lambda_j(u^l) = \sigma = \lambda_j(u^r) \tag{7.3.16b}$$

if the jth characteristic field is linearly degenerate. Here $\lambda_0 = -\infty$ and $\lambda_{N+1} = +\infty$.

Condition (7.3.16a) expresses that at a point of discontinuity there are $N+1$ incoming characteristics, whose speeds of propagation are the eigenvalues $\lambda_1(u^r)$, $\ldots$, $\lambda_j(u^r)$, $\lambda_j(u^l)$, $\ldots$, $\lambda_N(u^l)$, see Fig. 7.3. As a consequence, $N+1$ scalar data are leaded to a point of discontinuity instead of N at a point of continuity. This is consistent with the fact that the speed of the discontinuity is itself an unknown.

By using the parametrization of Theorem 7.2, we define $\mathscr{S}_j^a(u^l)$ as the subset of $\mathscr{S}_j(u^l)$ made of states that can be connected on the right to u^l by a j–shock that satisfies the Lax condition (7.3.16), see Fig. 7.2.

Theorem 7.4. *If the jth characteristic field is genuinely nonlinear, with the normalization* (7.1.5)*, the curve $\mathscr{S}_j^a(u^l)$ consists of the states $\Psi_j(\varepsilon)$ with $\varepsilon \in \mathbb{R}_-^*$ small enough. If the jth characteristic field is linearly degenerate, the curve $\mathscr{S}_j^a(u^l)$ coincides with the whole curve $\mathscr{S}_j(u^l)$.*

Proof. Assume that the jth characteristic field is genuinely nonlinear. By (7.1.5) and (7.3.10) we have that

$$\Psi_j(\varepsilon) = u^l + \varepsilon\, r_j(u^l) + O(\varepsilon^2) \quad , \quad \sigma\left(u^l, \Psi_j(\varepsilon)\right) = \lambda_j(u^l) + \frac{\varepsilon}{2} + O(\varepsilon^2)$$

so that by (7.1.5)

$$\lambda_j\left(\Psi_j(\varepsilon)\right) = \lambda_j(u^l) + \varepsilon\, \nabla\lambda_j(u^l)\cdot r_j(u^l) + O(\varepsilon^2) = \lambda_j(u^l) + \varepsilon + O(\varepsilon^2)\,.$$

Therefore, the conditions in (7.3.16a) are equivalent to

$$\lambda_j(u^l) + \varepsilon + O(\varepsilon^2) < \lambda_j(u^l) + \frac{\varepsilon}{2} + O(\varepsilon^2) < \lambda_j(u^l) \tag{7.3.17a}$$

$$\lambda_{j-1}(u^l) < \lambda_j(u^l) + \frac{\varepsilon}{2} + O(\varepsilon^2) < \lambda_{j+1}(u^l) + \varepsilon + O(\varepsilon^2)\,. \tag{7.3.17b}$$

Clearly, (7.3.17b) is satisfied iff $|\varepsilon|$ is sufficiently small because, by hypothesis, $\lambda_{j-1}(u^l) < \lambda_j(u^l) < \lambda_{j+1}(u^l)$, whereas (7.3.17a) is satisfied iff ε is sufficiently small and negative.

If the jth characteristic field is linearly degenerate, the result of Theorem 7.3 ensures that the whole curve $\mathscr{S}_j(u^l)$ satisfies the Lax condition (7.3.16b). □

If the jth characteristic field is genuinely nonlinear, then from (7.3.7) and (7.3.10a) it follows that the curves $\mathscr{R}_j(u^l)$ and $\mathscr{S}_j^a(u^l)$ are osculatory at u^l, i.e., they have a tangency of second–order at u^l. Hence, the composite function $\omega_j\,:\,\mathbb{R}\times\mathbb{R}^N \to \mathbb{R}^N$ defined for $|\varepsilon|$ small enough by

$$\omega_j(\varepsilon, u^l) = \begin{cases} \Psi_j(\varepsilon) & \text{if } \varepsilon \in \mathbb{R}_- \\ \Phi_j(\varepsilon) & \text{if } \varepsilon \in \mathbb{R}_+^* \end{cases} \tag{7.3.18}$$

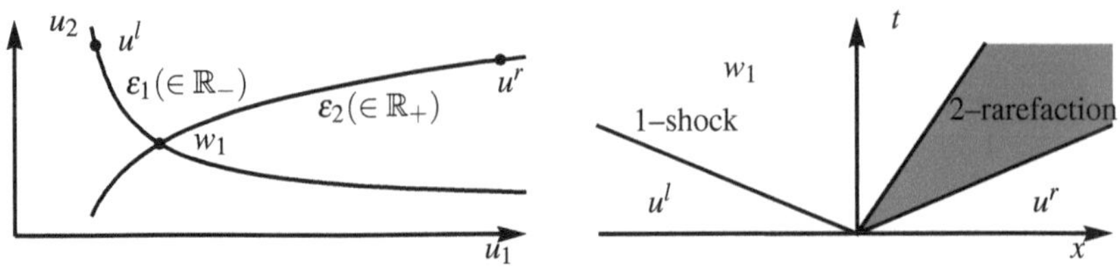

Fig. 7.4 Representation in the (u_1,u_2)–phase plane and in the (x,t)–plane, respectively on the left and on the right, of a solution to the Riemann problem (7.3.19) in the case $N = 2$.

is of class $\mathbf{C}^2$. Furthermore, by Theorem 7.1 and Theorem 7.4, the set $\mathscr{S}_j^a(u^l) \cup \mathscr{R}_j(u^l) = \{\omega_j(\varepsilon,u^l) : |\varepsilon| \text{ small enough}\}$ is exactly the set of all neighboring states u that can be connected to u^l either by a j–rarefaction wave or by a j–shock satisfying the Lax condition (7.3.16a), see Fig. 7.2. Analogously, when the jth characteristic field is linearly degenerate, we define $\omega_j(\varepsilon,u^l) = \Psi_j(\varepsilon)$ and obtain that the set $\{\omega_j(\varepsilon,u^l) : |\varepsilon| \text{ small enough}\}$ is the set of all neighboring states u that can be connected to u^l by a j–contact discontinuity.

For sufficiently weak j–shocks associated to genuinely nonlinear characteristic fields, the Lax condition (7.3.16) is equivalent to the entropy condition (7.1.11), see [11, Theorem 4.3.7] or [8, Chap. VI] for the proof, and to the **Liu's condition** [9]

$$\sigma(u^l,u^r) \leq \sigma(u^l,u) \qquad \text{for all } u \in \mathscr{S}_j(u^l) \text{ between } u^l \text{ and } u^r.$$

7.3.3 *General Solutions*

In sections 7.3.1 and 7.3.2, we have constructed particular weak solutions for strictly hyperbolic Riemann problems

$$\partial_t u + \partial_x f(u) = 0_N \tag{7.3.19a}$$

$$u(0,x) = \begin{cases} u^l & \text{if } x < 0 \\ u^r & \text{if } x \geq 0 \end{cases} \tag{7.3.19b}$$

in the special cases where the characteristic fields are either genuinely nonlinear or linearly degenerate, the initial data u^l and u^r are sufficiently close and u^r belongs to $\mathscr{W}(u^l) = \cup_{j=1}^N \left(\mathscr{S}_j(u^l) \cup \mathscr{R}_j(u^l)\right)$. In this section we generalize these results to the cases where u^r not necessarily belongs to $\mathscr{W}(u^l)$.

As in the scalar case, the solution to the Riemann problem (7.3.19) will be the juxtaposition of constant states, of rarefaction waves and of discontinuities (shock waves or contact discontinuities). On the other hand, differently from the one–dimensional case, for the multi–dimensional case, in general, the initial discontinuity breaks up into N discontinuities, see Fig. 7.4, right.

Theorem 7.5. *Assume that system* (7.3.19a) *is strictly hyperbolic and that, for any* $j = 1,\ldots,N$, *the* j*th characteristic field is either genuinely nonlinear or linearly* *degenerate. Then for all* $u^l \in \mathbb{R}^N$ *there exists a neighborhood* Ω *of* u^l *such that for all* $u^r \in \Omega$*, the Riemann problem* (7.3.19) *has a unique weak solution that consists of at most* $(N+1)$ *constant states*

$$w_0 = u^l \ , \quad w_j = \omega_j(\varepsilon_j, w_{j-1}) \ , \quad j = 1,\dots,N-1 \ , \quad w_N = \omega_N(\varepsilon_N, w_{N-1}) = u^r \ ,$$

for some $\varepsilon_j \in \mathbb{R}$*,* $j = 1,\dots,N$*, separated by up to N elementary waves, i.e., the states* w_{j-1} *and* w_j *are connected with either a contact discontinuity (if the jth characteristic field is linearly degenerate) or else, when* $\varepsilon_j \in \mathbb{R}^*_+$*, a rarefaction wave and, when* $\varepsilon_j < 0$*, a shock wave satisfying the entropy inequality* (7.1.12) *and the Lax condition* (7.3.16).

Proof. Define the function ω from a neighborhood of the origin in $\mathbb{R}^N$ into a neighborhood of u^l by

$$\omega(\varepsilon_1,\dots,\varepsilon_N) = \omega_N\left(\varepsilon_N, \omega_{N-1}\left(\varepsilon_{N-1},\dots,\omega_1(\varepsilon_1,u^l)\dots\right)\right) ,$$

where ω_j is defined by (7.3.18). We want to prove that for any u^r located in a neighborhood of u^l, there exists a solution $(\varepsilon_1,\dots,\varepsilon_N)^T \in \mathbb{R}^N$ for the equation

$$\omega(\varepsilon_1,\dots,\varepsilon_N) = u^r \ . \tag{7.3.20}$$

In fact, Eq. (7.3.20) implies that u^l can be connected to the right to $w_1 = \omega_1(\varepsilon_1, u^l)$ by a 1–elementary wave, then w_1 to $w_2 = \omega_2(\varepsilon_2, w_1)$ by a 2–elementary wave, ..., and w_{N-1} to $u^r = \omega_N(\varepsilon_N, w_{N-1})$ by a N–elementary wave, see Fig. 7.4. We begin observing that ω is a $\mathbf{C}^2$ function and that $\omega(0,\dots,0) = u^l$. Furthermore, by (7.3.7) and (7.3.10a) it follows by induction that

$$\omega(\varepsilon_1,\dots,\varepsilon_N) = u^l + \sum_{j=1}^{N} \varepsilon_j \, r_j(u^l) + O(\varepsilon^2) \ .$$

Therefore, for all $\eta = (\eta_1,\dots,\eta_N)^T \in \mathbb{R}^N$, the derivative of ω at the origin along η is given by $\nabla\omega(0)\cdot\eta = \sum_{j=1}^N \eta_j \, r_j(u^l)$. Since the vectors $r_j(u^l)$, $j = 1,\dots,N$, are linearly independent, the linear operator $\nabla\omega(0)$ is invertible. By the local inversion theorem, there exists a neighborhood Ω of u^l such that, for all $u^r \in \Omega$, the Eq. (7.3.20) has a unique solution $(\varepsilon_1,\dots,\varepsilon_N) \in \mathbb{R}^N$. □

Let us note that the solution to (7.3.20), if it exists, can have vanishing components. For example $\varepsilon_k = 0$ means that there is not a k–elementary wave, that is that $w_{k+1} = w_k$. Furthermore, the proof gives an equivalent of ε. In fact

$$0 = \omega(\varepsilon_1,\dots,\varepsilon_N) - u^r = u^l - u^r + \sum_{j=1}^{N} \varepsilon_j \, r_j(u^l) + O(\varepsilon^2)$$

and by using the normalization (7.2.2), we obtain that $\varepsilon_j = l_j(u^l)\cdot(u^r - u^l) + O(\varepsilon^2)$. In particular, the intermediate states are given by

$$w_k = u^l + \sum_{j=1}^{k} l_j(u^l) \cdot (u^r - u^l)\, r_j(u^l) + O(\varepsilon^2)\,.$$

Remark 7.9. Note that a weak solution consisting of combinations of rarefactions, shocks and contact discontinuities may not exist even for a strictly hyperbolic genuinely nonlinear system of conservation laws (7.3.19), see for instance [6, 10] and the references therein.

The existence of a solution to the Riemann problem (7.3.19) for a system that is not genuinely nonlinear or linearly degenerate is considered in [9, 13, 14]. For the existence of a solution to the Riemann problem (7.3.19) for non–strictly hyperbolic system we defer to [12] and the references at the end of Chap. 17. □

References

1. Boillat, G.: Chocs caractéristiques (Characteristic shocks)(1972)
2. Bressan, A.: Hyperbolic systems of conservation laws. Oxford Lecture Series in Mathematics and its Applications, vol. 20. Oxford University Press, Oxford (2000)
3. Courant, R., Hilbert, D.: Methods of Mathematical Physics, vol. 2 (Set of Volumes). Wiley India Pvt. Ltd. (2008)
4. Godlewski, E., Raviart, P.A.: Numerical approximation of hyperbolic systems of conservation laws. Applied Mathematical Sciences, vol. 118. Springer, New York (1996)
5. Holden, H., Risebro, N.H.: Front tracking for hyperbolic conservation laws. Applied Mathematical Sciences, vol. 152. Springer, New York (2002)
6. Keyfitz, B.L., Kranzer, H.C.: Spaces of Weighted Measures for Conservation Laws with Singular Shock Solutions. Journal of Differential Equations 118(2), 420–451 (1995)
7. Lax, P.D.: Hyperbolic Systems of Conservation Laws II. In: Sarnak, P., Majda, A. (eds.) Selected Papers, vol. I, pp. 233–262. Springer, New York (2005)
8. Lefloch, P.G.: Hyperbolic systems of conservation laws. The theory of classical and nonclassical shock waves. Lectures in Mathematics ETH Zürich. Birkhäuser Verlag, Basel (2002)
9. Liu, T.-P.: The Riemann problem for general systems of conservation laws. Journal of Differential Equations 18(1), 218–234 (1975)
10. Serre, D.: Richness and the classification of quasilinear hyperbolic systems. Institute for Mathematics and Its Applications 29, 315 (1991)
11. Serre, D.: Systems of conservation laws. 1 & 2. Translated from the 1996 French original by I. N. Sneddon. Cambridge University Press, Cambridge (1999)
12. Smoller, J.: Shock Waves and Reaction-Diffusion Equations. Grundlehren der mathematischen Wissenschaften, vol. 258. Springer (1983)
13. Wendroff, B.: The Riemann problem for materials with nonconvex equations of state: II: General Flow. Journal of Mathematical Analysis and Applications 38(3), 640–658 (1972)
14. Wendroff, B.: The Riemann problem for materials with nonconvex equations of state I: Isentropic flow. Journal of Mathematical Analysis and Applications 38(2), 454–466 (1972)

Chapter 8
One–Dimensional Systems of Balance Laws (Weakly Coupled)

Abstract. This chapter is concerned with nonlinear systems of balance laws in one space dimension weakly coupled, that is the coupling occurs only through source terms. We prove the well–posedness of the Cauchy problem in the space of functions with bounded total variation.

8.1 Introduction

Let $\mathscr{U}$ be a closed connected subset of $\mathbb{R}^N$ with non–empty interior and, for simplicity, we also require that $0 \in \mathscr{U}$. We focus on the global existence and stability of solutions to the Cauchy problem for a nonlinear hyperbolic system of balance laws

$$\partial_t u + \partial_x f(u) = G(t,u) \,, \qquad u(t=0) = \overline{u} \,, \tag{8.1.1}$$

where $u : \mathbb{R}_+^* \times \mathbb{R} \to \mathscr{U}$ denotes the unknown vector function of the **conserved quantities** $u_1, \ldots, u_N$, $f : \mathscr{U} \to \mathbb{R}^N$ is the **flux function**, $G : \mathbb{R}_+^* \times \mathbf{L^1}(\mathbb{R};\mathbb{R}^N) \to \mathbf{L^1}(\mathbb{R};\mathbb{R}^N)$ is the **source term** and $\overline{u} : \mathbb{R} \to \mathscr{U}$ is the **initial datum**. As will be proved in Theorem 8.1, on bounded time intervals the solutions to (8.1.1) turn out to be Lipschitz functions of the initial datum $\overline{u}$ with respect to the $\mathbf{L^1}$–norm, with a Lipschitz constant depending on the total variation of the initial datum.

Remark 8.1. Observe that the set $\mathscr{U}$ is not necessarily compact. Furthermore, G may be a (possibly) *non–local* operator, but in this case the non–locality in the source term will be referred to the space variable. For sources non–local in time, i.e. **systems with memory**, see [5]; while for the more general case, with a flux of the form $f = f(t,x,u)$ and a source of the form $G = G(t,x,u)$ and non–local both in time and space, see [9]. □

First, we introduce the following notation

$$|u| = \max_{i=1,\ldots,N} |u_i|$$

M.D. Rosini: *Macroscopic Models for Vehicular Flows & Crowd Dynamics*, UCS, pp. 111–120.
DOI: 10.1007/978-3-319-00155-5_8

$$
\begin{aligned}
|u|_{\mathbf{L^p}(\mathbb{R};\mathbb{R}^N)} &= \max_{i=1,\ldots,N} \|u_i\|_{\mathbf{L^p}(\mathbb{R};\mathbb{R})} \\
\mathrm{TV}(u) &= \max_{i=1,\ldots,N} \left\{ \mathrm{TV}\left(u_i(x)\right) \right\}
\end{aligned} \tag{8.1.2}
$$

and, for any positive constant M, we introduce the sets

$$
\mathscr{D} = \left\{ u \in \mathbf{L^1}(\mathbb{R};\mathscr{U}) \;:\; \mathrm{TV}(u) < +\infty \right\} \tag{8.1.3a}
$$

$$
\mathscr{D}_M = \left\{ u \in \mathbf{L^1}(\mathbb{R};\mathscr{U}) \;:\; \mathrm{TV}(u) \leq M \right\}. \tag{8.1.3b}
$$

Definition 8.1. A function u in $\mathbf{C^0}\left([0,T];\mathscr{D}\right)$ is an **entropy weak solution** of the Cauchy problem (8.1.1) if for any convex continuous function $U : \mathbb{R} \to \mathbb{R}$

$$
\begin{aligned}
\int_0^T \int_{\mathbb{R}} \left(U(u)\, \partial_t \varphi + F(u)\, \partial_x \varphi + \nabla U(u) \cdot G(t,u)\, \varphi \right) \mathrm{d}x\, \mathrm{d}t \\
+ \int_{\mathbb{R}} U\left(\bar{u}(x)\right)\, \varphi(0,x)\, \mathrm{d}x \geq 0
\end{aligned} \tag{8.1.4}
$$

holds for all $\varphi \in \mathbf{C_c^\infty}\left(]-\infty,T[\, \times \mathbb{R}; \mathbb{R}_+^*\right)$, being $\nabla F(v) = Df(v)^T\, \nabla U(v)$.

As usual when considering hyperbolic balance laws, the proof exploits the **fractional step** (or **operator splitting**) method, see Sect. 8.4. This procedure works under the hypothesis of well–posedness of both the convective part

$$
\partial_t u + \partial_x f(u) = 0_N \tag{8.1.5}
$$

and of the dynamical system

$$
\partial_t u = G(t,u) \tag{8.1.6}
$$

associated to the source part, together with a geometric compatibility condition.

On the conservative part (8.1.5) we assume that (8.1.1) is **weakly coupled**, i.e.

(F) $f : \mathscr{U} \to \mathbb{R}^N$ is a $\mathbf{C^1}$ function and $f(u) = \left(f_1(u_1), \ldots, f_N(u_N)\right)^T$.

Observe that the system (8.1.1) is then hyperbolic, whereas the strict hyperbolicity of Df is not ensured by condition **(F)**.

On the source term G we assume that

(G) $G : \mathbb{R}_+^* \times \mathscr{D} \to \mathbf{L^1}(\mathbb{R};\mathbb{R}^N)$ and for all compact subsets K of $\mathscr{U}$ there exist functions $l, a, b \in \mathbf{L^\infty_{loc}}\left(\mathbb{R}_+^*;\mathbb{R}_+\right)$ and a constant $C \in \mathbb{R}_+$ such that for all $u, u' \in \mathbf{L^1} \cap \mathbf{BV}(\mathbb{R};K)$ and for all $t \in \mathbb{R}_+^*$ the following estimates hold:

$$
\left|G(t,u) - G(t,u')\right|_{\mathbf{L^1}(\mathbb{R};\mathbb{R}^N)} \leq l(t)\, \left|u - u'\right|_{\mathbf{L^1}(\mathbb{R};\mathbb{R}^N)} \tag{8.1.7a}
$$

$$
\left|G(t,u)\right|_{\mathbf{L^\infty}(\mathbb{R};\mathbb{R}^N)} \leq a(t) + b(t)\, |u|_{\mathbf{L^\infty}(\mathbb{R};\mathbb{R}^N)} \tag{8.1.7b}
$$

$$
\mathrm{TV}\left(G(t,u)\right) \leq C\, \left(1 + \mathrm{TV}(u)\right). \tag{8.1.7c}
$$

Above, we did not indicate explicitly the dependence of l, a, b and C on the compact set K to simplify the notations. Because of **(G)**, for any compact real interval I we have that $G : I \times \mathscr{D} \to \mathbf{L^1} \cap \mathbf{BV}(\mathbb{R};\mathbb{R}^N)$. Observe that condition **(G)** includes the possibility of local terms, [7, 8].

Finally, the geometric compatibility between (8.1.5) and (8.1.6) is ensured by

(I) $\mathscr{U}$ is invariant with respect to both (8.1.5) and (8.1.6).

Here, invariance is understood as follows, see [10].

Definition 8.2. A set $\mathscr{U}$ is **invariant** for (8.1.5), respectively (8.1.6), if any admissible initial datum $\bar{u}$ taking values in $\mathscr{U}$ leads to a solution u to (8.1.5), respectively (8.1.6), taking values in $\mathscr{U}$.

Above, by **admissible initial data** $\bar{u}$ we mean that the Cauchy problem for (8.1.5), respectively (8.1.6), with initial datum $\bar{u}$ admits an entropy weak solution, respectively Carathéodory solution (see [13, Sect. 8.1]), defined for all times $t \in \mathbb{R}_+^*$. In both cases, $\mathscr{U}$ needs neither be convex nor compact. See [14] as a general reference for invariant domains for conservation laws. Concerning ordinary differential equations, a classical condition implying invariance is the so called Nagumo condition, see [2, Sect. 4].

8.2 The Convective Part

We adapt the approximation algorithm introduced in Chap. 5 for the scalar case to approximate the Cauchy problem associated to the system (8.1.5) satisfying the condition **(F)**. Our algorithm needs the introduction of a grid. By **(I)**, there exist some closed possibly unbounded intervals $\mathscr{I}_1, \ldots, \mathscr{I}_N \subseteq \mathbb{R}$ such that $\mathscr{U} = \prod_{i=1}^N \mathscr{I}_i$. For any fixed $\varepsilon \in \mathbb{R}_+$ sufficiently small, introduce in $\mathscr{U}$ an **ε–grid** $\mathscr{G}^\varepsilon = \prod_{i=1}^N \mathscr{I}_i^\varepsilon$, where each $\mathscr{I}_i^\varepsilon$ is a finite subset of $\mathscr{I}_i$ such that

1. $\bigcup_{u \in \mathscr{I}_i^\varepsilon}]u-\varepsilon, u+\varepsilon[\supseteq (\mathscr{I}_i \cap [-1/\varepsilon, 1/\varepsilon])$;
2. there exists $\delta^\varepsilon \in \mathbb{R}_+$ such that $\min_{u_i', u_i'' \in \mathscr{I}_i^\varepsilon,\, u_i' \neq u_i''} \left| u_i' - u_i'' \right| > \delta^\varepsilon$;
3. $\min \mathscr{I}_i^\varepsilon = \begin{cases} -1/\varepsilon & \text{if } \inf \mathscr{I}_i = -\infty \\ \min \mathscr{I}_i & \text{if } \inf \mathscr{I}_i \neq -\infty \end{cases}, \quad \max \mathscr{I}_i^\varepsilon = \begin{cases} 1/\varepsilon & \text{if } \sup \mathscr{I}_i = +\infty \\ \max \mathscr{I}_i & \text{if } \sup \mathscr{I}_i \neq +\infty \,. \end{cases}$

Correspondingly to $\mathscr{D}$ and $\mathscr{D}_M$ defined in (8.1.3), let

$$\begin{aligned} \mathscr{D}\left(\mathscr{G}^\varepsilon\right) &= \left\{ u \in \mathscr{D} \;:\; u \in \mathbf{PC}_c\left(\mathbb{R};\mathbb{R}^N\right) \text{ and } u\left(\mathbb{R}\right) \subseteq \mathscr{G}^\varepsilon \right\} \\ \mathscr{D}_M\left(\mathscr{G}^\varepsilon\right) &= \mathscr{D}\left(\mathscr{G}^\varepsilon\right) \cap \mathscr{D}_M \,. \end{aligned}$$

Introduce a piecewise linear continuous function $f^\varepsilon \;:\; \mathscr{U} \to \mathbb{R}^N$ that coincides with the function f on $\mathscr{G}^\varepsilon$ and such that ∇f^ε is continuous on $\mathscr{U} \setminus \mathscr{G}^\varepsilon$.

For any fixed $u^l, u^r \in \mathcal{G}^\varepsilon$, by **(F)** a weak (although possibly non–entropic) solution to the Riemann problem

$$\partial_t u + \partial_x f^\varepsilon(u) = 0_N \quad , \qquad u(0,x) = \begin{cases} u^l & \text{if } x < 0 \\ u^r & \text{if } x \geq 0 \end{cases} \tag{8.2.1}$$

is the vector function $u = (u_1, \ldots, u_N)^T : \mathbb{R}_+^* \times \mathbb{R} \to \mathcal{U}$, where each $u_i : \mathbb{R}_+^* \times \mathbb{R} \to \mathcal{I}_i$, is the entropy weak solution to the scalar Riemann problem

$$\partial_t u_i + \partial_x f_i^\varepsilon(u_i) = 0_N \quad , \qquad u_i(0,x) = \begin{cases} u_i^l & \text{if } x < 0 \\ u_i^r & \text{if } x \geq 0 \,. \end{cases}$$

By Corollary 4.1, each entropy weak solution to the above Riemann problem attains values in $\mathcal{I}_i^\varepsilon$. As a consequence, the weak solution to the Riemann problem (8.2.1) takes values in $\mathcal{G}^\varepsilon$. Finally, by Proposition 5.1, we have the following result. Below, we write the x–jump of a function $h : \mathbb{R}_+ \times \mathbb{R} \to \mathbb{R}$ at (t, x_α) as $\Delta h_\alpha(t) = h(t, x_\alpha+) - h(t, x_\alpha-)$.

Proposition 8.1. *Assume that* **(F)** *holds. Then, for any $\varepsilon \in \mathbb{R}_+$ and for any ε–grid $\mathcal{G}^\varepsilon$, the system* (8.1.5) *generates an operator*

$$\begin{aligned} S^\varepsilon \ : \ \mathbb{R}_+^* \times \mathcal{D}(\mathcal{G}^\varepsilon) &\to \mathcal{D}(\mathcal{G}^\varepsilon) \\ (t,u) &\mapsto S_t^\varepsilon u \end{aligned}$$

such that the map $[t \mapsto S_t^\varepsilon \bar{u}]$ is a weak solution to the Cauchy problem for (8.1.5) *with initial datum $\bar{u} \in \mathcal{D}(\mathcal{G}^\varepsilon)$. Moreover S^ε has the following properties:*

1. *S^ε is a semigroup, i.e. $S_0^\varepsilon = \mathrm{Id}_N$ and $S_{t_1}^\varepsilon \circ S_{t_2}^\varepsilon = S_{t_1+t_2}^\varepsilon$ for all $t_1, t_2 \in \mathbb{R}_+^*$.*
2. *The map $[(t,x) \mapsto S_t^\varepsilon \bar{u}(x)]$ is piecewise constant with discontinuities along finitely many polygonal lines and with finitely many interaction points.*
3. *For $u \in \mathcal{D}(\mathcal{G}^\varepsilon)$, both maps $\left[t \mapsto \left|S_t^\varepsilon u\right|_{\mathbf{L}^\infty(\mathbb{R};\mathbb{R}^N)}\right]$ and $[t \mapsto \mathrm{TV}(S_t^\varepsilon u)]$ are non–increasing.*
4. *Let $U : \mathcal{U} \to \mathbb{R}$ be a convex entropy for* (8.1.5) *with entropy flux $F : \mathcal{U} \to \mathbb{R}$, $U^\varepsilon = U\left(S_t^\varepsilon u\right)$, $F^\varepsilon = F\left(S_t^\varepsilon u\right)$ and let $x = x_\alpha(t)$ be the support of the αth discontinuity in $S_t^\varepsilon u$. For any $M \in \mathbb{R}_+$ there exists a positive constant C independent from ε, such that for all $u \in \mathcal{D}_M(\mathcal{G}^\varepsilon)$ and $t \in \mathbb{R}_+^*$*

$$\sum_\alpha \left(x'_\alpha \, \Delta U_\alpha^\varepsilon - \Delta F_\alpha^\varepsilon\right) \geq -C\,\varepsilon \,.$$

In other words, the orbits of S^ε are weak solutions that satisfy the entropy jump condition (7.1.12) only approximately. The proof of this result follows by a slight modification of the construction in [3], see [6].

8.3 The Non–local Source Term

In this section we pass to the source term and consider the Eq. (8.1.6).

Proposition 8.2. *Let G satisfy* **(G)** *and* **(I)** *hold. Then, the Cauchy problem for* (8.1.6) *generates a map*

$$\begin{aligned} \Sigma \ : \ \left\{(t_1,t_2)\in\mathbb{R}_+^*\times\mathbb{R}_+^* \ : \ t_1\le t_2\right\}\times\mathscr{D} &\to \mathscr{D} \\ (t_1,t_2,u) &\mapsto \Sigma_{t_1,t_2}u \end{aligned}$$

such that for all $t_o\in\mathbb{R}_+^*$, $\bar u\in\mathscr{D}$ *and* $t\ge t_o$ *the map* $[t\mapsto\Sigma_{t_o,t}\bar u]$ *is the solution to* (8.1.6) *with initial datum* $\bar u$ *assigned at time* t_o.

For every $R\in\mathbb{R}_+$ *and* $T>t_o\ge 0$ *there exists a compact set* $K\subset\mathscr{U}$ *such that for every* $\bar u\in\mathscr{D}$ *with* $\left|\bar u\right|_{\mathbf{L}^\infty(\mathbb{R};\mathbb{R}^N)}\le R$ *the solutions to* (8.1.6) *with initial datum* $\bar u$ *at time* t_o *attain values in K for* $t\in[t_o,T]$.
Moreover the following estimates hold for all $t\in[t_o,T]$ *and u, u′ as above:*

$$\left|\Sigma_{t_o,t}u-\Sigma_{t_o,t}u'\right|_{\mathbf{L}^1(\mathbb{R};\mathbb{R}^N)}\le\left|u-u'\right|_{\mathbf{L}^1(\mathbb{R};\mathbb{R}^N)}\exp\left(\left|l\right|_{\mathbf{L}^1([t_o,t];\mathbb{R})}\right) \tag{8.3.1}$$

$$\left|\Sigma_{t_o,t}u\right|_{\mathbf{L}^\infty(\mathbb{R};\mathbb{R}^N)}\le\left(\left|a\right|_{\mathbf{L}^1([t_o,t];\mathbb{R})}+\left|u\right|_{\mathbf{L}^\infty(\mathbb{R};\mathbb{R}^N)}\right)\exp\left(\left|b\right|_{\mathbf{L}^1([t_o,t];\mathbb{R})}\right) \tag{8.3.2}$$

$$TV(\Sigma_{t_o,t}u)\le\left(C\,(t-t_o)+TV(u)\right)\exp\left(C\,(t-t_o)\right). \tag{8.3.3}$$

Finally, if $\mathscr{U}$ *is compact, then there exists a positive constant* $\hat{\mathscr{C}}$ *such that for all* $t>t_o$ *we have* $\left|\Sigma_{t_o,t}u\right|_{\mathbf{L}^\infty(\mathbb{R})}\le\hat{\mathscr{C}}$ *and* $TV(\Sigma_{t_o,t}u)\le\hat{\mathscr{C}}$.

Proof. The existence follows from Banach Fixed Point Theorem in $\mathbf{L}^1$ by (8.1.7a), (8.1.7b) and **(I)**.

Consider an initial datum $\bar u$ satisfying $\left|\bar u\right|_{\mathbf{L}^\infty(\mathbb{R};\mathbb{R}^N)}\le R$. From the standard theory of differential equations it follows that there exists a compact set $K\subset\mathscr{U}$ such that all solutions to (8.1.6) with initial datum $\bar u$ as above attain values in K for all $(t,x)\in[t_o,T]\times\mathbb{R}$. Clearly the functions l, a, b and the constants C, L depend on K; K refers to R, t_o and T.

The Lipschitz bound (8.3.1) follows from a direct application of the Gronwall Lemma 2.1, noting that from (8.1.7a)

$$\begin{aligned} &\left|\Sigma_{t_o,t}u-\Sigma_{t_o,t}u'\right|_{\mathbf{L}^1(\mathbb{R};\mathbb{R}^N)} = \\ &= \left|u-u'+\int_{t_o}^t\left(G(\tau,\Sigma_{t_o,\tau}u)-G(\tau,\Sigma_{t_o,\tau}u')\right)\,d\tau\right|_{\mathbf{L}^1(\mathbb{R};\mathbb{R}^N)} \\ &\le \left|u-u'\right|_{\mathbf{L}^1(\mathbb{R};\mathbb{R}^N)}+\int_{t_o}^t l(\tau)\,\left|\Sigma_{t_o,\tau}u-\Sigma_{t_o,\tau}u'\right|_{\mathbf{L}^1(\mathbb{R};\mathbb{R}^N)}\,d\tau\,. \end{aligned}$$

Consider now (8.3.2). From the sublinearity (8.1.7b) it follows that

$$\left|\Sigma_{t_o,t}u\right|_{\mathbf{L}^\infty(\mathbb{R};\mathbb{R}^N)}=\left|u+\int_{t_o}^t G\left(\tau,\Sigma_{t_o,\tau}u\right)\,d\tau\right|_{\mathbf{L}^\infty(\mathbb{R};\mathbb{R}^N)}$$

$$
\begin{aligned}
&\leq |u|_{\mathbf{L}^\infty(\mathbb{R};\mathbb{R}^N)} + \int_{t_o}^t \left| G\left(\tau, \Sigma_{t_o,\tau} u\right) \right|_{\mathbf{L}^\infty(\mathbb{R};\mathbb{R}^N)} \, \mathrm{d}\tau \\
&\leq |u|_{\mathbf{L}^\infty(\mathbb{R};\mathbb{R}^N)} + \int_{t_o}^t \left(a(\tau) + b(\tau) \, |\Sigma_{t_o,\tau} u|_{\mathbf{L}^\infty(\mathbb{R};\mathbb{R}^N)} \right) \, \mathrm{d}\tau \, .
\end{aligned}
$$

Then (8.3.2) follows again by the Gronwall Lemma 2.1. In particular it follows that the compact set K is contained in the ball of center 0 and radius $\left(|u|_{\mathbf{L}^\infty(\mathbb{R};\mathbb{R}^N)} + |a|_{\mathbf{L}^1([t_o,T];\mathbb{R})}\right) \exp(|b|_{\mathbf{L}^1([t_o,T];\mathbb{R})})$.

To prove (8.3.3) it is sufficient to apply once again the Gronwall Lemma 2.1 to the estimate

$$
\begin{aligned}
\mathrm{TV}(\Sigma_{t_o,t} u) &\leq \mathrm{TV}(u) + \int_{t_o}^t \mathrm{TV}\left(G(\tau, \Sigma_{t_o,\tau} u)\right) \, \mathrm{d}\tau \\
&\leq \mathrm{TV}(u) + C \int_{t_o}^t \left(1 + \mathrm{TV}(\Sigma_{t_o,\tau} u)\right) \, \mathrm{d}\tau \, ,
\end{aligned}
$$

which follows from (8.1.7c).

Finally, the case of a compact $\mathscr{U}$ follows immediately from **(I)**. □

Introduce the operator $\Pi_\varepsilon : \mathbf{L}^1(\mathbb{R};\mathbb{R}^N) \to \mathbf{PC}_c(\mathbb{R};\mathbb{R}^N)$ defined

$$
\Pi_\varepsilon(u)(x) = \frac{1}{\varepsilon} \sum_{k=-\varepsilon^{-2}-1}^{\varepsilon^{-2}-1} \left(\int_{k\,\varepsilon}^{(k+1)\,\varepsilon} u(\xi) \, \mathrm{d}\xi \right) \chi_{]k\,\varepsilon,(k+1)\,\varepsilon]}(x) \, , \quad x \in \mathbb{R} \, .
$$

Lemma 8.1. *The operator Π_ε is linear with norm 1 both in $\mathbf{L}^\infty(\mathbb{R};\mathbb{R}^N)$ and in $\mathbf{L}^1(\mathbb{R};\mathbb{R}^N)$. Furthermore, for any $u \in \mathbf{L}^1 \cap \mathbf{BV}(\mathbb{R};\mathbb{R}^N)$ we have $\mathrm{TV}(\Pi_\varepsilon(u)) \leq \mathrm{TV}(u)$ and $\lim_{\varepsilon \searrow 0} \left|\Pi_\varepsilon(u) - u\right|_{\mathbf{L}^1(\mathbb{R};\mathbb{R}^N)} = 0$.*

Proof. Linearity and estimates on the norms in $\mathbf{L}^\infty$ and $\mathbf{L}^1$ are immediate. For $u \in \mathbf{L}^1(\mathbb{R};\mathbb{R}^N)$ introduce $u_\varepsilon = u \, \chi_{[-\varepsilon-\varepsilon^{-1},\varepsilon^{-1}]}$. Then $\Pi_\varepsilon(u_\varepsilon) = \Pi_\varepsilon(u)$ and $\mathrm{TV}(u_\varepsilon) \leq \mathrm{TV}(u)$. Therefore

$$
\begin{aligned}
\mathrm{TV}\left(\Pi_\varepsilon(u)\right) &= \mathrm{TV}\left(\Pi_\varepsilon(u_\varepsilon)\right) = \left|\Pi_\varepsilon(u_\varepsilon)(-\varepsilon^{-1})\right| + \left|\Pi_\varepsilon(u_\varepsilon)(\varepsilon^{-1})\right| \\
&\quad + \sum_{k=-\varepsilon^{-2}}^{\varepsilon^{-2}-1} \left|\Pi_\varepsilon(u_\varepsilon)\left((k+1)\,\varepsilon\right) - \Pi_\varepsilon(u_\varepsilon)(k\,\varepsilon)\right| \\
&= \frac{1}{\varepsilon} \sum_{k=-1-\varepsilon^{-2}}^{\varepsilon^{-2}} \left| \int_{k\,\varepsilon}^{(k+1)\,\varepsilon} u_\varepsilon(\xi) \, \mathrm{d}\xi - \int_{(k-1)\,\varepsilon}^{k\,\varepsilon} u_\varepsilon(\xi) \, \mathrm{d}\xi \right| \\
&\leq \frac{1}{\varepsilon} \sum_{k=-1-\varepsilon^{-2}}^{\varepsilon^{-2}} \int_{k\,\varepsilon}^{(k+1)\,\varepsilon} \left|u_\varepsilon(\xi) - u_\varepsilon(\xi-\varepsilon)\right| \, \mathrm{d}\xi \\
&= \frac{1}{\varepsilon} \int_{-\varepsilon-\varepsilon^{-1}}^{\varepsilon+\varepsilon^{-1}} \left|u_\varepsilon(\xi) - u_\varepsilon(\xi-\varepsilon)\right| \, \mathrm{d}\xi \leq \frac{1}{\varepsilon} \int_{\mathbb{R}} \left|u_\varepsilon(\xi) - u_\varepsilon(\xi-\varepsilon)\right| \, \mathrm{d}\xi \\
&\leq \mathrm{TV}(u_\varepsilon) \leq \mathrm{TV}(u)
\end{aligned}
$$

where in the last line we used Lemma 2.6. Finally, the $\mathbf{L}^1$ convergence of $\Pi_\varepsilon(u)$ to u is proved in [8]. □

We introduce now the operator $\Sigma^\varepsilon = \Pi_\varepsilon \circ \Sigma$. By convexity of $\mathscr{U}$ and because of the Lemma 8.1, we have $\Sigma^\varepsilon : \{(t_1,t_2) \in \mathbb{R}_+^* \times \mathbb{R}_+^* : t_1 \le t_2\} \times \mathscr{D} \to \mathscr{D}$.

Corollary 8.1. *Let G satisfy* **(G)** *and* **(I)** *hold. For every $R \in \mathbb{R}_+$ and $T > t_o \geq 0$, there exists a compact set $K \subset \mathscr{U}$ such that for every $u \in \mathscr{D}$ with $|u|_{\mathbf{L}^\infty(\mathbb{R};\mathbb{R}^N)} \le R$ the solutions to the Cauchy problem for* (8.1.6) *with initial datum u at time t_o attain values in K for $t \in [t_o,T]$.*

Furthermore, for all $t \in [t_o,T]$ and u,u' as above

$$\left|\Sigma^\varepsilon_{t_o,t}u - \Sigma^\varepsilon_{t_o,t}u'\right|_{\mathbf{L}^1(\mathbb{R};\mathbb{R}^N)} \le \left|u-u'\right|_{\mathbf{L}^1(\mathbb{R};\mathbb{R}^N)} \exp\left(|l|_{\mathbf{L}^1([t_o,t];\mathbb{R})}\right) \tag{8.3.4a}$$

$$\left|\Sigma^\varepsilon_{t_o,t}u\right|_{\mathbf{L}^\infty(\mathbb{R};\mathbb{R}^N)} \le \left(|a|_{\mathbf{L}^1([t_o,t];\mathbb{R})} + |u|_{\mathbf{L}^\infty(\mathbb{R};\mathbb{R}^N)}\right) \exp\left(|b|_{\mathbf{L}^1([t_o,t];\mathbb{R})}\right) \tag{8.3.4b}$$

$$TV(\Sigma^\varepsilon_{t_o,t}u) \le \left(C\,(t-t_o) + TV(u)\right)\,\exp\left(C\,(t-t_o)\right)\,. \tag{8.3.4c}$$

Finally, if $\mathscr{U}$ is compact, then there exists a positive constant $\hat{\mathscr{C}}$ such that for all $t > t_o$

$$\left|\Sigma^\varepsilon_{t_o,t}u\right|_{\mathbf{L}^\infty(\mathbb{R};\mathbb{R}^N)} \le \hat{\mathscr{C}} \quad \text{and} \quad TV(\Sigma^\varepsilon_{t_o,t}u) \le \hat{\mathscr{C}}\,.$$

Proof. The estimates (8.3.4) immediately follow by Proposition 8.2 and Lemma 8.1. By using the estimates (8.3.4) it is easy to complete the proof. □

8.4 Operator Splitting

An approximate solution to (8.1.1) with an initial datum $u \in \mathscr{D}(\mathscr{G}^\varepsilon)$ is constructed through the following operator splitting scheme, see [12]. Let $h \in \mathbb{N}$ and define recursively

$$F^\varepsilon_{0,t}u = \begin{cases} u & \text{if } t = 0 \\ S^\varepsilon_t u & \text{if } t \in \left]0,\varepsilon\right[\\ \Sigma^\varepsilon_{0,\varepsilon} S^\varepsilon_\varepsilon u & \text{if } t = \varepsilon \\ S^\varepsilon_{t-h\,\varepsilon}\left(\bigcirc_{i=0}^{h-1} \Sigma^\varepsilon_{i\,\varepsilon,(i+1)\,\varepsilon} S^\varepsilon_\varepsilon\right) u & \text{if } t \in \left]h\,\varepsilon,(h+1)\,\varepsilon\right[\\ \left(\bigcirc_{i=0}^{h} \Sigma^\varepsilon_{i\,\varepsilon,(i+1)\,\varepsilon} S^\varepsilon_\varepsilon\right) u & \text{if } t = (h+1)\,\varepsilon\,. \end{cases} \tag{8.4.1}$$

Concerning the ε–grids, refine them recursively at each convective step: start with an initial datum $u \in \mathscr{D}(\mathscr{G}^\varepsilon)$. $F^\varepsilon_{0,t}u$ attains values in the same ε–grid $\mathscr{G}^\varepsilon$ for all $t \in [0,\varepsilon[$. At time $t = \varepsilon$ we apply $\Sigma^\varepsilon_{0,\varepsilon}$ and at the same time pass to another ε–grid $\mathscr{G}^\varepsilon_1$ that contains $\Sigma^\varepsilon_{0,\varepsilon}(\mathscr{G}^\varepsilon)$. Note that $\mathscr{G}^\varepsilon_1$ exists because $\Sigma^\varepsilon_{0,\varepsilon}(\mathscr{G}^\varepsilon)$ is a finite subset of $\mathscr{U}$.

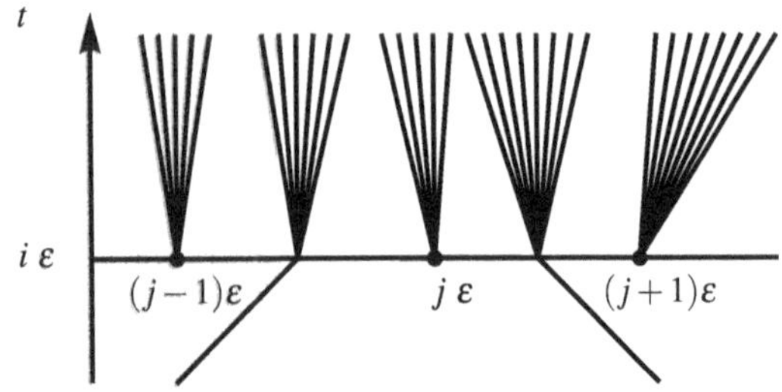

Fig. 8.1 The operator splitting procedure (8.4.1). The dots at time $t = i\,\varepsilon$ denote the starting points of the discontinuities generated by $\Sigma^{\varepsilon}_{(i-1)\,\varepsilon,i\,\varepsilon}$. Observe that S^{ε}_t may split the discontinuities at time $t = i\,\varepsilon$ in (at most) N discontinuities for $t \in \left]i\,\varepsilon,(i+1)\,\varepsilon\right[$.

Recursively, if $F^{\varepsilon}_{0,h\,\varepsilon}u$ attains values in $\mathscr{G}^{\varepsilon}_h$, then $F^{\varepsilon}_{0,t}u$ is valued in the same grid for all $t \in \left[h\,\varepsilon,(h+1)\,\varepsilon\right[$. Applying $\Sigma^{\varepsilon}_{h\,\varepsilon,(h+1)\,\varepsilon}$ at time $t = (h+1)\,\varepsilon$ we pass to another ε–grid $\mathscr{G}^{\varepsilon}_{h+1}$, see Fig. 8.1. Note that the total number of gridpoints is bounded on any strip $[0,T] \times \mathbb{R}$ by [6, Lemma 4.4].

Lemma 8.2. *The operator F^{ε} is well defined on $\{(t_1,t_2) \in \mathbb{R}^*_+ \times \mathbb{R}^*_+ \,:\, t_1 \le t_2\} \times \mathscr{D}(\mathscr{G}^{\varepsilon})$ and for all $u \in \mathscr{D}(\mathscr{G}^{\varepsilon})$, the total number of discontinuities in $F^{\varepsilon}_{0,t}u$ is finite on any strip $[0,T] \times \mathbb{R}$. Moreover, for all $R,T \in \mathbb{R}_+$, $t \in [0,T]$ and $u \in \mathscr{D}(\mathscr{G}^{\varepsilon})$ with $|u|_{\mathbf{L}^\infty(\mathbb{R};\mathbb{R}^N)} \le R$, by using the functions and constants defined in Proposition 8.2, the operator F^{ε} satisfies the following estimates:*

$$\left|F^{\varepsilon}_{0,t}u\right|_{\mathbf{L}^\infty(\mathbb{R};\mathbb{R}^N)} \le \left(|a|_{\mathbf{L}^1([0,t];\mathbb{R})} + |u|_{\mathbf{L}^\infty(\mathbb{R};\mathbb{R}^N)}\right)\,\exp\left(|b|_{\mathbf{L}^1([0,t];\mathbb{R})}\right)$$

$$\mathrm{TV}\left(F^{\varepsilon}_{0,t}u\right) \le \left(C\,t + \mathrm{TV}(u)\right)\,\exp\left(|c|_{\mathbf{L}^1([0,t];\mathbb{R})}\right).$$

Finally, if $\mathscr{U}$ is compact, then there exists a positive constant $\hat{\mathscr{C}}$ such that for all $t \in \mathbb{R}_+$ we have $\left|F^{\varepsilon}_{0,t}u\right|_{\mathbf{L}^\infty(\mathbb{R};\mathbb{R}^N)} \le \hat{\mathscr{C}}$ and $\mathrm{TV}(F^{\varepsilon}_{0,t}u) \le \hat{\mathscr{C}}$.

Proof. Observe first that all compositions in (8.4.1) are possible thanks to (8.3.3) and to the above choice of the grids. The grid $\mathscr{G}^{\varepsilon}_h$ ensures that in each interval where the semigroup S^{ε} is used, the total number of interaction points is finite.

The first estimate follows from 3. in Proposition 8.1 and (8.3.4b) in Corollary 8.1. Similarly, to prove the second estimate we use again 3. in Proposition 8.1 and (8.3.4c) in Corollary 8.1. □

8.5 Well Posedness of the Cauchy Problem

The construction in Sect. 8.4 of the process F allows to extend to the present non–local setting the results of the Standard Riemann Semigroup (SRS) theory, see [4], as extended to balance laws, see [1, 8, 11]. In particular, we underline that the characterization provided in [8, (6) and (7), Theorem 1.2] as *viscosity solutions*, see also [4, Sect. 9.2], still holds here. Similarly, the *tangent vectors* to the trajectories are given by the sum of the SRS and Euler polygonals, as in [8, (4), Theorem 1.2]. The corresponding proofs are slight modifications of the cited results in [8].

Theorem 8.1. *Let* (8.1.1) *satisfy assumptions* **(F)**, **(G)** *and* **(I)**. *Then, there exists a unique process*

$$F : \left\{(t_1,t_2) \in \mathbb{R}_+^* \times \mathbb{R}_+^* : t_1 \le t_2\right\} \times \mathscr{D} \to \mathscr{D}$$
$$(t_1,t_2,\overline{u}) \mapsto F_{t_1,t_2}\overline{u}$$

with the following properties:

1. $F_{t_2,t_3} \circ F_{t_1,t_2} = F_{t_1,t_3}$ *for all* $0 \le t_1 \le t_2 \le t_3$ *and* $F_{t,t} = \mathrm{Id}_N$ *for all* $t \in \mathbb{R}_+^*$;
2. *for all* $\overline{u} \in \mathscr{D}$, *the function* $u(t) = F_{0,t}\overline{u}$ *is an entropy weak solution to* (8.1.1) *for* $t \in \mathbb{R}_+^*$.

Moreover, for every $T, M \in \mathbb{R}_+$ *there exist constants* $\mathscr{L}, \mathscr{C} \in \mathbb{R}_+$ *such that:*

3. *fix* $\overline{u}$, $\overline{u}'$ *in* $\mathscr{D}_M$ *and let* u, u' *be the corresponding solutions to* (8.1.1) *yielded by* F; *then, for all* $t, t' \in [0,T]$
$$\left|u(t) - u'(t')\right|_{\mathbf{L}^1(\mathbb{R};\mathbb{R}^N)} \le \mathscr{L}\left(\left|\overline{u} - \overline{u}'\right|_{\mathbf{L}^1(\mathbb{R};\mathbb{R}^N)} + \left|t - t'\right|\right) ; \tag{8.5.1}$$
4. *for any initial datum* $\overline{u} \in \mathscr{D}_M$ *the solution* $u = u(t)$ *yielded by* F *satisfies for all* $t \in [0,T]$
$$\begin{aligned} \left|u(t)\right|_{\mathbf{L}^\infty(\mathbb{R};\mathbb{R}^N)} &\le \left|\overline{u}\right|_{\mathbf{L}^\infty(\mathbb{R};\mathbb{R}^N)} \exp(\mathscr{C}\, t) \\ \mathrm{TV}\big(u(t)\big) &\le \mathrm{TV}(\overline{u}) \exp(\mathscr{C}\, t) ; \end{aligned} \tag{8.5.2}$$
5. *if* $\mathscr{U}$ *is compact, then for any* $t \in [0,T]$ *we have* $\mathrm{TV}\big(u(t)\big) \le \hat{\mathscr{C}}\ \mathrm{TV}(\overline{u})$ *and* $\left|u(t)\right|_{\mathbf{L}^\infty(\mathbb{R};\mathbb{R}^N)} \le \hat{\mathscr{C}}\ \left|\overline{u}\right|_{\mathbf{L}^\infty(\mathbb{R};\mathbb{R}^N)}$, *where the constant* $\hat{\mathscr{C}}$ *is independent from* T.

Proof. The proof is based on the classical arguments used, for instance, in [1, 3, 4, 6, 8, 11]. Approximate any initial datum $\overline{u} \in \mathscr{D}$ with $\overline{u}^k \in \mathscr{D}(\mathscr{G}^{2^{-k}})$ and construct the sequence of approximate solutions $F_{0,t}^{2^{-k}}\overline{u}^k$. This is a Cauchy sequence in $\mathbf{L}^1$ and it converges to an entropy weak solution of (8.1.1).

The Lipschitz continuous dependence (8.5.1) is established following the same technique in [3, 6], thanks to the key estimates (8.3.4a) and (8.3.4b). The bounds (8.5.2) are obtained passing to the limit in the estimates provided by Lemma 8.2. The case of a compact $\mathscr{U}$ is dealt with as in [6, (vi),Theorem 2.2]. □

References

1. Amadori, D., Guerra, G.: Global weak solutions for systems of balance laws. Appl. Math. Lett. 12(6), 123–127 (1999)
2. Aubin, J.P., Cellina, A.: Differential inclusions. Grundlehren der Mathematischen Wissenschaften, vol. 264. Springer, Berlin (1984)

3. Bianchini, S.: The semigroup generated by a Temple class system with non-convex flux function. Differential Integral Equations 13(10-12), 1529–1550 (2000)
4. Bressan, A.: Hyperbolic systems of conservation laws. Oxford Lecture Series in Mathematics and its Applications, vol. 20. Oxford University Press, Oxford (2000)
5. Cleopatra, C.: Systems of hyperbolic conservation laws with memory. Journal of Hyperbolic Differential Equations 04 (2007)
6. Colombo, R.M., Corli, A.: On a class of hyperbolic balance laws. J. Hyperbolic Differ. Equ. 1(4), 725–745 (2004)
7. Colombo, R.M., Corli, A., Rosini, M.D.: Non local balance laws in traffic models and crystal growth. ZAMM Z. Angew. Math. Mech. 87(6), 449–461 (2007)
8. Colombo, R.M., Guerra, G.: Hyperbolic Balance Laws with a Non Local Source. Communications in Partial Differential Equations 32(12), 1917–1939 (2007)
9. Colombo, R.M., Mercier, M., Rosini, M.D.: Stability and total variation estimates on general scalar balance laws. Commun. Math. Sci. 7(1), 37–65 (2009)
10. Colombo, R.M., Rosini, M.D.: Well posedness of balance laws with boundary. J. Math. Anal. Appl. 311(2), 683–702 (2005)
11. Crasta, G., Piccoli, B.: Viscosity solutions and uniqueness for systems of inhomogeneous balance laws. Discrete Contin. Dynam. Systems 3(4), 477–502 (1997)
12. Dafermos, C.M.: Hyperbolic conservation laws in continuum physics, 2nd edn. Grundlehren der Mathematischen Wissenschaften, vol. 325. Springer, Berlin (2005)
13. Deimling, K.: Ordinary differential equations in Banach spaces. Lecture Notes in Mathematics, vol. 596. Springer, Berlin (1977)
14. Hoff, D.: Invariant regions for systems of conservation laws. Trans. Amer. Math. Soc. 289(2), 591–610 (1985)

Part II
Models for Vehicular Traffic

Chapter 9
Vehicular Traffic

Abstract. This chapter introduces the main theoretical definitions and models encountered in the study of vehicular traffic. After a brief non–exhaustive overview and classification of the models for traffic, we describe the fundamental traffic variables and their relations.

9.1 Introduction

Transportation problem have plagued man much before the advent of cars. However, in recent years, traffic congestion has become especially acute in cities worldwide: too many vehicles on too few roads! Traffic jams and congested roads are a daily problem. The increasing demand for mobility is also a major challenge. Rising levels of traffic bring increased safety, health, environmental and economic concerns. The resulting costs can be measured as incremental delay, vehicle operating costs (fuel and wear), accidents, pollution emissions and stress of the drivers. Smart traffic management systems can reduce congestion and the related costs by optimizing the use of transport resources and infrastructures of the transport system as a whole, bringing more efficiency in the areas of traffic fluidity and transport services reliability.

The problems of mobility in cities are clear. However, the possible solutions are counterintuitive and still in their infancy, highlighting the essential role of the research. Simply building more roads is not the solution. Traffic engineers have long been familiar with the **fundamental law of highway congestion** in which latent demand expands to fill the gap created whenever highway capacity is improved [65, page 85]. A celebrate example is the so called **Braess' paradox for vehicular traffic** [10]: an extension of the road network may cause a redistribution of the traffic that results in longer individual travel times! Roughly speaking, this paradox can be explained saying that uncoordinated individual drivers pursuing their personally optimal strategies do not always achieve the most beneficial state to the traffic as a whole. Having this in mind, sections of roads were closed in Stuttgart

M.D. Rosini: *Macroscopic Models for Vehicular Flows & Crowd Dynamics*, UCS, pp. 123–138.
DOI: 10.1007/978-3-319-00155-5_9

(Germany) [46] and New York City (USA) [74], with the result that the traffic conditions were improved.

A wide range of traffic flow theories and models have been developed to answer research questions such as what causes congestion, what determines the time and location of traffic breakdown, how does the congestion propagate through the network, etc., which are essential for an effective traffic flow planning. These models of control and simulation of vehicular traffic flow are classified in two main categories: mathematical models and computational models. In the next two sections we briefly describe the main general characteristics of these two approaches. However, a special attention is devoted only to macroscopic mathematical models, since they represent the main object of the present book. In particular, the last two sections of this chapter concern with the introduction and description of the macroscopic traffic variables together with their relations.

9.2 Mathematical Models

Mathematical models are classified based on the following:

- Level of details (microscopic, mesoscopic, macroscopic).
- Scale of the independent variables (continuous, discrete, semi–discrete).
- Representation of the processes (deterministic, stochastic).
- Operationalisation (analytical, simulation).
- Scale of application (networks, stretches, links, and intersections).

Table 9.1 presents a non–exhaustive overview of the main models classified according to the above criteria, see [2, 9, 12, 24, 30, 38, 45, 49, 52, 54] for more details. Let us discuss in particular the first criterium. According to the level of details the models can be distinguished in:

Microscopic models: they give a detailed description of the traffic flow and describe *individually* each vehicle and its interaction with the other vehicles and with the environment. Higher level of details, such as the functioning of specific parts and processes of vehicles, driving tasks, driver behavior (how he changes the gear, applies brakes, changes lane and the corresponding reaction times), characterizes the so called **submicroscopic models**.

Mesoscopic models: they specify the behavior of each individual in probabilistic terms. To this end, traffic is represented by small groups of vehicles. Only the interactions of each group is considered, overlooking the interaction of the individual vehicles.

Macroscopic models: they describe traffic as a flow without distinguishing its constituent vehicles. The traffic streaming is represented in terms of flow, density and velocity.

The correct level of details that should be considered is largely dependent on the envisaged model application.

Table 9.1 Overview of traffic flow models.
DI: dimension (other than time and space): velocity v, desired velocity d, lateral position l (lanes), and other o.
SC: scale (**c**ontinuous, **d**iscrete, and semi–**d**iscrete);
RE: process representation (**d**eterministic, **s**tochastic);
OP: operationalisation (**a**nalytical, **s**imulation);
AR: area of application (**c**ross–section, **s**ingle **l**ane stretches, **m**ulti–**l**ane stretches, **a**ggregate **l**ane stretches, **d**iscontinuities, motorway **n**etwork, and **u**rban network).

Detail Level	Model Name / Ref.	DI				SC	RE	OP	AR						
		v	d	l	o				c	sl	ml	al	d	n	u
Microscopic	MIXIC [4]	*		*	*	d	s	s			*				
	SIMONE [55]	*	*	*	*	d	s	s			*		*		
	PELOPS [51]	*		*	*	d	s	s			*				
	car-following models [49, 54, 71]	*	*	*		c	d,s	a,s		*	*				
	FOSIM [69]	*	*	*	*	d	s	s			*		*		
	cellular automata [21, 56, 57]	*	*			d	s	s						*	*
	INTEGRATION [1]	*				d	d	s						*	
Mesoscopic	headway distr. models [35]			*	*	c	s	a	*						
	reduced gas-kinetic model [62]	*				c	d	a				*			
	improved gas-kinetic model [59]	*	*			c	d	a				*			
	multi–lane gas-kinetic model [29]	*	*		*	c	d	a			*		*		
	multi–class gas-kinetic model [37]	*	*	*		c	d	a				*			
	multi–class multi–lane model [36]	*	*	*		c	d	a			*	*	*		
	cluster models [8]	*	*			c	d	a				*			
Macroscopic	LWR model [50, 64]	*				c	d	a				*			
	Payne–type models [60]	*				c	d	a				*			
	Helbing–type models [29, 29]	*			*	c	d	a				*			
	Cell-Transmission Model [14–16]	*				d	d	s						*	
	METANET [47]	*				d	d	s						*	
	semi–discrete model [66]					sd	s	a				*	*		
	FREFLO [61]	*				d	d	s						*	
	MASTER [68]	*				d	d	a			*				

Let us compare macroscopic and microscopic models in more detail. The number of developed macroscopic models is far less than that of microscopic ones. In [2], the authors identified 58 microscopic simulation models already in 1997.

In a microscopic simulation model each vehicle is described by its own equation of motion. The dynamics are described by a system of ordinary second–order differential equations of the form

$$x_i''(t) = a_i\left(t, x_i(t), x_{i+1}(t), v_i(t), v_{i+1}(t)\right) , \qquad i = 1, \ldots, N , \tag{9.2.1}$$

where N is the total number of vehicles, t is the time, $x_i = x_i(t)$ is the position along the road of the ith vehicle, $v_i = x_i'(t)$ is its velocity and $a_i = x_i''(t)$ is its acceleration.

Each vehicle is commonly assumed to be influenced by its heading vehicle only. It is immediately seen that the size of system (9.2.1) grows proportionally to the number of simulated vehicles and, consequently, so do also the computer time and memory required to solve it. For this reason, microscopic models are most suitable for off–line traffic simulations, for instance to perform detailed studies of geometric design and vehicle equipment, or to gain insight into flow quantities that are difficult to determine empirically. However, their application in on–line traffic control is limited due to the large computation times and the absence of an explicit model input–output relation. In addition, due to a large number of sometimes unobservable parameters that characterizes microscopic models, only few microscopic simulation models have been extensively calibrated and validated. Furthermore, from the analytical point of view, it is often difficult to investigate the relevant global features of the system via microscopic models, also in connection with control and optimization problems. Finally, microscopic simulation tools do not provide insight into the macroscopic mechanisms of traffic flow.

On the other hand, macroscopic models are computationally less demanding, thereby allowing on–line traffic simulations of traffic flow in networks. Moreover, solutions to macroscopic models are usually formulated in a closed analytical form, making them excellently applicable for model based estimation, optimization, prediction, and control approaches. In macroscopic models, the number of parameters is relatively small and, more importantly, comparably easy to observe and measure. Therefore, in general, macroscopic models are relatively easy to calibrate: essentially, speed–density relations derived from observations are required. Therefore, calibration and validation of macroscopic models require less effort than calibration of microscopic models. Furthermore, macroscopic models are deemed to describe macroscopic characteristics of vehicular traffic more accurately by means of mathematical analysis and manipulation. For this reason they are very suited for analyzing and reproducing macroscopic characteristics of vehicular traffic, such as shock waves, and queue lengths.

Since macroscopic models represent the main object of the book, we further describe in more details the properties of these models. Macroscopic models are all based on the fluid dynamic. But there are essential differences between vehicular traffics and fluids:

(D.1) A fluid particle is isotropic and responds to stimuli from the surrounding particles, while a vehicle is anisotropic and responds only to frontal stimuli.
(D.2) The law of conservation of momentum holds for fluids but not for vehicular traffic.
(D.3) The number of vehicles is far less than that of particles.
(D.4) A vehicle is a living system since his driver has a personality (e.g., aggressive or timid) and this remains unchanged by motion (e.g., a slow vehicle is unaffected by its interaction with faster vehicles passing it, or queueing behind it).
(D.5) Vehicles with negative speed can not arise.

According to Aw and Rascle [5], a "good" macroscopic traffic flow model needs to fulfill the following list of requirements:

(R.1) The system must be hyperbolic.
(R.2) The solution of any Riemann problem with arbitrary bounded non–negative Riemann data in a suitable region of the phase plane must remain non–negative and bounded from above.
(R.3) The speeds of propagation of the waves of any solution to any Riemann problem must be at most equal to the average speed.
(R.4) Braking must produce shock waves, whose propagation speed must be negative or non–negative, whereas accelerating produces rarefaction waves, which in any case satisfy (R.3).
(R.5) Near the vacuum, the solution to the Riemann problem must be very sensitive to the data.

Condition (R.1) corresponds to the assumption that the cars are neither created nor destroyed. Condition (R.2) represents a minimal requirement that any reasonable model has to satisfy. Condition (R.3) is actually under an interesting debate [31, 32, 76]. The existence of characteristics faster than traffic produces counterintuitive predictions that still need to be verified through empirical observations. Condition (R.4) is suggested by direct observations of real traffic. Finally, condition (R.5) says that there must be no continuous dependence of the solution with respect to the initial datum corresponding to the vacuum.

9.3 Computational Models

Since their introduction in the 1960s, supercomputers underwent fast development, see Fig. 9.1. According to `http://www.top500.org`, the actual world's top three supercomputers are: Titan based on Cray XK7 architecture (ORNL, USA), Sequoia based on IBM Blue Gene/Q architecture (LLNL, USA) and Fujitsu K computer (RIKEN AICS, Japan). Their new systems, such as grid computing, computer cluster and multi–core processors combined with centralization, enable very highly calculation–intensive tasks. All of these systems are designed and optimized to obtain high performances by developing new methods and algorithms, that enable simulations of complex real–world processes on extremely large scale as well as the development of the so called computational models, based on computational processes and algorithms.

In this section we briefly introduce the most celebrated meta–heuristic computational models used in vehicular traffic controlling and forecasting: ant colony optimization algorithm [20], particle swarm intelligent algorithm [6, 44], genetic algorithm [63], fuzzy and neural network algorithm [41].

Ant Colony Optimization (ACO). ACO algorithms are one of the most successful technique for approximate optimization. The method takes inspiration from

Fig. 9.1 IBM Blue Gene/Q system available for open science at the Interdisciplinary Centre for Mathematical and Computational Modelling, University of Warsaw.

how an ant colony is able to coordinate as a whole to locate and collect food without a central control communicating with the individual ants of the system. Even though Dorigo [19] originally introduced the first ACO algorithms to optimize discrete problems, the application of ACO to continuous optimization problems, such as those related to vehicular traffic, was feasible after its extension to Continuous ACO (**CACO**) [7, 53, 72]. In the corresponding algorithms, the behavior of drivers is simulated by assuming that they choose the route that results to be optimal according several objectives such as, for instance, traveling the shortest and quickest route to their destination with minimum route changes. From the point of view of managing the road network as a whole, the goal is rather to minimize the average travel time by maximizing the traffic flow in the network, which may mean that some drivers does not travel according to their personal optimal route or speed. For example, in [23] the drivers select optimum route between origin and destination according the path length and traffic congestion of roads. Another application [18] consists in optimally routing a fleet of vehicles taking into account the number of tours first and the total travel time in a time dependent context. Further models that use CACO algorithms are **SVRCACO** [34] to simulate traffic flow along inter–urban motorways, **SuRJE** [33] that optimizes traffic light turn and time sequences, [28, 48] that deal with centralized vehicle routing and [13] that deals with decentralized vehicle routing.

Particle Swarm Optimization (PSO). PSO is a population based stochastic optimization technique inspired by the movement of swarms. In [40] the PSO algorithm is combined with wavelet networks for predicting urban traffic conditions. We recall that wavelet networks [77] are based on the combination of wavelet theory [22, 27] and neural networks [39]. In [67], the problem of early traffic incidents detection is solved by using artificial neural network and the PSO algorithm for the learning phase of neural network. In [79], the PSO algorithm is used to solve the time dependent routing problem. In [78], a traffic coordination control system for urban area network based upon multi–agent technology was optimized utilizing PSO.

Genetic Algorithm (GA). GAs belong to the larger class of evolutionary algorithms, which generate solutions to optimization problems using techniques inspired by natural evolution, such as crossover and mutation. In [75] is proposed an optimization model for the coordination of vehicular traffic signals at the intersections based on a GA. [25] presents a genetic algorithm approach for a traffic light optimization problem. [43] describes a traffic control simulation based on the exchange of messages between local intersections, which incorporate dynamically assembled cellular automata. A GA is employed to determine parameters governing the messaging and cellular behavior. A self–adaptive string length GA is presented in [11] to solve the urban rerouting problem, so as to meet the inconstant intersection number in different potential routes.

Fuzzy and Neural Network Algorithm (FNNA). FNNA is inspired by biological neural networks. Fuzzy approach is mostly applied to predict, control and direct vehicular traffic by using traffic lights. [17] devises an adaptive and cooperative multi–agent fuzzy system for a decentralized traffic signal control based on FNNA. To achieve this goal, every intersection has three levels of control: its own traffic situation, correlated intersection recommendations and a knowledge base which provides its traffic pattern. [58] describes an intelligent system architecture for urban traffic control which integrates a neural network and an expert system on silicon. The intelligent decision making system consists of a back–propagation based neural network for adaptive learning and a rule–based fuzzy expert system for decision making. In [73] is developed the **FNM model** to predict the traffic flows in an urban street network. It consists of two modules: a gate network GN and an expert network EN. GN categories input data using a fuzzy approach and EN uses neural network to associate input and output data. The authors of [3] propose a hybrid modeling approach which combines FNNA and a simple statistical approach in order to provide a one hour forecast of urban traffic flow rates. [42] presents a two–stage method to control four–phased intersection signal timing control by using the FNNA.

The increasing application of swarm intelligence algorithms in engineering, biology, chemistry, physics, etc., results from their numerous powerful properties, such as their ability to cope with systems characterized by huge number of variables and objectives, with a rapidly changing topology and complexity. Swarm intelligence however is a new field of research, still in its infancy, and thus much work remains to be done. Analytic proof of the swarm–based algorithm performance remains topic of ongoing research. Therefore, a major problem with the application of swarm intelligence algorithms is the amount of trust that must be placed onto the suggested solution.

The main shortcoming of swarm intelligence algorithms is that they do not take into account the most fundamental random element of the system: the human factor. In general, the persecution of an optimal solution, could cause a complete break in the efficiency of the system, or the proposed solution can not be even completely assured in its safety and applicability. For instance, a single delay could disturb the most efficient system completely and make it very inefficient or dangerous. Furthermore, such algorithms always converge to one solution and do not even take into account the possible existence of other optimal solutions. These drawbacks are consequences of the lack of a well established theory able to analyze all the aspects that may characterize the solutions proposed by computational models, such as their stability and uniqueness.

In respect to the application to vehicular traffic, the different behavior of each driver may affect the optimal strategy, but this is not taken into account by swarm intelligence algorithms. Another problem may arise if at the same time many drivers have the same starting point, the same destination and they use the same algorithm. In this case the proposed optimal solution fails, as all these driver will select the same roads at the same time lowering down the efficiency of the system.

It is also questionable the replacement of human controlled network with intelligent systems. The human knowledge should confirm the solution and help to reduce the risk of dangerous decision being made only by blindly following a computational model. These models should be used as tools to take the final decision by considering their results as suggestions rather than final word of any critical action.

In conclusion, the combination of different approaches and different mathematical and computational models seems to be the most effective way to work with problems characterized by too many aspects, variables, objectives and levels of description, such as the vehicular traffic, to be described by a unique comprehensive model. For instance, the optimal solutions proposed by computational models could select the most common routes of the drivers and the mathematical models can be applied to make simulations of traffic along these routes and detect the dangerous intersections that need more attentions and more accurate considerations.

9.4 The Fundamental Macroscopic Traffic Variables

Consider a long homogeneous multi–lane highway. What traffic variables could an observer easily measure? Imagine that we can keep track of each vehicle moving in one direction. As a first insight, the dimensions of the vehicles could be overlooked and the position of the ith vehicle at time t might refer, for instance, to the center of it and denoted by $X_i(t) = \big(x_i(t), \ell_i(t)\big) \in \mathbb{R} \times \mathbb{N}$, where the first component gives the position along the lane that is given by the second component. Then it is possible to provide a space–time diagram for a graphical description of the traffic. In

Fig. 9.2, position is measured along the horizontal axis, and the time is measured along the vertical axis. Each trajectory line represents a unique vehicle. Vehicles following each other have parallel trajectories, and trajectories cross when one vehicle passes another. Trajectories parallel to the vertical axis correspond to stopped vehicles waiting, for instance, in a queue. If we introduce the **velocity** v defined as

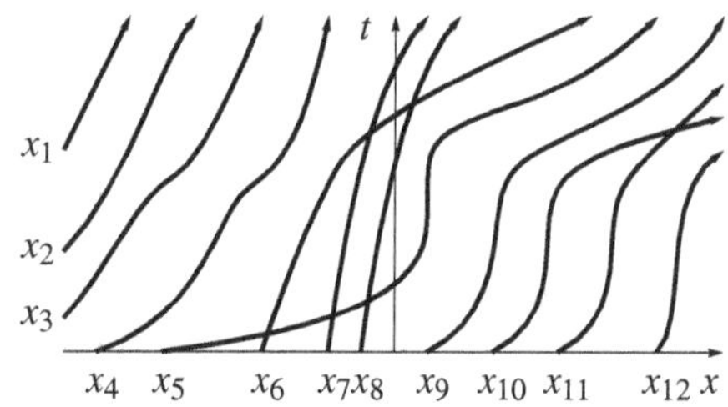

Fig. 9.2 Space–time diagram for the positions of, only, 12 vehicles. The fastest vehicle is the 5th one (maybe a Bugatti?) that overtake the vehicles 6, 7 and 8. These last two move relatively slowly and follow each other (maybe two trucks?).

the distance covered per unit time, then the velocity at time t of the ith vehicle is $v = x_i'(t)$.

In many real situations the number of vehicles is so large that it is almost impossible to keep track of each of them. For this reason, it might be more reasonable to associate to each lane ℓ, position x and time t a unique velocity $v = v(t,x,\ell)$, called **velocity field**. This would be the velocity measured along the lane ℓ by an observer in x at time t. The case $v(t,x,\ell) = 0$ corresponds to have along the lane ℓ no vehicle in x at time t. A velocity field $v = v(t,x,\ell)$ is well defined iff at each t, x and ℓ there is *one* velocity, in other words, iff overtaking is allowed only among vehicles moving along different lanes. From the mathematical point of view, the velocity of the ith car and the velocity field satisfy the equation

$$v\left(t, X_i(t)\right) = x_i'(t) .$$

An observer in x could also measure the number of vehicles that pass and compute, for instance, the average number of vehicles passing per unit time. This quantity is called **traffic flow** (or **flow–rate**) and denoted f. Another standard traffic measurement is the number of cars at a fixed time per unit space. This quantity is called **traffic density** and denoted by ρ. It can be computed, for instance, by taking photographes. Vehicles not completely in a given region at a fixed time can be taken into account by using estimates of fractional vehicles or by counting a vehicle only if its center is in the region.

Example 9.1. The simplest possible traffic situation occurs when vehicles have the same length L, the same velocity v and are equally spaced along a one lane highway, see Fig. 9.3. We can parameterize the road by a single coordinate x and we can assume that the traffic moves in the direction of increasing x. Since each vehicle moves with the same speed, the distance between vehicles remains constant. Hence also the traffic density does not change. If d is the distance between vehicles, then

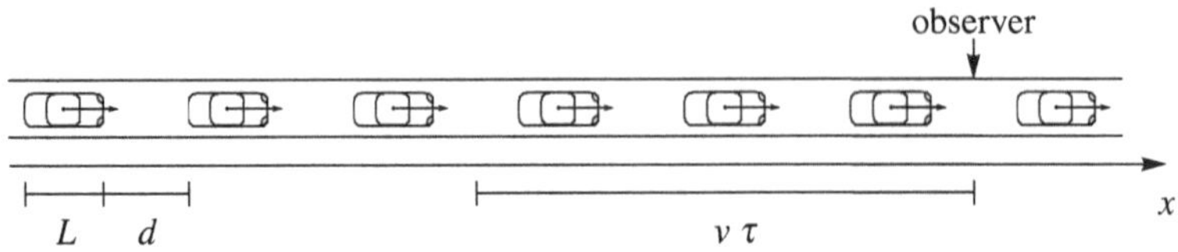

Fig. 9.3 Vehicles with the same length L and velocity v move equally spaced along a one lane highway

the density is

$$\rho = \frac{1}{L+d} \,. \tag{9.4.1}$$

The maximal density ρ_m is achieved when there is a traffic jam, the vehicles stand still bumper to bumper, namely $d = 0$. Thus

$$\rho_m = \frac{1}{L} \,.$$

What is the flow? After τ hours, an observer in x see at time t the vehicle that was at time $t-\tau$ in $x-\tau\, v$. Thus, the number of vehicles that pass the observer in τ hours is the number of vehicles in the interval $[x-\tau\, v, x]$ at time $t-\tau$. Since the density ρ is constant, this number is $\big(x-(x-\tau\, v)\big)\ \rho = \tau\, v\, \rho$. Thus, by (9.4.1) the flow is

$$f = \frac{v}{L+d} \,.$$

The above expression for the flux might suggest that the maximal flow, also called **road capacity**, would occur iff $d = 0$, namely iff vehicles are bumper to bumper and $\rho = \rho_m$. Clearly this is not safe, but furthermore, the observations of real traffics show that as the density approach the maximal one, the drivers slow their vehicles and $v = 0$ when $\rho = \rho_m$. Therefore, the case $d = 0$ corresponds rather to a minimum traffic, namely zero. □

9.5 Relations between the Fundamental Traffic Variables

In this section we derive the relations among the three fundamental traffic variables: velocity, density and flow. For the case considered in the Example 9.1 we have

$$f(t,x) = \rho(t,x)\ v(t,x) \,. \tag{9.5.1}$$

Although this law has been derived from an oversimplified case, we will prove that it is always valid. Consider the number of vehicles that pass $x = x_o$ in a very small time Δt, namely, during the time interval $[t_o, t_o + \Delta t]$. If the density and the velocity are continuous function of t and x, in this small time interval these quantities can be approximated by constants. Consequently, the number of passing cars is approximately $v(t,x)\,\Delta t\,\rho(t,x)$ and (9.5.1) is still valid.

By (9.5.1), we can choose the two fundamental variables to be the density ρ and the velocity v and express the flux as $f(\rho,v)=\rho\ v$. Suppose that we know at time $t=t_o$ the density and the velocity along a one lane highway, namely $\rho(t_o,x)$ and $v(t_o,x)$ are given. Can we predict the densities and velocities at future times? Suppose that a traffic control device records the number of vehicles $N=N(x,t,h)$ in the segment $[x,x+h]$ of the road at time t. If some vehicle is located at the boundary of this interval, the traffic control device accounts for that by allowing N to vary in all $\mathbb{R}_+^*$. If N is large, h is large compared with the average length of the cars and small compared with the length of the road, the density satisfies

$$\rho(t,x)=\lim_{h\searrow 0}\frac{N(x,t,h)}{h}\,.$$

By definition $N(x,t,0)=0$, therefore $\rho(t,x)=\partial_h N(x,t,0)$ and

$$N(x,t,h)=\int_x^{x+h}\rho(t,y)\ \mathrm{d}y\ . \tag{9.5.2}$$

We wish to determine how the number of cars changes in a space interval $[a,b]$ of the road. If no entries or exits are present in $[a,b]$, then this number can change only as cars are entering the interval from the left endpoint $x=a$, or leaving the interval at the right endpoint $x=b$. Consequently, for any $T>t_o$, the number of cars in $[a,b]$ at time $t=T$ is

$$\underbrace{N(a,T,b-a)}_{\substack{\text{cars in } [a,b]\\ \text{at time } t=T}}=\underbrace{N(a,t_o,b-a)}_{\substack{\text{cars in } [a,b]\\ \text{at time } t=t_o}}+\underbrace{\int_{t_o}^{T}\rho(t,a)\,v(t,a)\,\mathrm{d}t}_{\substack{\text{cars entering in } x=a\\ \text{for } t\in[t_o,T]}}-\underbrace{\int_{t_o}^{T}\rho(t,b)\,v(t,b)\,\mathrm{d}t}_{\substack{\text{cars exiting in } x=b\\ \text{for } t\in[t_o,T]}}$$

and by (9.5.2) we have

$$\int_a^b\big(\rho(T,x)-\rho(t_o,x)\big)\ \mathrm{d}x=\int_{t_o}^{T}\big(\rho(t,a)\,v(t,a)-\rho(t,b)\,v(t,b)\big)\ \mathrm{d}t\ ,$$

or equivalently

$$\int_{t_o}^{T}\int_a^b\Big[\partial_t\rho(t,x)-\partial_x\big(\rho(t,x)\,v(t,x)\big)\Big]\ \mathrm{d}x\ \mathrm{d}t=0\,.$$

As a consequence, since a, b, T and t_o are arbitrary, we deduce

$$\partial_t\rho+\partial_x(\rho\ v)=0\,. \tag{9.5.3}$$

This equation is a scalar conservation law in one space dimension and expresses the fact that the cars are neither created nor destroyed, namely the conservation of the number of cars, see Chap. 3.

Equation (9.5.3) represents an equation with two unknown variables, the density and the velocity; therefore it is not enough to determine the densities and velocities at future times. To get a complete description of traffic dynamics, it is necessary to add a further independent equation. Different approaches to this problem are available in the literature and can be essentially divided into two types:

Equilibrium traffic models: They assume that the velocity is a function of the density and consider beside (9.5.3) a constitutive equation of the form $v = v(\rho)$.

Non–equilibrium traffic models: They add to (9.5.3) another partial differential equation with density and velocity as its variables.

The word "equilibrium" has in this context an obvious reason: all traffic dynamics in the theory of equilibrium traffic models occur along the equilibrium curve $\{(\rho, v(\rho)) \, : \, \rho \in [0, \rho_m]\}$. Non–equilibrium traffic models attempt to relax this restriction by adding a partial differential equation that describes the rate of change of the velocity, so that non–equilibrium states can also occur. We recall that often equilibrium traffic models are referred to as **first–order traffic models**, and the non–equilibrium traffic models as **second–order traffic models**.

We underline that the only accurate physical law in traffic flow theory is the conservation of vehicles, represented by (9.5.3); all other assumptions result from a coarse approximations of empirical observations. However, as vehicle traffic is influenced by decision–making and psychological effects, nobody would expect that traffic models could reach an accuracy comparable to that attained in other domains of science (e.g. Newtonian physics or thermodynamics). Nevertheless, they can have sufficient descriptive power of the specific application–purpose and help to understand non–trivial properties of traffic flows, to predict and optimize them. In general, models do not even aim at giving an accurate reproduction of reality. The strength of models lies in the simplification of reality by means of good approximations able to reproduce empirical facts.

The LWR model proposed by Lighthill, Whitham [50] and Richards [64] is the first model to describe traffic flow. It is an equilibrium traffic model and takes the Greenshields [26] expression as constitutive equation, see Chap. 10 for more details and further examples of equilibrium traffic models. The PW model proposed by Payne [60] and Whitham [70], see Sect. 14.2, is the first non–equilibrium traffic model. We defer to Chap. 14 for the description of the PW model and of other non–equilibrium traffic models.

References

1. Aerde, M.V.: Modelling of integrated traffic networks using the integration simulation model. In: Ontario Ministry of Transportation Research and Development Branch TDS, Ontario Ministry of Transportation, Research and Development Branch (1990)

2. Algers, S., Bernauer, E., Boero, M., Breheret, L., Di Taranto, C., Dougherty, M., Fox, K., Gabard, J.F.: Review of Micro-Simulation Models. Analysis 3 (August 1997)
3. Annunziato, M., Moretti, F., Pizzuti, S.: Urban Traffic Flow Forecasting Using Neural-Statistic Hybrid Modeling. In: Snášel, V., Abraham, A., Corchado, E.S. (eds.) SOCO Models in Industrial & Environmental Appl. AISC, vol. 188, pp. 183–190. Springer, Heidelberg (2013)
4. Arem, B.: TNO Beleidsstudies en Advies. The Microscopic traffic simulation model, MIXIC 1.1. TNO Institute for Policy Studies (1994)
5. Aw, A., Rascle, M.: Resurrection of "second order" models of traffic flow. SIAM J. Appl. Math. 60(3), 916–938 (2000) (electronic)
6. Beni, G.: Swarm intelligence. In: Meyers, R.A. (ed.) Computational Complexity, pp. 3150–3169. Springer, New York (2012)
7. Bilchev, G., Parmee, I.C.: The Ant Colony Metaphor for Searching Continuous Design Spaces. In: Fogarty, T.C. (ed.) AISB-WS 1995. LNCS, vol. 993, pp. 25–39. Springer, Heidelberg (1995)
8. Botma, H.: Traffic-flow models: state of the art report. In: Institute for Road Safety Research SWOV, The Netherlands, (1981)
9. Brackstone, M., McDonald, M.: Car-following: a historical review. Transportation Research Part F: Traffic Psychology and Behaviour 2(4), 181–196 (1999)
10. Braess, D., Nagurney, A., Wakolbinger, T.: On a paradox of traffic planning. Transportation Science 39(4), 446–450 (2005)
11. Cao, L., Shi, Z., Bao, P.: Self-adaptive Length Genetic Algorithm for Urban Rerouting Problem. In: Jiao, L., Wang, L., Gao, X.-b., Liu, J., Wu, F. (eds.) ICNC 2006. LNCS, vol. 4221, pp. 726–729. Springer, Heidelberg (2006)
12. Chowdhury, D., Santen, L., Schadschneider, A.: Statistical physics of vehicular traffic and some related systems. Physics Report 329, 199–329 (2000)
13. Claes, R., Holvoet, T., Weyns, D.: A decentralized approach for anticipatory vehicle routing using delegate multiagent systems (2011)
14. Daganzo, C.F.: The cell transmission model: A dynamic representation of highway traffic consistent with the hydrodynamic theory. Transportation Research Part B: Methodological 28(4), 269–287 (1994)
15. Daganzo, C.F.: The cell transmission model, part II: Network traffic. Transportation Research Part B: Methodological 29(2), 79–93 (1995)
16. Daganzo, C.F.: The lagged cell-transmission model. In: 14th Int. Symp. on Transportation and Traffic Theory, pp. 81–104 (1999)
17. Daneshfar, F., RavanJamJah, J., Mansoori, F., Bevrani, H., Azami, B.Z.: Adaptive fuzzy urban traffic flow control using a cooperative multi-agent system based on two stage fuzzy clustering. In: IEEE 69th Vehicular Technology Conference, VTC Spring 2009, pp. 1–5 (April 2009)
18. Donati, A., Montemanni, R., Casagrande, N., Rizzoli, A., Gambardella, L.: Time dependent vehicle routing problem with a multi ant colony system. European Journal of Operational Research 185(3), 1174–1191 (2008)
19. Dorigo, M.: Optimization, learning and natural algorithms. Ph. D. Thesis, Politecnico di Milano, Italy (1992)
20. Dorigo, M., Stützle, M.D.T.: Ant Colony Optimization. Bradford Bks. Mit Press (2004)
21. Esser, J., Schreckenberg, M.: Microscopic simulation of urban traffic based on cellular automata. International Journal of Modern Physics 8(5), 1025–1036 (1997)
22. Feichtinger, H.G., Strohmer, H.G.F.T.: Gabor Analysis and Algorithms: Theory and Applications. In: Applied and Numerical Harmonic Analysis Series, Birkhäuser (1998)

23. Foroughi, R., Montazer, G., Sabzevari, R.: Design of a new urban traffic control system using modified ant colony optimization approach. Iranian Journal of Science and Technology Transaction B: Engineering 32, 167–173 (2008)
24. Garavello, M., Piccoli, B.: Traffic flow on networks. AIMS Series on Applied Mathematics, vol. 1. American Institute of Mathematical Sciences (AIMS), Springfield, MO (2006) (Conservation laws models)
25. Gora, P.: A Genetic Algorithm Approach to Optimization of Vehicular Traffic in Cities by Means of Configuring Traffic Lights. In: Ryżko, D., Rybiński, H., Gawrysiak, P., Kryszkiewicz, M. (eds.) Emerging Intelligent Technologies in Industry. SCI, vol. 369, pp. 1–10. Springer, Heidelberg (2011)
26. Greenshields, B.D.: A study of traffic capacity. Highway Research Board 14, 448–477 (1934)
27. Haar, A.: Zur Theorie der orthogonalen Funktionensysteme. Mathematische Annalen 69(3), 331–371 (1910)
28. Hallam, N., Hartley, M., Blanchfield, P., Kendall, G.: Optimisation in a road traffic system using collaborative search (2004)
29. Helbing, D.: Modeling Multi-Lane Traffic Flow with Queuing Effects. Arxiv Preprint Condmat9806050 242(1-2), 175–194 (1998)
30. Helbing, D.: Traffic and related self-driven many-particle systems. Reviews of Modern Physics (2001)
31. Helbing, D.: Reply to comment on Şon the controversy around DaganzoŠs requiem for and Aw-RascleŠs resurrection of second-order traffic flow models by H.M. Zhang. The European Physical Journal B - Condensed Matter and Complex Systems 69, 569–570 (2009), 10.1140/epjb/e2009-00193-4
32. Helbing, D., Johansson, A.F.: On the controversy around DaganzoŞs requiem for and Aw-RascleŠs resurrection of second-order traffic flow models. The European Physical Journal B - Condensed Matter and Complex Systems 69, 549–562 (2009)
33. Hoar, R., Penner, J., Jacob, C.: Evolutionary swarm traffic: if ant roads had traffic lights. In: Proceedings of the Congress Evolutionary Computation, CEC 2002, vol. 2, pp. 1910–1915. IEEE Computer Society, Washington, DC (2002)
34. Hong, W.-C., Pai, P.-F., Yang, S.-L., Lai, C.-Y.: Continuous Ant Colony Optimization in a SVR Urban Traffic Forecasting Model. In: Sandoval, F., Prieto, A.G., Cabestany, J., Graña, M. (eds.) IWANN 2007. LNCS, vol. 4507, pp. 765–773. Springer, Heidelberg (2007)
35. Hoogendoorn, S., Bovy, P.: New Estimation Technique for Vehicle-Type-Specific Headway Distributions. Transportation Research Record 1646(1), 18–28 (1998)
36. Hoogendoorn, S.P.: Multiclass continuum modelling of multilane traffic flow. TRAIL thesis series. Delft University Press (1999)
37. Hoogendoorn, S.P., Bovy, P.H.L.: Continuum modeling of multiclass traffic flow. Transportation Research Part B: Methodological 34(2), 123–146 (2000)
38. Hoogendoorn, S.P., Bovy, P.H.L.: State-of-the-art of Vehicular Traffic Flow Modelling, pp. 283–303. Delft University of Technology, Delft (2001)
39. Hopfield, J.J.: Neural networks and physical systems with emergent collective computational abilities. Proceedings of the National Academy of Sciences of the United States of America 79(8), 2554–2558 (1982)
40. Huang, Y.: Short-term traffic flow forecasting based on wavelet network model combined with pso. In: 2008 International Conference on Intelligent Computation Technology and Automation (ICICTA), vol. 1, pp. 249–253 (October 2008)
41. Jang, J.-S.R.: Chuen-Tsai Sun. Neuro-fuzzy modeling and control. Proceedings of the IEEE 83(3), 378–406 (1995)

42. Kaedi, M., Movahhedinia, N., Jamshidi, K.: Traffic signal timing using two-dimensional correlation, neuro-fuzzy and queuing based neural networks. Neural Computing and Applications 17, 193–200 (2008)
43. Kelly, M.: Decentralised urban traffic control using genetic algorithm and cellular automata. In: IEEE Congress on Evolutionary Computation, CEC 2008, IEEE World Congress on Computational Intelligence, pp. 945–950 (June 2008)
44. Kennedy, J., Eberhart, R.: Particle swarm optimization. In: Proceedings., IEEE International Conference on Neural Networks, vol. 4, pp. 1942–1948 (November/December 1995)
45. Kerner, B.S.: The physics of traffic: empirical freeway pattern features, engineering applications, and theory. In: Understanding Complex Systems. Springer (2004)
46. Knödel, W.: Graphentheoretische Methoden und ihre Anwendungen. In: Econometrics and Operations Research. Springer (1969)
47. Kotsialos, A., Papageorgiou, M., Diakaki, C., Pavlis, Y., Middelham, F.: Traffic flow modeling of large-scale motorway networks using the macroscopic modeling tool METANET. IEEE Transactions on Intelligent Transportation Systems 3(4), 282–292 (2002)
48. Kromer, P., Martinovic, J., Radecky, M., Tomis, R., Snasel, V.: Ant colony inspired algorithm for adaptive traffic routing (2011)
49. Leutzbach, W.: Introduction to the theory of traffic flow. Springer (1988)
50. Lighthill, M.J., Whitham, G.B.: On kinematic waves. II. A theory of traffic flow on long crowded roads. Proc. Roy. Soc. London. Ser. A 229, 317–345 (1955)
51. Ludmann, J.: Beeinflussung des Verkehrsablaufs auf Strassen: Analyse mit dem fahrzeugorientierten Verkehrssimulationsprogramm PELOPS. In: Schriftenreihe Automobiltechnik. Forschungsges. Kraftfahrwesen (1998)
52. Maerivoet, S., De Moor, B.: Cellular automata models of road traffic. Physics Reports 419(1), 1–64 (2005)
53. Mathur, M., Karale, S.B., Priye, S., Jayaraman, V.K., Kulkarni, B.D.: Ant Colony Approach to Continuous Function Optimization. Ind. Eng. Chem. Res. 39(10), 3814–3822 (2000)
54. May, A.D.: Traffic flow fundamentals. Prentice-Hall (1990)
55. Minderhoud, M.M.: Supported driving: impacts on motorway traffic flow. TRAIL Thesis Series. Delft University Press (1999)
56. Nagel, K., Schleicher, A.: Microscopic traffic modeling on parallel high performance computers. Report // Angewandte Mathematik und Informatik, Universität zu Köln. Univ. (1993)
57. Ning, W., Brilon, W.: Cellular Automata for Highway Traffic Flow Simulation. homepageruhrunibochumde (1992)
58. Patel, M.I., Ranganathan, N.: An intelligent system architecture for urban traffic control applications. In: Eighth IEEE Symposium on Parallel and Distributed Processing, pp. 10–17 (October 1996)
59. Paveri-Fontana, S.: On Boltzmann-like treatmants for traffic flow: A critical review of the basic model and an alternative proposal for dilute traffic analysis. Transportation Research 9(18), 225–235 (1975)
60. Payne, H.J.: Models of freeway traffic and control. Math. Models Publ. Sys. Simul. Council Proc. (28), 51–61 (1971)
61. Payne, H.J.: Freflo: A Macroscopic Simulation Model of Freeway Traffic. Transportation Research Record 722(722), 68–77 (1979)
62. Prigogine, I., Herman, R.: Kinetic theory of vehicular traffic. American Elsevier Pub. Co. (1971)

63. Rechenberg, I.: Evolutionsstrategie – Optimierung technisher Systeme nach Prinzipien der biologischen Evolution, Frommann-Holzboog, Stuttgart, GER (1973)
64. Richards, P.I.: Shock waves on the highway. Operations Res. 4, 42–51 (1956)
65. Small, K.A., Winston, C., Evans, C.A.: Road work: a new highway pricing and investment policy. Brookings Institution (1989)
66. Smulders, S.: Control of freeway traffic flow by variable speed signs. Transportation Research Part B: Methodological 24(2), 111–132 (1990)
67. Srinivasan, D., Loo, W.H., Cheu, R.L.: Traffic incident detection using particle swarm optimization. In: Proceedings of the 2003 IEEE Swarm Intelligence Symposium, SIS 2003, pp. 144–151 (April 2003)
68. Treiber, M., Hennecke, A., Helbing, D.: Derivation, Properties, and Simulation of a Gas-Kinetic-Based, Non-Local Traffic Model. Physical Review E 59(1), 239–253 (1999)
69. Vermijs, R.G.M.M.: Kalibratie en validatie van het mikrosimulatiemodel FOSIM voor symmetrische weefvakken. LVV rapport. Technische Universiteit Delft, Faculteit der Civiele Techniek, Vakgroep Verkeer (1991)
70. Whitham, G.B.: Linear and nonlinear waves. Pure and Applied Mathematics. Wiley-Interscience [John Wiley & Sons], New York (1974)
71. Wiedemann, M.: Simulation Des Schwingungsverhaltens Spanender Werkzeugmaschinen. Iwg Forschungsberichte, Band 64. Springer Verlag (1995)
72. Wodrich, M., Bilchev, G.: Cooperative Distributed Search: The Ant's Way. Control and Cybernetics 26(3), 413–446 (1997)
73. Yin, H., Wong, S.C., Xu, J., Wong, C.K.: Urban traffic flow prediction using a fuzzy-neural approach. Transportation Research Part C: Emerging Technologies 10(2), 85–98 (2002)
74. Youn, H., Gastner, M.T., Jeong, H.: Price of anarchy in transportation networks: efficiency and optimality control. Physical Review Letters 101(12), 128701 (2008)
75. Zang, L., Jia, L., Meng, X.: An optimization control algorithm of traffic signals for urban arterials. In: IITA International Conference on Control, Automation and Systems Engineering, CASE 2009, pp. 338–341 (July 2009)
76. Zhang, H.M.: Comment on Şon the controversy around DaganzoŠs requiem for and Aw-RascleŠs resurrection of second-order traffic flow models" by D. Helbing and A.F. Johansson. The European Physical Journal B - Condensed Matter and Complex Systems 69, 563–568 (2009)
77. Zhang, Q., Benveniste, A.: Approximation by nonlinear wavelet networks. In: International Conference on Acoustics, Speech, and Signal Processing, ICASSP 1991, vol. 5, pp. 3417–3420 (April 1991)
78. Zhao, J., Tang, D., Geng, X., Jia, L.: Urban Arterial Traffic Coordination Control System. In: Wang, F.L., Deng, H., Gao, Y., Lei, J. (eds.) AICI 2010, Part II. LNCS, vol. 6320, pp. 275–283. Springer, Heidelberg (2010)
79. Zhu, Q., Qian, L., Li, Y., Zhu, S.: An improved particle swarm optimization algorithm for vehicle routing problem with time windows. In: IEEE Congress on Evolutionary Computation, CEC 2006, pp. 1386–1390 (2006)

Chapter 10
Equilibrium Traffic Models

Abstract. This chapter deals with equilibrium traffic models for traffic flows. In the first section we introduce the basic assumptions common to all equilibrium traffic models; then, in the second section, we show how to construct the solution to the related Riemann problems. In the final section we discuss the drawbacks of the equilibrium traffic models.

10.1 Introduction

In this section we describe equilibrium models for vehicular traffic. The first equilibrium traffic model is the LWR model, independently proposed by Lighthill, Whitham [15] and Richards [21], and represents the starting point for the modeling of vehicular flows. Its basic assumptions are:

(ET.1) There is only one class of vehicles moving along a unique homogeneous lane and overtaking is not allowed.
(ET.2) Cars do not enter or exit the road.
(ET.3) The (average) speed of the cars depends on the (average) density alone.

Remark 10.1. The first assumption **(ET.1)** is realistic, for instance, for traffic in a narrow tunnel. A generalization to the multi–class can be found in Sect. 11.5 and the generalization to the multi–lane case with overtaking can be found in Sect. 11.6. The second assumption **(ET.2)** corresponds to choose a section of the road without exit or entrance gates. A generalization to the case with an entrance is considered in Sect. 11.3. The last assumption **(ET.3)** is rather controversial. Indeed it means that a unique velocity corresponds to a certain density. But the analysis of real data in congested situations contradicts this assumption, see Fig. 10.1. Furthermore, the last assumption implies that density change causes an immediate speed variation. Non–equilibrium traffic models overcomes this shortcoming, typical of all equilibrium traffic models, by taking also the velocity as a fundamental variable; see, for instance, the non–equilibrium traffic models presented in Chap. 14. □

M.D. Rosini: *Macroscopic Models for Vehicular Flows & Crowd Dynamics*, UCS, pp. 139–148.
DOI: 10.1007/978-3-319-00155-5_10

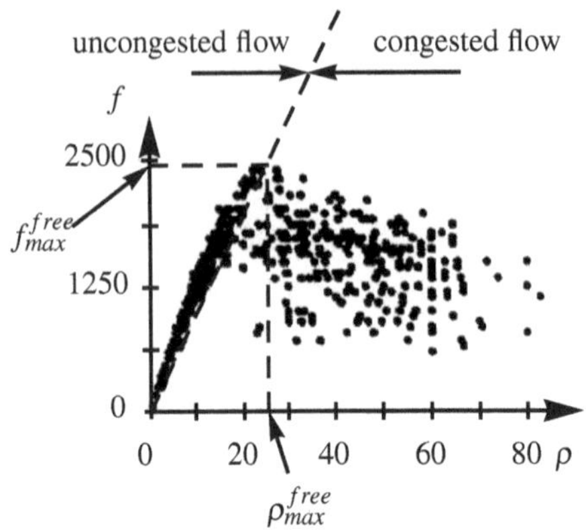

Fig. 10.1 Empirical construction of the fundamental diagram [13]. Here f is the flow (vehicles/h), ρ is the density (vehicles/km), f_{max}^{free} is the road capacity and is the maximal flow corresponding to the density ρ_{max}^{free}, which divide the stable from the unstable regime.

By assumption **(ET.1)**, we can parameterize the road by a single coordinate x and we can assume that the traffic moves in the direction of increasing x. Assumption **(ET.2)** ensures that the number of cars is conserved or, equivalently, that (9.5.3) holds. Assumption **(ET.3)** implies that the velocity in x at time t is given by $v = v\left(\rho(t,x)\right)$. In conclusion, from the assumptions **(ET.1–3)** we deduce the conservation law

$$\partial_t \rho + \partial_x \left(\rho\, v(\rho)\right) = 0 \tag{10.1.1}$$

that represents the general form of the **equilibrium traffic models**. We recall that here ρ is the (mean) density and $v = v(\rho)$, the corresponding preferred (mean) velocity, is a non–negative Lipschitz function for ρ between 0 and some positive maximal density ρ_m, which corresponds to a total traffic jam, so $1/\rho_m$ is the (mean) vehicle length. In standard situations, it is reasonable to require that v is a non–increasing function of the density and that $v(\rho_m) = 0$. As a consequence, $v_m = v(0)$ is the maximal velocity. Furthermore, the unique characteristic speed $\lambda(\rho) = v(\rho) + \rho\, v'(\rho)$ is a quantity not greater than the average speed $v(\rho)$. In particular, the model results to be anisotropic and condition (D.1) of Sect. 9.1 is satisfied. The choice of such function $[\rho \mapsto v(\rho)]$ depends on the behavior the model is trying to mimic. It must be either taken as a phenomenological relation extracted from empirical data or derived from more microscopic considerations. Several expressions for the velocity $v = v(\rho)$ have been proposed in the literature; among the others, see Fig. 10.2, we mention the following:

$$\text{Greenshields [10]}: \quad v(\rho) = v_m \left(1 - \frac{\rho}{\rho_m}\right) \tag{10.1.2a}$$

$$\text{Greenberg [9]}: \quad v(\rho) = \lambda\, \ln\left(\frac{\rho_m}{\rho}\right), \qquad \lambda \in \mathbb{R}_+ \tag{10.1.2b}$$

$$\text{Underwood [9]}: \quad v(\rho) = v_m\, \exp(-\rho) \tag{10.1.2c}$$

$$\text{Facchi [7]}: \quad v(\rho) = \begin{cases} v_m & \text{if } \rho \in [0,\rho_*[\\ v_m\left(1-\dfrac{\rho-\rho_*}{\rho_m-\rho_*}\right) & \text{if } \rho \in [\rho_*,\rho_m] \end{cases} \tag{10.1.2d}$$

$$\text{Chen [1]}: \quad v(\rho) = \begin{cases} v_1(\rho) & \text{if } \rho \in [0,\rho_*[\\ v_2(\rho) & \text{if } \rho \in [\rho_*,\rho_m]\,, \end{cases} \tag{10.1.2e}$$

where v_i introduced in (10.1.2e) is such that $(-1)^i\left(2\,v_i'(\rho)+\rho\,v_i''(\rho)\right)>0$, $i=1,2$.

In [15, 21] the chosen constitutive equation is the most straightforward expression (10.1.2a), which corresponds to the **LWR model**. It is well known that the LWR model results from the Hopf equation (3.3.1a) through the simple scaling and translation of variables

$$u(t,x) = \rho\left(\frac{\rho_m\,t}{2\,v_m}, -x\right) - \frac{\rho_m}{2}\,.$$

The velocity (10.1.2b) is not bounded at zero density and, consequently, it appears reasonable only for large densities, i.e. near $\rho=\rho_m$. Observe also that (10.1.2c) admits any positive density value; in other words, the maximal density is infinity. The choice (10.1.2d) accounts for a maximum speed limit imposed and respected by the drivers along the highway. Finally, the flux related to (10.1.2e) is defined by two quadratic functions that correspond to two different regimes.

A key role is played by the flux function $f(\rho)=\rho\;v(\rho)$. By definition $f(0)=f(\rho_m)=0$. In the **flux–density diagrams**, also called **fundamental diagrams**, exist two regimes, that means it is possible to have two speeds at the same traffic flow. By this the traffic flow is classified in an **unstable regime** $[0,\rho_{max}^{free}[$ and a **stable regime** $]\rho_{max}^{free},\rho_m]$, see Fig. 10.2, that correspond, respectively, to **light** and **heavy traffic**. The maximal flow f_{max}^{free} is called **road capacity** and is reached at the density ρ_{max}^{free}, namely $f_{max}^{free}=f(\rho_{max}^{free})$. The above relations (10.1.2a), (10.1.2b), (10.1.2c) would be called **univariate models**, because both regimes are computed with the same formula. Moreover, the fluxes related to (10.1.2a), (10.1.2b) and (10.1.2d) are concave but not that ones related to (10.1.2c) and (10.1.2e).

We want to underline that the assumption of a non–increasing speed law has only a technical reason. In fact, it allows for an easy description of the anisotropic property of the resulting model. However, this assumption is not in general satisfied in real situations and is not even necessary to guarantee the anisotropic property of the model [12]. For instance, consider a traffic traveling on an unfamiliar road at night. In this situation it is easier to drive quickly if there are other cars ahead on the road. On the other hand, when there is an empty road ahead, the speed is limited by the distance the headlights can illuminate. Thus, in this case a reasonable speed law might be

$$v(\rho) = \begin{cases} v_0 & \text{if } \rho \in [0,\rho_*[\\ v_1(\rho) & \text{if } \rho \in [\rho_*,\rho_m]\,, \end{cases} \tag{10.1.2f}$$

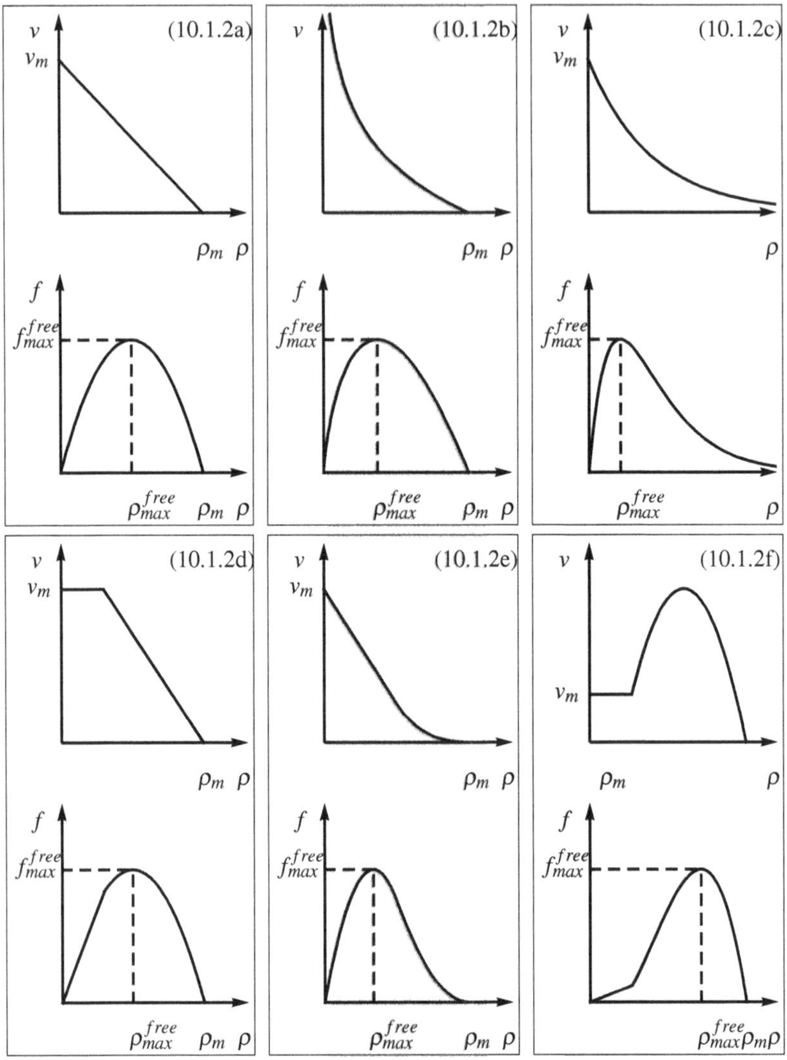

Fig. 10.2 In each box, above, the speed–density diagram given by (10.1.2) and below, the related fundamental diagram.

where v_0 is a positive constant and $[\rho \mapsto v_1(\rho)]$ is a $\mathbf{C}^2$ concave function, roughly speaking, as in the last box of Fig. 10.2. The construction of the resulting non–entropy solution can be found in [14], where the anisotropic property is considered.

Also the continuity assumption for the flux $[\rho \mapsto f(\rho)]$ is in general not satisfied. Edie [5] was among the first to point out that traffic behaved differently at different two density regimes, and introduced the idea of a two–regime model leading to a discontinuous fundamental diagram. The discontinuous fundamental diagrams have also been observed from empirical works [4, 17].

10.2 Riemann Problems

This section is devoted to study the Riemann problem for the equilibrium traffic model (10.1.1)

$$\partial_t \rho + \partial_x \left(\rho\, v(\rho)\right) = 0 \quad , \qquad \rho(0,x) = \begin{cases} \rho^l & \text{if } x < 0 \\ \rho^r & \text{if } x \geq 0\,. \end{cases} \tag{10.2.1}$$

We first consider the case of a strictly concave flux such as, for instance, that ones related to (10.1.2a) and (10.1.2b). Then we generalize the result by assuming conditions on the flux satisfied, for instance, by the fluxes related to (10.1.2c) and (10.1.2e).

CASE 1: As a first step, we assume that $[\rho \mapsto f(\rho) = \rho\, v(\rho)]$ is a strictly concave $\mathbf{C}^2$ function. In this case, if $\rho^l < \rho^r$, then $f'(\rho^l) > f'(\rho^r)$ and, in accordance with (4.2.1), the entropy weak solution to (10.2.1) is the shock wave

$$\rho(t,x) = \begin{cases} \rho^l & \text{if } x < \sigma\, t \\ \rho^r & \text{if } x \geq \sigma\, t\,, \end{cases} \tag{10.2.2a}$$

where the speed of propagation of the discontinuity is

$$\sigma = \frac{f(\rho^l) - f(\rho^r)}{\rho^l - \rho^r}\,. \tag{10.2.2b}$$

Therefore, the speed of the wave is positive iff $f(\rho^l) - f(\rho^r) > 0$, and is negative iff $f(\rho^l) - f(\rho^r) < 0$. On the other hand, if $\rho^l > \rho^r$, then $f'(\rho^l) < f'(\rho^r)$ and, in accordance with (4.4.5), the entropy weak solution to (10.2.1) is the rarefaction wave

$$\rho(t,x) = \begin{cases} \rho^l & \text{if } x < f'(\rho^l)\, t \\ (f')^{-1}\,(x/t) & \text{if } f'(\rho^l)\, t \leq x < f'(\rho^r)\, t \\ \rho^r & \text{if } x \geq f'(\rho^r)\, t\,. \end{cases} \tag{10.2.3}$$

CASE 2: Let us now consider the case of a not concave flux f. More precisely, we assume that:

C.1 f is a $\mathbf{C}^2$ function.
C.2 There exist $0 < \hat{\rho} < \rho_s < 1$ such that the following holds: f is strictly increasing on $[0, \hat{\rho}]$ and strictly decreasing on $[\hat{\rho}, \rho_m[$; f is strictly concave on $[0, \rho_s[$ and strictly convex in $]\rho_s, \rho_m[$; $f'(\rho_m) < 0$.
C.3 $f(0) = f(\rho_m) = 0$.

Let us now illustrate the corresponding solutions to the Riemann problem (10.2.1). Recalling the construction of Sect. 4.6, we define ρ_T to be the point such that the line from $\left(\rho_T, f(\rho_T)\right)$ to $(\rho_m, 0)$ is tangent to the graph of f at ρ_m, see Fig. 10.3; in formulas:

$$\frac{f(\rho_m) - f(\rho_T)}{\rho_m - \rho_T} = f'(\rho_m)\,.$$

Then, similarly, define two functions $\alpha_1 : [\rho_T, \rho_s] \to [\rho_s, \rho_m]$ and $\alpha_2 : [\rho_s, \rho_m] \to$

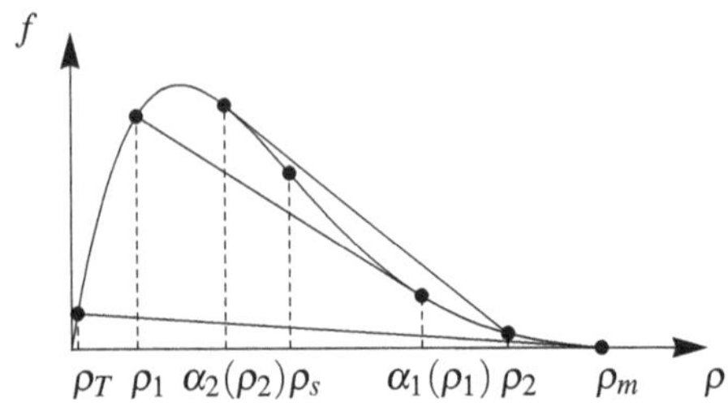

Fig. 10.3 A flux function f satisfying **C**.1–3 and definition of functions α_1 and α_2

$[\hat{\rho}, \rho_s]$ in the following way. The secant from $\left(\alpha_i(\rho), f\left(\alpha_i(\rho)\right)\right)$ to $(\rho, f(\rho))$ is tangent to the graph of f at $\alpha_i(\rho)$, see Fig. 10.3. Observe that by definition $\alpha_1(\rho_s) = \alpha_2(\rho_s) = \rho_s$ and $\alpha_1(\rho_T) = \rho_m$. Now, we fix $\rho^l \in [0, \rho_m]$ and distinguish four cases:

Case $\rho^l \in [0, \rho_T]$: in this case, the solution to Riemann problems with data (ρ^l, ρ^r) is the same as in the case of a strictly concave flux.

Case $\rho^l \in [\rho_T, \rho_s]$: in this case, if $\rho^r \leq \alpha_1(\rho^l)$, then the solution is the same as before, i.e. either a rarefaction ($\rho^r < \rho^l$) or a shock ($\rho^r > \rho^l$). If $\rho^r > \alpha_1(\rho^l)$, then the solution is formed by a shock from ρ^l to $\alpha_1(\rho^l)$ followed by rarefaction from $\alpha_1(\rho^l)$ to ρ^r.

Case $\rho^l = \rho_s$: in this case, the solution is always a rarefaction.

Case $\rho^l \in [\rho_s, \rho_m]$: in this case, if $\rho^r > \rho^l$, then the solution is a rarefaction; if $\alpha_2(\rho^l) \leq \rho^r < \rho^l$, then the solution is a shock; finally, if $\rho^r < \alpha_2(\rho^l)$, then the solution is formed by a shock from ρ^l to $\alpha_2(\rho^l)$ followed by a rarefaction from $\alpha_2(\rho^l)$ to ρ^r.

Again, we can relax **C**.2 at the prize of a more involved treatment.

Example 10.1. Let us consider the case when there is a traffic jam ahead $x = 0$ and a light traffic moves towards it with uniform density $\rho = \rho_l (\ll \rho_m)$. The corresponding model is given by the Riemann problem (10.2.1) with $\rho^l = \rho_l$ and $\rho^r = \rho_m$. For simplicity, consider the **normalized LWR model**, i.e. take the velocity given by (10.1.2a) and normalized with $\rho_m = 1$ and $v_m = 1$. Then the solution is given by (10.2.2), namely

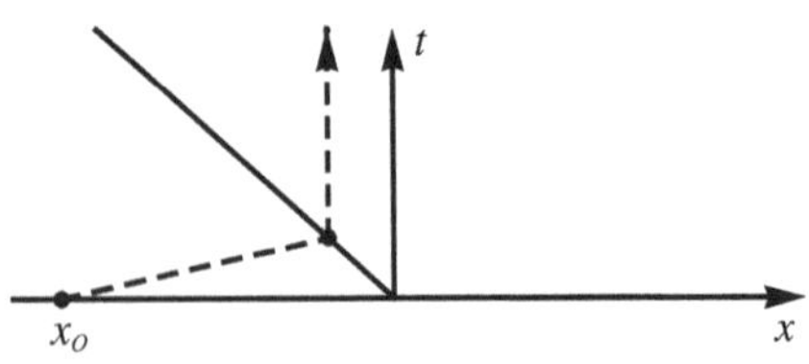

Fig. 10.4 The solution (10.2.4), representing a queue starting from $x = 0$ and growing with velocity $v = -\rho_l$. The dashed line represents the trajectory of a driver given by (10.2.5).

$$\rho(t,x) = \begin{cases} \rho_l & \text{if } x < -\rho_l\, t \\ 1 & \text{if } x \geq -\rho_l\, t\,. \end{cases} \tag{10.2.4}$$

According with the experience, the cars reach the traffic jam and immediately stop creating a queue that starts from $x=0$ and grows with velocity $v=-\rho_l$. In particular, if a driver has position $x = x_o \in \mathbb{R}_-$ at time $t = 0$, then it moves with velocity $v = 1-\rho_l$ and reaches the queue at time $t = -x_o$. Therefore, with reference to the driver that at time $t=0$ is in $x=x_o \in \mathbb{R}_-$, see Fig. 10.5, we have

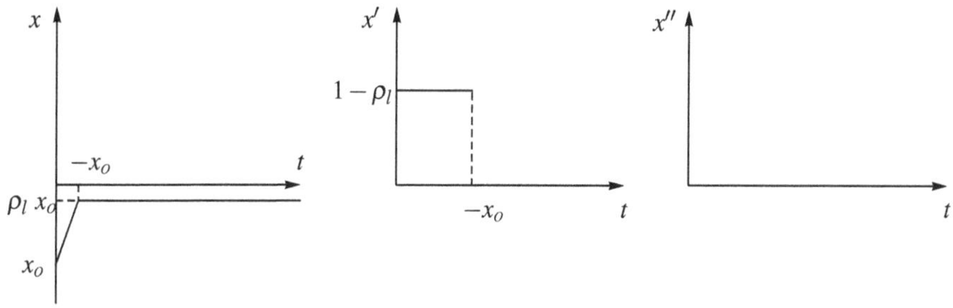

Fig. 10.5 From the left to the right, respectively, the position, velocity and acceleration of the driver that at time $t=0$ is in $x=x_o \in \mathbb{R}_-$ and given by (10.2.5)

$$\text{position:} \qquad x(t) = \begin{cases} x_o + (1-\rho_l)\, t & \text{if } t < -x_o \\ \rho_l\, x_o & \text{if } t \geq -x_o, \end{cases} \tag{10.2.5a}$$

$$\text{velocity:} \qquad x'(t) = \begin{cases} 1-\rho_l & \text{if } t < -x_o \\ 0 & \text{if } t \geq -x_o, \end{cases} \tag{10.2.5b}$$

$$\text{acceleration:} \qquad x''(t) \equiv 0\,. \tag{10.2.5c}$$

Observe that $[t \mapsto x(t)]$ is a $\mathbf{C^0}$ function. □

Example 10.2. Consider the situation when the traffic light sited in $x=0$ turns to green at time $t=0$ after a long queue forms behind it, see Fig. 10.6. Then the corresponding model is given by the Riemann problem (10.2.1) with $\rho^l = \rho_m$ and $\rho^r = 0$. Again, for simplicity, consider the normalized LWR model. Then by (10.2.3) the solution is

$$\rho(t,x) = \begin{cases} 1 & \text{if } x < -t \\ \frac{1}{2}\left(1 - \frac{x}{t}\right) & \text{if } -t \leq x < t \\ 0 & \text{if } x \geq t\,. \end{cases} \tag{10.2.6}$$

According with the experience, the cars closer to the traffic light start to move before the others. Furthermore, when we emerge from a traffic jam, we experience a gradual decrease in the density of cars ahead to us. More in detail, consider a driver with position $x = x_o \in \mathbb{R}_-$ at time $t = 0$. After the light turns to green, he has to

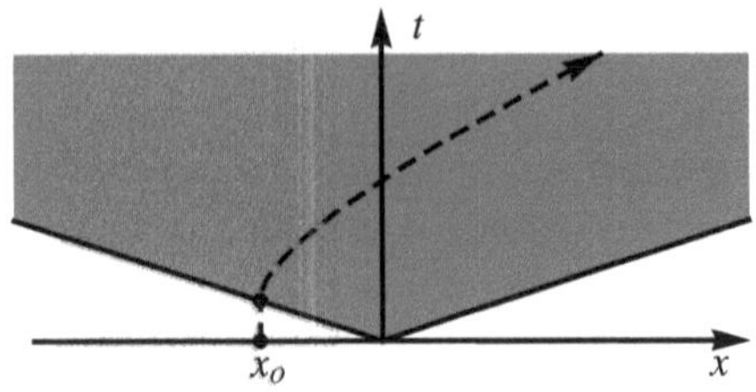

Fig. 10.6 The solution (10.2.6), representing a queue behind a streetlight sited in $x = 0$ that turns to green at time $t = 0$. The shaded area represents the rarefaction, while the dashed line is the trajectory of a driver given by (10.2.7).

wait until time $t = -x_o$ before he can start to move. Then, its position $[t \mapsto x(t)]$ is computed by solving the ordinary differential equation

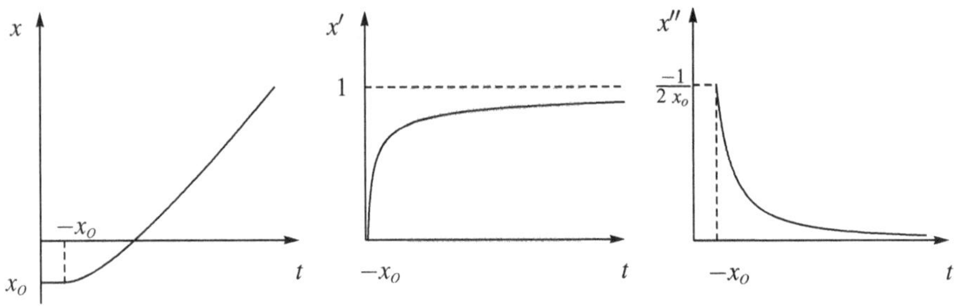

Fig. 10.7 From the left to the right, respectively, the position, velocity and acceleration of the driver that at time $t = 0$ is in $x = x_o \in \mathbb{R}_-$ and given by (10.2.7)

$$x'(t) = \frac{1}{2}\left(1 + \frac{x(t)}{t}\right) \quad , \qquad x(-x_o) = x_o \,.$$

Therefore, with reference to the driver that at time $t = 0$ is in $x = x_o \in \mathbb{R}_-$, see Fig. 10.7, we have

$$\text{position:} \qquad x(t) = \begin{cases} x_o & \text{if } t < -x_o \\ t - 2\sqrt{-x_o\, t} & \text{if } t \geq -x_o, \end{cases} \tag{10.2.7a}$$

$$\text{velocity:} \qquad x'(t) = \begin{cases} 0 & \text{if } t < -x_o \\ 1 - \sqrt{\dfrac{-x_o}{t}} & \text{if } t \geq -x_o, \end{cases} \tag{10.2.7b}$$

$$\text{acceleration:} \qquad x''(t) = \begin{cases} 0 & \text{if } t < -x_o \\ \dfrac{1}{2\,t}\sqrt{\dfrac{-x_o}{t}} & \text{if } t \geq -x_o \,. \end{cases} \tag{10.2.7c}$$

Observe that $[t \mapsto x(t)]$ is a $\mathbf{C}^1$ function. □

10.3 The Drawbacks of the Equilibrium Traffic Models

Equilibrium traffic models have some shortcomings, given in the following list [2, 16, 20]:

- They take into account only the desired velocity of each vehicle, but do not recognize the distribution of desired velocities across vehicles. For this reason, equilibrium traffic models can not predict the observed linear in time dispersion of a platoon of vehicles, since the variation within each vehicle causes a platoon to disentangle only with the square root of time, see [18].
- They contain stationary speed–density relations, implying that the mean velocity adapts instantaneously to the traffic density rather than considering some delay, which is obviously in contrast with the real observations.
- They do not predict instabilities of the **stop–and–go** (**start–stop**) type that are observed in real–life traffic [6].
- A unique velocity corresponds to a certain density, in contrast with experimental observations, see Fig. 10.1.
- In real–life traffic flow, **hysteresis** phenomena have been observed [22], showing a generally retarded behavior displayed by a platoon of vehicles after emerging from a kinematic disturbance as compared to the behavior of the same vehicles approaching the disturbance, resulting in a retardation in the recovery of the conditions prior to the disturbance. These hysteresis phenomena are not described by equilibrium traffic models.
- Equilibrium traffic models are not able to predict the occurrence of localized structures and **phantom–jams** [8, 11], i.e. they do not describe the amplification of small disturbances in heavy traffic.
- The kinematic wave theory of equilibrium traffic models produces discontinuous solutions irrespective of the smoothness of initial conditions, due to the dominating convective term in (10.1.1). These are in contradiction with smooth shocks observed in real–life traffic.

Non–equilibrium traffic models aim to overcome the above drawbacks of the equilibrium traffic models, see Chap. 14 for more details. Other approaches are proposed by Newell [19] and Daganzo et al. [3]. In particular, Newell's model recovers traffic hysteresis by imposing the fact that deceleration flows and acceleration flows follow distinctively different paths in the speed–density plane.

On a positive note, it should be emphasized that the shock structure is irrelevant for some applications, and in those cases, equilibrium traffic models should be sufficient. For instance, to model freeway traffic networks numerically, one has to discretize freeway links into sections that are large compared with the width of a shock. As a consequence, even if a numerical model could capture the shock structure, the improved accuracy could be lost in the subsequent aggregation of the data within each freeway section. This additional computational effort would have been wasted.

References

1. Chen, W., Wong, S.C., Shu, C.W., Zhang, P.: Front tracking algorithm for the Lighthill-Whitham-Richards traffic flow model with a piecewise quadratic, continuous, non-smooth and non-concave fundamental diagram. Int. J. Numer. Anal. Model. 6(4), 562–585 (2009)
2. Daganzo, C.F.: Requiem for high-order fluid approximations of traffic flow. Trans. Res. 29B(4), 277–287 (1995)
3. Daganzo, C.F., Cassidy, M.J., Bertini, R.L.: Possible explanations of phase transitions in highway traffic. Transportation Research Part A: Policy and Practice 33(5), 365–379 (1999)
4. Drake, J.S., Schofer, J.L., May, A., May, A.D.: A statistical analysis of speed density hypotheses. Report (Expressway Surveillance Project (Ill.)). Expressway Surveillance Project (1965)
5. Edie, L.C.: Car-following and steady-state theory for non-congested traffic. Tunnel traffic capacity study. Port of New York Authority, Tunnels and Bridges Dept., Project and Planning Division (1960)
6. Edie, L.C., Foote, R.S.: Traffic flow in tunnels, vol. 37, pp. 334–344 (1958)
7. Facchi, G.: Crowd Dynamics: On a New Continuum Model for Crowd Dynamics. Master's thesis, University of Brescia (2008)
8. Flynn, M.R., Kasimov, A.R., Nave, J.C., Rosales, R.R., Seibold, B.: Self-sustained nonlinear waves in traffic flow. Phys. Rev. E 79, 056113 (2009)
9. Greenberg, H.: An Analysis of Traffic Flow. Operations Research 7(1), 79–85 (1959)
10. Greenshields, B.D.: A study of traffic capacity. Highway Research Board 14, 448–477 (1934)
11. Helbing, D.: Traffic and related self-driven many-particle systems. Reviews of Modern Physics (2001)
12. Zhang, H.M.: Anisotropic property revisited–does it hold in multi-lane traffic? Transportation Research Part B: Methodological 37(6), 561–577 (2003)
13. Kerner, B.S.: Complexity of Synchronized Flow and Related Problems for Basic Assumptions of Traffic Flow Theories. Networks and Spatial Economics 1(1), 35–76 (2001)
14. Leveque, R.J.: Some traffic flow models illustrating interesting hyperbolic behavior. In: Minisymposium on Traffic Flow. SIAM Annual Meeting (2001)
15. Lighthill, M.J., Whitham, G.B.: On kinematic waves. II. A theory of traffic flow on long crowded roads. Proc. Roy. Soc. London. Ser. A 229, 317–345 (1955)
16. Liu, G., Lyrintzis, A.S., Michalopoulos, P.G.: Improved High-Order Model for Freeway Traffic Flow. Transportation Research Record 1644(1), 37–46 (1998)
17. May, A.D., Keller, H.: Non-integer car-following model. Highway Research Record (199), 19–35 (1967)
18. Newell, G.F.: The Flow of Highway Traffic Through a Sequence of Synchronized Traffic Signals. Operations Research 8(3), 390–405 (1960)
19. Newell, G.F.: Instability in dense highway traffic: a review. Organisation for Economic Co-operation and Development 1, 73–85 (1965)
20. Papageorgiou, M.: Some remarks on macroscopic traffic flow modelling. Transportation Research Part A: Policy and Practice 32(5), 323–329 (1998)
21. Richards, P.I.: Shock waves on the highway. Operations Res. 4, 42–51 (1956)
22. Treiterer, J., Myers, J.A.: The hysteresis phenomenon in traffic flow, vol. 33, pp. 13–38. Elsevier (1974)

Chapter 11
Generalizations of Equilibrium Traffic Models

Abstract. In this chapter we review macroscopic models for vehicular traffic resulting from a generalization of equilibrium traffic models presented in the previous chapter.

11.1 Introduction

The equilibrium traffic models introduced in Chap. 10 can only be used to describe very simple situations, when we are dealing with a very long one lane highway, with neither entrances nor exits and overtaking is not allowed. However, they are basic bricks for more complicated and sophisticated models. Some of these models will be presented in the following sections. In more detail, in the following five sections we concentrate our attention to model, respectively, a road with an entrance and a time depending constraint [6], two merging roads [6], a traffic circle [3], a multi–population traffic [1] and a multi–lane road [4, 9–11].

11.2 Highway with an Entrance and Constraints

The evolution of traffic flowing along a highway with a given entry, say sited at $x = 0$, can be described by Eq. (10.1.1) supplemented with an initial datum $\overline{\rho} = \overline{\rho}(x)$ and, depending on the situation at hand, also by a time dependent inflow q_b at the entry, namely $f\left(\rho(t,0)\right) = q_b(t)$. Furthermore, often traffic flow is subject to various restrictions, such as traffic lights, toll gates, construction sites, or the effects of accidents. All these situations amount to limit the flow at specific locations along the road and, hence, can be described adding suitable unilateral constraints on the flow, such as $f\left(\rho(t,x_c)\right) \leq q_c(t)$, $x = x_c$ being the location of the traffic light (or toll gate, or accident, ...) and $q_c = q_c(t)$ being the possibly time dependent maximal flow through $x = x_c$ allowed by the traffic light (or toll gate, or accident, ...). The resulting system is then [6]

M.D. Rosini: *Macroscopic Models for Vehicular Flows & Crowd Dynamics*, UCS, pp. 149–160.
DOI: 10.1007/978-3-319-00155-5_11

$$\begin{aligned}
&\partial_t \rho + \partial_x f(\rho) = 0 && (t,x) \in \mathbb{R}_+^2 && (11.2.1a)\\
&\rho(0,x) = \overline{\rho}(x) && x \in \mathbb{R}_+ && (11.2.1b)\\
&f\big(\rho(t,0)\big) = q_b(t) && t \in \mathbb{R}_+ && (11.2.1c)\\
&f\big(\rho(t,x_c)\big) \le q_c(t) && t \in \mathbb{R}_+ \,. && (11.2.1d)
\end{aligned}$$

A typical solution to (11.2.1) represented in the (x,t)–plane is depicted in Fig. 11.1.

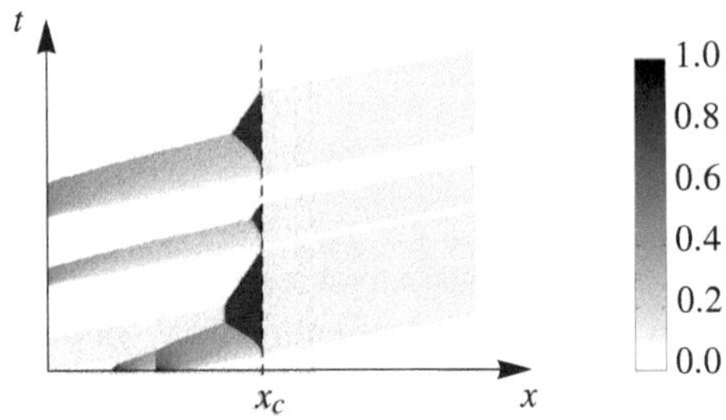

Fig. 11.1 A typical solution to (11.2.1), portrayed on the (x,t)–plane, with two square waves in the initial datum, three waves as inflow from $x = 0$ and a constraint at $x = x_c$

The theory developed in Sect. 6.5 directly applies to the present case. In particular, Theorem 6.3 ensures the well posedness of (11.2.1).

We stress that, in our intentions, the standardized situation (11.2.1) and the corresponding well posedness results serve as a model situation. In other words, problem (11.2.1), Theorem 6.3 and Corollary 6.1 constitute the basic bricks for a framework where more complex models can be constructed, used and managed. Indeed, formally different and more complicate situations can be described through a sort of juxtaposition of several simpler problems of the type (11.2.1). The analytical results of Sect. 6.5 can be then easily applied to each of the simpler problems, giving then information about the full situation. We provide examples of these constructions in Chap. 13.

11.3 Merging Roads

Consider two one–way roads $\boxed{1}$ and $\boxed{2}$ entering through a junction, say at $x = 0$, into a one–way road $\boxed{3}$, see Fig. 11.2. Traffic along road $\boxed{i}$ can then be described

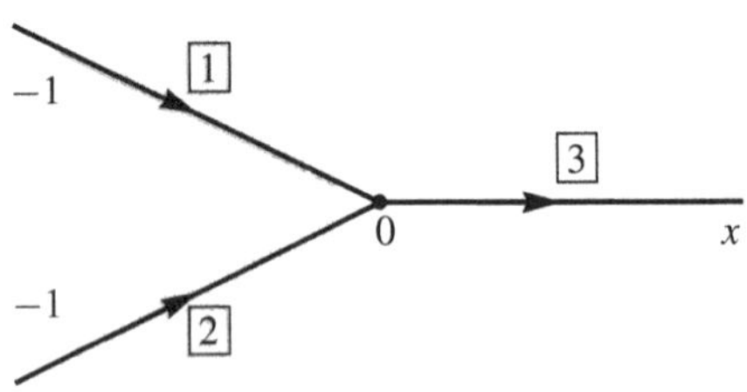

Fig. 11.2 Two roads merging into a third one, as considered in (11.3.1)

by an equilibrium traffic model of the type (11.2.1) with density $\rho = \rho_i$ and flow $f = f_i(\rho_i)$, $i = 1,2,3$. The junction is regulated so that the flow from road $\boxed{i}$ into $\boxed{3}$ is bounded by $q_{ci}(t)$, $i = 1,2$. We are thus lead to consider the problem [6]

$$\begin{array}{rlr} \text{incoming roads:} & \partial_t \rho_i + \partial_x f_i(\rho_i) = 0 & (t,x) \in \mathbb{R}_+ \times [-1,0] \\ \text{outgoing road:} & \partial_t \rho_3 + \partial_x f_3(\rho_3) = 0 & (t,x) \in \mathbb{R}_+ \times \mathbb{R}_+ \\ \text{initial data:} & \rho_i(0,x) = 0 & x \in [-1,0] \\ \text{initial datum:} & \rho_3(0,x) = 0 & x \in \mathbb{R}_+ \\ \text{inflows:} & f_i\left(\rho_i(t,-1)\right) = q_{bi}(t) & t \in \mathbb{R}_+ \\ \text{conservation:} & f_3\left(\rho_3(t,0)\right) = \sum_{i=1}^{2} f_i\left(\rho_i(t,0)\right) & t \in \mathbb{R}_+ \\ \text{traffic lights:} & f_i\left(\rho_i(t,0)\right) \leq q_{ci}(t) & t \in \mathbb{R}_+\,, \end{array} \tag{11.3.1a}$$

where $i = 1,2$. By solution to (11.3.1a) we mean the obvious adaptation of Definition 6.1 and Definition 6.3. We require that the regulation at the junction is *reasonable*, i.e.

$$q_{c1}(t) + q_{c2}(t) \leq f_{max3}^{free} \qquad \text{for all } t \in \mathbb{R}_+^*\,, \tag{11.3.1b}$$

where f_{max3}^{free} is the capacity of road $\boxed{3}$, consistently with condition **(F)** of Sect. 6.2. In other words, (11.3.1b) ensures that the total outflow from roads $\boxed{1}$ and $\boxed{2}$ is regulated so that it never exceeds the capacity of road $\boxed{3}$. Observe that this condition is consistent with the formation of queues along the roads $\boxed{1}$ and $\boxed{2}$ before $x = 0$. Stating and proving the well posedness of (11.3.1) amount to a repeated application of the arguments in Theorem 6.1 and Theorem 6.3.

Corollary 11.1. *Let f_1, f_2, f_3 satisfy **(F)**, q_{b1}, q_{b2} satisfy **(B)** and q_{c1}, q_{c2} satisfy **(C)**. Then, there exists a unique solution*

$$\rho(t,x) = \begin{cases} \rho_i(t,x) & (t,x) \in \mathbb{R}_+ \times [-1,0[,\ i = 1,2 \\ \rho_3(t,x) & (t,x) \in \mathbb{R}_+ \times \mathbb{R}_+^* \end{cases}$$

to (11.3.1). *Finally, if also q'_{bi} and q'_{ci} satisfy, respectively, **(B)** and **(C)** for $i = 1,2$, then the following Lipschitz estimates on the corresponding solutions ρ, ρ' hold, for every $t \in \mathbb{R}_+$:*

$$\left\|\rho_i(t) - \rho_i'(t)\right\|_{\mathbf{L}^1([-1,0];\mathbb{R})} \leq \left\|q_{bi} - q'_{bi}\right\|_{\mathbf{L}^1([0,t];\mathbb{R})} + 2\left\|q_{ci} - q'_{ci}\right\|_{\mathbf{L}^1([0,t];\mathbb{R})}, \quad i = 1,2$$

$$\left\|\rho_3(t) - \rho_3'(t)\right\|_{\mathbf{L}^1(\mathbb{R}_+;\mathbb{R})} \leq \sum_{i=1}^{2} \left(\left\|q_{bi} - q'_{bi}\right\|_{\mathbf{L}^1([0,t];\mathbb{R})} + 2\left\|q_{ci} - q'_{ci}\right\|_{\mathbf{L}^1([0,t];\mathbb{R})}\right).$$

Proof. First, apply Theorem 6.3 and solve the two independent problems

$$\begin{array}{rr} \partial_t \rho_1 + \partial_x f_1(\rho_1) = 0 & (t,x) \in \mathbb{R}_+ \times [-1,0] \\ \rho_1(0,x) = 0 & x \in [-1,0] \\ f\left(\rho_1(t,-1)\right) = q_{b1}(t) & t \in \mathbb{R}_+ \\ f_1\left(\rho_1(t,0)\right) \leq q_{c1}(t) & t \in \mathbb{R}_+ \end{array}$$

and

$$
\begin{aligned}
\partial_t \rho_2 + \partial_x f_2(\rho_2) &= 0 && (t,x) \in \mathbb{R}_+ \times [-1,0] \\
\rho_2(0,x) &= 0 && x \in [-1,0] \\
f\left(\rho_2(t,-1)\right) &= q_{b2}(t) && t \in \mathbb{R}_+ \\
f_2\left(\rho_2(t,0)\right) &\leq q_{c2}(t) && t \in \mathbb{R}_+ \, .
\end{aligned}
$$

Then, add the traces of the two flows from the incoming roads at $x = 0$, and set $q_{b3}(t) = f_1\left(\rho_1(t,0-)\right) + f_2\left(\rho_2(t,0-)\right)$. Finally solve the problem

$$
\begin{aligned}
\partial_t \rho_3 + \partial_x f_3(\rho_3) &= 0 && (t,x) \in \mathbb{R}_+ \times \mathbb{R}_+ \\
\rho_3(0,x) &= 0 && x \in \mathbb{R}_+ \\
f\left(\rho_3(t,0)\right) &= q_{b3}(t) && t \in \mathbb{R}_+
\end{aligned}
$$

by means of Theorem 6.1. □

Obviously, the insertion of junctions or other constraints along road $\boxed{3}$ simply amounts to further applications of Theorem 6.1 and Theorem 6.3.

Refer to [8, Chap. 4] for a thorough treatment of junctions through the LWR model.

11.4 Traffic Circle

In this section we consider a simple traffic circle formed by four roads, $R_{2\ell}$, $\ell = 1,\ldots,4$, reached by two incoming roads, R_1, R_5, and two outgoing roads, R_3, R_7, see Fig. 11.3, left. Each road is parameterized by $[a_\ell, b_\ell]$, $\ell = 1,\ldots,8$, with traffic moving in the direction of increasing x and described by an equilibrium traffic model. Let q_{b1} and q_{b5} be the flows from the incoming roads R_1 and R_5 respectively. Assume that the traffic is low, in the sense that the number of cars reaching the circle is less then the capacity f_{max}^{free} of the circle itself. If the network is initially empty, then we obtain the following problem [3]

$$
\begin{aligned}
&\partial_t \rho_\ell + \partial_x f(\rho_\ell) = 0 && (t,x) \in \mathbb{R}_+ \times [a_\ell, b_\ell] && (11.4.1a) \\
&\rho_\ell(0,x) = 0 && x \in [a_\ell, b_\ell] && (11.4.1b) \\
&f\left(\rho_1(t,a_1)\right) = q_{b1}(t) && t \in \mathbb{R}_+ && (11.4.1c) \\
&f\left(\rho_5(t,a_5)\right) = q_{b5}(t) && t \in \mathbb{R}_+ && (11.4.1d) \\
&q_{1\ell}(t) + q_{5\ell}(t) \leq f_{max}^{free} && t \in \mathbb{R}_+ && (11.4.1e) \\
&f\left(\rho(t,a_2)\right) = f\left(\rho(t,b_1)\right) + f\left(\rho(t,b_8)\right) && t \in \mathbb{R}_+ && (11.4.1f) \\
&f\left(\rho(t,b_2)\right) = f\left(\rho(t,a_3)\right) + f\left(\rho(t,a_4)\right) && t \in \mathbb{R}_+ && (11.4.1g) \\
&f\left(\rho(t,a_6)\right) = f\left(\rho(t,b_4)\right) + f\left(\rho(t,b_5)\right) && t \in \mathbb{R}_+ && (11.4.1h)
\end{aligned}
$$

$$f\left(\rho(t,b_6)\right) = f\left(\rho(t,a_7)\right) + f\left(\rho(t,a_8)\right) \qquad t \in \mathbb{R}_+ \tag{11.4.1i}$$

for $\ell = 1,\ldots,8$. The only conservation through the node is not sufficient to ensure the uniqueness of the solution and consequently the above problem is not well posed. In fact, (11.4.1) determines the dynamics on each arc, while that on nodes needs to be defined. Thus we introduce two additional rules:

(A) fluxes distribute on outgoing roads according to fixed coefficients;
(B) the through flux is maximized (respecting rule **(A)**).

The ŕfirst rule expresses driver preferences, while the second one is an "entropy" type condition. Furthermore, at each junction with two incoming roads, it is also necessary to introduce the priority among the incoming roads. The rule **(A)** can be explicitly expressed as it follows. Given a road R_i, let $\alpha_{ij} \in]0,1[$ be the probability that the traffic from R_i decide to move trough the road R_j. This means that if N is the quantity of traffic incoming from the road R_i, then $\alpha_{ij}\ N$ traffic decides to move trough the road R_j. Since the roads of the circle are just intermediate between the incoming roads and the outgoing roads, it is sufficient to assign the distribution coefficients describing how the traffic coming from the incoming roads choose to exit to outgoing roads. Thus we assume to have fixed the parameters $\alpha_{13}, \alpha_{57} \in]0,1[$. In particular, if N vehicles reach the circle from road R_1, then $\alpha_{13}\ N$ decide to drive to road R_3 and $(1-\alpha_{13})\ N$ to drive to R_7.

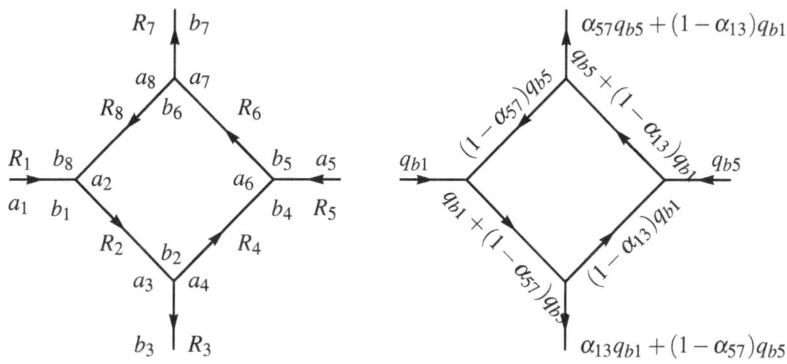

Fig. 11.3 Left: A traffic circle formed by four roads, $R_{2\ell}$, $\ell = 1,\ldots,4$, reached by two incoming roads, R_1, R_5, and two outgoing roads, R_3, R_7. Right: The equilibrium for a circle corresponding to the asymptotic behavior described in Theorem 11.1.

Theorem 11.1. *Consider the above model for a traffic circle. If q_{b1} and q_{b5} are constants, then there exist time dependent coefficients $\alpha_{13}, \alpha_{57} : \mathbb{R}_+^* \to]0,1[$ and $T \in \mathbb{R}_+$ such that the solution is constant for every $t \geq T$.*

Proof. By (11.4.1b), (11.4.1c), (11.4.1d), firstly the vehicles from roads R_1 and R_5 reach the roads R_2 and R_6 respectively and therefore the coefficients have to satisfy

$$\alpha_{23} = \alpha_{13} \qquad \alpha_{24} = 1 - \alpha_{13} \qquad \alpha_{67} = \alpha_{57} \qquad \alpha_{68} = 1 - \alpha_{57}\,.$$

Then, also the vehicles from road R_1 reach the road R_7 and vehicles from road R_5 reach road R_3. Therefore we should modify in time the coefficients and finally set

$$\alpha_{23} = \frac{\alpha_{13}\, q_{b1} + (1-\alpha_{57})\, q_{b5}}{q_{b1} + (1-\alpha_{57})\, q_{b5}} \qquad \alpha_{24} = \frac{(1-\alpha_{13})\, q_{b1}}{q_{b1} + (1-\alpha_{57})\, q_{b5}}$$
$$\alpha_{67} = \frac{\alpha_{57}\, q_{b5} + (1-\alpha_{13})\, q_{b1}}{q_{b5} + (1-\alpha_{13})\, q_{b1}} \qquad \alpha_{68} = \frac{(1-\alpha_{57})\, q_{b5}}{q_{b5} + (1-\alpha_{13})\, q_{b1}}\,.$$

With this choice of time dependent coefficients, we can construct the solution by wave front tracking algorithm and find $T \in \mathbb{R}_+$ such that the solution is given by the fluxes indicated in Fig. 11.3, right, for every $t \geq T$. □

Remark 11.1. Piccoli and Chitour [3] considered also traffic circles in the situation of heavy traffic and multi–lane traffic circles with no interactions. They also provide an interesting comparison between traffic lights and traffic circles. According with their analysis of the related asymptotic behaviors, traffic circles result to be the best solution with an appropriate choice of the right of way rules. Further, they suggest to put traffic lights at crossing of the circles, working only in the case of really heavy traffic, to avoid that the circle itself is stuck. □

11.5 Multi–population

In this section we present the model proposed by Benzoni and Colombo [1]. It accounts for various qualitative features of the evolution of the traffic along an highway characterized by n heterogeneous drivers / vehicle behaviors. The model reads

$$\partial_t \rho_\ell + \partial_x \left(\rho_\ell\, v_\ell(\rho)\right) = 0 \qquad \ell = 1,\dots,n \tag{11.5.1a}$$
$$v_\ell(\rho) = V_\ell\, \psi(\rho) \qquad \ell = 1,\dots,n\,, \tag{11.5.1b}$$

where, for all ℓ, ρ_ℓ is concentration of vehicles belonging to the ℓ–th class, the average speed v_ℓ is a function of the **total density** $\rho = \sum_{\ell=1}^{n} \rho_\ell$ and its maximum value is V_ℓ. For notational symplicity, we rescale the maximal total density to 1, so that the model (11.5.1) is defined on the simplex

$$\mathscr{S} = \left\{ (\rho_1,\dots,\rho_n)^T \in (\mathbb{R}_+^*)^n \,:\, \sum_{\ell=1}^{n} \rho_\ell \leq 1 \right\}.$$

Remark 11.2. Equation (11.5.1b) expresses that drivers belonging to different classes adjust their speed to the local total density in the same way and independently of the traffic conditions. Roughly speaking, this corresponds to assume that the speed depends only on the vehicle performance. In general, each class may be characterized by a specific density–speed relation. But then, none significant analytical result could be reached since even the hypotheses that would ensure that (11.5.1a) is hyperbolic would be unclear. We also observe that the model proposed by Zhang and Jin [14] consider two classes, namely slow and fast vehicles, and assumes that they have the same kinematic behavior. Such a model results then from (11.5.1a), (11.5.1b) by taking $n = 2$ and $V_1 = V_2$. □

In order to simplify the following computations and avoid overloading the section, we consider the case $n = 2$, $V_1 > V_2 > 0$ and take (11.5.1b) of the form (10.1.2a), namely

$$\psi(\rho) = 1 - \rho\,, \tag{11.5.1c}$$

deferring the reader to [1, 2, 14, 15] for a study of more general cases.

Proposition 11.1. *The system* (11.5.1) *is strictly hyperbolic in* $\mathscr{S} \setminus \{u_U\}$, *where*

$$u_U = \left(\frac{V_1 - V_2}{2V_1 - V_2}, 0\right)^T$$

is an umbilical point. Furthermore the first characteristic field is genuinely nonlinear in all $\mathscr{S}$ *and the second characteristic field is genuinely nonlinear in* $\mathscr{S} \setminus \{(\rho_1, \rho_2) \,:\, \rho_1 + \rho_2 = 1\}$.

Proof. The model (11.5.1) can be written in the conservative form (7.1.2)

$$\partial_t u + \partial_x f(u) = 0_2$$

with $u = (\rho_1, \rho_2)^T$ and $f(u) = \big(\rho_1\, v_1(\rho), \rho_2\, v_2(\rho)\big)^T$. The jacobian matrix of f is

$$a(u) = \begin{pmatrix} \big(\psi(\rho) - \rho_1\big)\, V_1 & -\rho_1\, V_1 \\ -\rho_2\, V_2 & \big(\psi(\rho) - \rho_2\big)\, V_2 \end{pmatrix}.$$

The solutions of the related characteristic polynomial

$$\begin{aligned} &\det\big[a(u) - \lambda\, \mathrm{Id}_2\big] = \\ =&\lambda^2 - \Big(\big(\psi(\rho) - \rho_1\big)\, V_1 + \big(\psi(\rho) - \rho_2\big)\, V_2\Big)\,\lambda + \psi(\rho)\, \psi(2\rho)\, V_1\, V_2 \end{aligned}$$

are the eigenvalues

$$\begin{aligned} \lambda_1(u) &= \frac{\big(\psi(\rho) - \rho_1\big)\, V_1 + \big(\psi(\rho) - \rho_2\big)\, V_2 - \Delta(u)}{2} \\ \lambda_2(u) &= \frac{\big(\psi(\rho) - \rho_1\big)\, V_1 + \big(\psi(\rho) - \rho_2\big)\, V_2 + \Delta(u)}{2}\,, \end{aligned}$$

where

$$\Delta(u) = \sqrt{\Big[\big(\psi(\rho) - \rho_1\big)\, V_1 - \big(\psi(\rho) - \rho_2\big)\, V_2\Big]^2 + 4\, \rho_1\, \rho_2\, V_1\, V_2}\,.$$

The eigenvalues are real in all $\mathscr{S}$ and coincide iff $u = u_U$. In fact, $\lambda_1(u) = \lambda_2(u)$ iff $\Delta(u) = 0$, namely

$$\begin{aligned} 4\,\rho_1\,\rho_2\,V_1\,V_2 &= 0 \\ \big(\psi(\rho)-\rho_1\big)\,V_1 - \big(\psi(\rho)-\rho_2\big)\,V_2 &= 0 \end{aligned}$$

and

$$\begin{aligned} \Big[\big(\psi(\rho)-\rho_1\big)\,V_1 - \big(\psi(\rho)-\rho_2\big)\,V_2\Big]_{\rho_1=0} = 0 \quad &\Longleftrightarrow \quad \rho_2 = \frac{V_1-V_2}{V_1-2V_2} \notin [0,1] \\ \Big[\big(\psi(\rho)-\rho_1\big)\,V_1 - \big(\psi(\rho)-\rho_2\big)\,V_2\Big]_{\rho_2=0} = 0 \quad &\Longleftrightarrow \quad \rho_1 = \frac{V_1-V_2}{2V_1-V_2} \in [0,1]\,. \end{aligned}$$

Observe that $\lambda_1(u_U) = \lambda_2(u_U) = V_1\,V_2\,(2V_1-V_2)^{-1}$. As corresponding eigenvectors we choose

$$\begin{aligned} r_1(u) &= \frac{(2V_1-V_2)}{2V_1\,(V_1-V_2)}\begin{bmatrix} \big(3\rho_1-\psi(\rho)\big)\,V_1 + \big(\psi(\rho)-\rho_2\big)\,V_2 + \Delta(u) \\ \big(\psi(\rho)-\rho_1\big)\,V_1 + \big(3\rho_2-\psi(\rho)\big)\,V_2 + \Delta(u) \end{bmatrix} \\ r_2(u) &= \frac{(2V_1-V_2)}{2V_1\,(V_1-V_2)}\begin{bmatrix} \big(\psi(\rho)+\rho_1\big)\,V_1 - \big(\psi(\rho)-\rho_2\big)\,V_2 + \Delta(u) \\ \big(\psi(\rho)-\rho_1\big)\,V_1 - \big(\psi(\rho)+\rho_2\big)\,V_2 - \Delta(u) \end{bmatrix}. \end{aligned}$$

The eigenvectors are independent in $\mathscr{S}\setminus\{u_U\}$, indeed

$$\det[r_1(u), r_2(u)] = -\frac{\Delta(u)\;(2V_1-V_2)^2\;\big(\Delta(u)+\rho_1\,V_1+\rho_2\,V_2\big)}{V_1^2\,(V_1-V_2)^2}\,.$$

Therefore, the system (11.5.1) is strictly hyperbolic in $\mathscr{S}\setminus\{u_U\}$ and u_U represents an umbilical point. Observe that $r_1(u_U) = r_2(u_U) = (1,0)^T$.

We consider now the genuinely nonlinearity of the characteristic fields. By direct computations we get

$$\begin{aligned} \nabla\lambda_1(u)\cdot r_1(u) &= -\frac{2V_1-V_2}{V_1}\,\big(\Upsilon_1(u)+\Upsilon_2(u)\big) = \\ = -\frac{2V_1-V_2}{V_1}&\left[\frac{(\rho_1\,V_1+\rho_2\,V_2)\Big(\big(\rho_1-\psi(\rho)\big)\,V_1+\big(\psi(\rho)-\rho_2\big)\,V_2+\Delta(u)\Big)}{\Delta(u)}\right. \\ &\left.+\frac{(V_1+V_2)\,\Delta(u)+\big(\rho_1-\psi(\rho)\big)\,V_1^2+(2-\rho_1-3\rho_2)\,V_1\,V_2-\big(\psi(\rho)-3\rho_2\big)\,V_2^2}{V_1-V_2}\right] \end{aligned}$$

$$\begin{aligned} &\nabla\lambda_2(u)\cdot r_2(u) = \\ &= -\frac{\psi(\rho)(2V_1-V_2)}{V_1}\left[\frac{(V_1-V_2)\,\big(\big(\psi(\rho)-\rho_1\big)\,V_1-\big(\psi(\rho)-\rho_2\big)\,V_2\big)}{\Delta(u)}+V_1+V_2\right]. \end{aligned}$$

Since

$$\lim_{u \to u_U} \Upsilon_1(u) = \frac{V_1\,(V_1 - V_2)}{2V_1 - V_2}\,,$$

we have that $[u \mapsto \nabla\lambda_1(u) \cdot r_1(u)]$ is well defined in all $\mathscr{S}$ and that the term $\Upsilon_1(u)$ is strictly positive in $\mathscr{S} \setminus \{0_2\}$. In $u = 0_2^T$ we have $\nabla\lambda_1(0_2) \cdot r_1(0_2) = -2V_2\, V_1^{-1}\,(2V_1 - V_2)$. Therefore, it is sufficient to prove that $\Upsilon_2(u) \geq 0$ in all $\mathscr{S}$ to get the genuine nonlinearity of the first vector field. If $\big(\psi(\rho) - \rho_1\big)\; V_1 \geq \big(\psi(\rho) - \rho_2\big)\; V_2$, then

$$\Upsilon_2(u) \geq \frac{2V_2\left[\big(\psi(\rho) - \rho_1\big)\; V_1 - \big(\psi(\rho) - \rho_2\big)\; V_2 + \rho_1\, V_1 + \rho_2\, V_2\right]}{V_1 - V_2} \geq 0\,,$$

and, if $\big(\psi(\rho) - \rho_1\big)\; V_1 \leq \big(\psi(\rho) - \rho_2\big)\; V_2$, then

$$\Upsilon_2(u) \geq \frac{2\left[V_1\,\Big(\big(\rho_1 - \psi(\rho)\big)\; V_1 - \big(\rho_2 - \psi(\rho)\big)\; V_2\Big) + V_2\;(\rho_1\, V_1 + \rho_2\, V_2)\right]}{V_1 - V_2} \geq 0\,.$$

In the above two estimates we used that $\Delta(u) \geq \Big|\big(\psi(\rho) - \rho_1\big)\, V_1 - \big(\psi(\rho) - \rho_2\big)\, V_2\Big|$. As a consequence, we have proved that $\lambda_1(u) \cdot r_1(u) < 0$ for all $u \in \mathscr{S}$. Finally, since

$$\lim_{u \to u_U} \nabla\lambda_2(u) \cdot r_2(u) = -V_1 - V_2$$

and for all $u \in \mathscr{S} \setminus \{u_U\}$

$$\frac{\big(\psi(\rho) - \rho_1\big)\, V_1 - \big(\psi(\rho) - \rho_2\big)\, V_2}{\Delta(u)} \geq -1\,,$$

we have that $\nabla\lambda_2(u) \cdot r_2(u) \leq -2V_2\; \psi(\rho)\; (2V_1 - V_2)\; V_1^{-1}$ for all $u \in \mathscr{S}$. Therefore $\nabla\lambda_2(u) \cdot r_2(u) < 0$ for all $u \in \mathscr{S} \setminus \big\{(\rho_1, \rho_2)\,:\; \rho_1 + \rho_2 = 1\big\}$. □

For the above proposition we can apply the results of Theorem 7.5 to the system (11.5.1) in $\mathscr{S} \setminus \Big(\{u_U\} \cup \big\{(\rho_1, \rho_2)\,:\; \rho_1 + \rho_2 = 1\big\}\Big)$. In particular, for any fixed $(\rho_1^l, \rho_2^l) \in \mathscr{S} \setminus \Big(\{u_U\} \cup \big\{(\rho_1, \rho_2)\,:\; \rho_1 + \rho_2 = 1\big\}\Big)$, there exists a neighborhood Ω of (ρ_1^l, ρ_2^l) such that for all $(\rho_1^r, \rho_2^r) \in \Omega$, the Riemann problem

$$\begin{aligned}
&\partial_t \rho_\ell + \partial_x \big(\rho_\ell\, v_\ell\, (\rho_1 + \rho_2)\big) = 0 && \ell = 1,2\\
&v_\ell(\rho) = V_\ell\;(1 - \rho) && \ell = 1,2\\
&\rho_\ell(0,x) = \begin{cases} \rho_\ell^l & \text{if } x < 0\\ \rho_\ell^r & \text{if } x \geq 0 \end{cases} && \ell = 1,2
\end{aligned}$$

has a unique weak solution that consists of at most three constant states separated by up to two elementary waves, namely rarefaction and shock waves.

11.6 Multi–lane Traffic Flow

Many macroscopic models for **multi–lane** traffic flows are based on systems of balance laws, see for instance [4, 9–11]. In the case of a road with n lanes, the corresponding model writes

$$\partial_t \rho_\ell + \partial_x \left[\rho_\ell \, v_\ell(\rho_\ell)\right] = G_\ell(t, \rho_1, \ldots, \rho_n) \qquad \ell = 1, \ldots, n \,, \tag{11.6.1}$$

where ρ_ℓ is the (mean) traffic density on the ℓ–th lane and $v_\ell = v_\ell(\rho_\ell)$ is the corresponding (mean) speed. Each $[\rho \mapsto v_\ell(\rho)]$ is a smooth decreasing function and satisfies $v_\ell(1) = 0$, the maximal traffic densities being all normalized to 1. The rightmost lane is labeled with $\ell = 1$, the leftmost with $\ell = n$, see Fig. 11.4. In each equation of the system (11.6.1), the convective part describes the intraŰ-lane dynamics, while the source models the inter–lane dynamics coupling between adjacent lanes. In the case of an homogeneous road with neither entries nor exits, we require that $\sum_{\ell=1}^n G_\ell(t, \rho_1, \ldots, \rho_n) = 0$ so that the total amount of vehicles $\sum_{\ell=1}^n (\int_{\mathbb{R}} \rho_\ell \, dx)$ is conserved.

For instance, as in [5], we postulate

$$\left[G_\ell(\rho)\right](x) = \int_{-\infty}^{x+h} g_\ell\left(\rho_1(y), \ldots, \rho_n(y)\right) \, e^{-k\,(x-y)} \, dy \,, \tag{11.6.2a}$$

for $h \in \mathbb{R}_+^*$ and $k \in \mathbb{R}_+$. With this choice for the source, a driver at position $\bar{x}$ decides whether to change lane or not evaluating an average of a function of the densities on $]-\infty, \bar{x}+h]$, the constant h represents then an anticipation term. The source term (11.6.2a) can be written as a convolution operator $G_\ell(\rho) = g_\ell(\rho) * \mathcal{K}$ for the kernel $\mathcal{K}(x) = e^{k\,x} \, \chi_{[-h,+\infty[}(x)$. For the functions g_ℓ we follow the *local* model

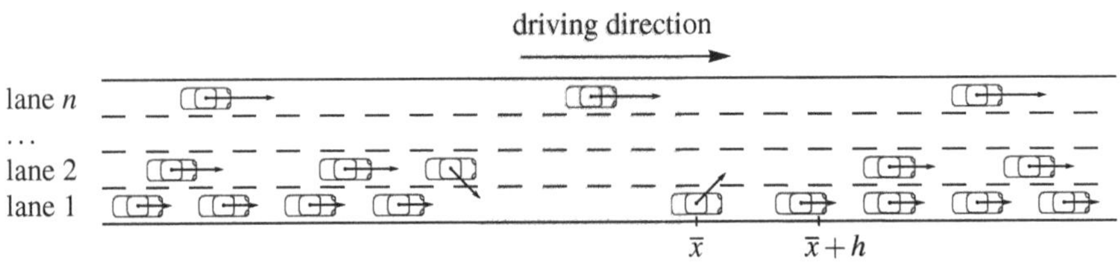

Fig. 11.4 A n–lane traffic flow

proposed in [10, Eq. (4.6)] and take

$$\begin{aligned} g_\ell(\rho_1, \ldots, \rho_n) = {} & \left[g_{\ell-1}^{\ell}(\rho_{\ell-1}, \rho_\ell)\rho_{\ell-1} - g_\ell^{\ell-1}(\rho_{\ell-1}, \rho_\ell, \rho_{\ell+1})\rho_\ell\right](1 - \delta_{1\ell}^K) \\ & - \left[g_\ell^{\ell+1}(\rho_\ell, \rho_{\ell+1})\rho_\ell - g_{\ell+1}^{\ell}(\rho_\ell, \rho_{\ell+1}, \rho_{\ell+2})\rho_{\ell+1}\right](1 - \delta_{\ell n}^K) \end{aligned} \tag{11.6.2b}$$

$$g_\ell^{\ell+1}(\rho_\ell, \rho_{\ell+1}) = P_L(\rho_{\ell+1}) \, \mathrm{i}_v(\rho_\ell) \tag{11.6.2c}$$

$$g_\ell^{\ell-1}(\rho_{\ell-1}, \rho_\ell, \rho_{\ell+1}) = P_R(\rho_{\ell-1}) \, \left(1 - P_L(\rho_{\ell+1})\right) \, \mathrm{i}_v(\rho_\ell) \,. \tag{11.6.2d}$$

Here i_v is the **interaction frequency** and P_R, P_L are the **lane changing probabilities**, respectively, to the right and to the left. Assume that i_v, P_R, P_L are smooth, positive and satisfy $\mathrm{i}_v(0) = \mathrm{i}_v(1) = 0$, $P_R(0) = P_L(0) = 1$, $P_R(1) = P_L(1) = 0$. The terms $g^{\ell}_{\ell-1}$ and $g^{\ell-1}_{\ell}$ model, respectively, the inflow into lane ℓ from lane $\ell - 1$ and the outflow from lane ℓ into lane $\ell - 1$; their expressions (11.6.2c), (11.6.2d) are deduced from a kinetic approximation. Remark that we recover formally the model in [10] replacing in (11.6.2a) the kernel $e^{-k\,(x-y)}$ with the delta function $\delta^D_0(x-y)$.

System (11.6.1), (11.6.2) is weakly coupled, that is the coupling occurs only through source terms. One then could try to prove the global existence of solutions by the methods developed in Chap. 8. We underline that both the vanishing viscosity method and the compensated compactness method apply only in the homogeneous case, see [13, §13.3] and [13, §13.2] respectively. Furthermore, in the non–homogenous case with *local* sources, system (11.6.1), (11.6.2) is not in general diagonally dominant, so the results in [7, 12] can not be applied. With the next proposition [5] we show that indeed the multi–lane traffic flow model presented in this section fits in the framework described in Chap. 8.

Proposition 11.2. *The system* (11.6.1) *with source* (11.6.2) *satisfies assumptions* ***(F)****,* ***(G)*** *and* ***(I)*** *given in Sect. 8.1 with* $\mathscr{U} = [0,1]^n$.

Proof. In this case, the homogeneous system (8.1.5) consists of n independent, scalar conservation laws, therefore, **(F)** holds.

Consider now **(G)**. Note that $g = (g_1, \ldots, g_n)^T$ is a Lipschitz function, so (8.1.7a) holds with $l = \mathscr{L}ip(g)\; e^{h\,k}/k$. Similarly, (8.1.7b) is satisfied with $a = 0$ and $b = \mathscr{L}ip(g)\; e^{h\,k}/k$ because $g(0) = 0$. A direct calculation proves then that $\mathrm{TV}\big(G(\rho)\big) \leq \mathscr{L}ip(g)\; e^{h\,k}\,\mathrm{TV}(\rho)/k$, whence (8.1.7c) holds.

Finally, the invariance of $\mathscr{U}$ with respect to the homogeneous system is immediate. Moreover, if $\rho_\ell = 0$ then $g_\ell \in \mathbb{R}^*_+$, while if $\rho_\ell = 1$ then $g_\ell = 0$, completing the proof of **(I)**. □

Hence, by the above proposition, Theorem 8.1 applies to the present case and the Cauchy problem for (11.6.1), (11.6.2) is well posed globally in time for all initial data in $\mathbf{L^1}$ with bounded total variation.

Models with source terms more general than (11.6.2a) can be considered as well. For instance one can allow the parameters h, k and the kernel $\mathscr{K}$ to depend on the lanes and on time by introducing functions $h_\ell(t), k_\ell(t), \mathscr{K}_\ell(t) \in \mathbf{L^1}(\mathbb{R};\mathbb{R})$. Condition **(G)** still holds under suitable assumptions on the time dependence.

References

1. Benzoni-Gavage, S., Colombo, R.M.: An n-populations model for traffic flow. European Journal of Applied Mathematics 14(05), 587–612 (2003)
2. Benzoni-Gavage, S., Colombo, R.M., Gwiazda, P.: Measure valued solutions to conservation laws motivated by traffic modelling. Proceedings of the Royal Society A: Mathematical, Physical and Engineering Science 462(2070), 1791–1803 (2070)

3. Chitour, Y., Piccoli, B.: Traffic circles and timing of traffic lights for cars flow. Discrete and Continuous Dynamical Systems Series B 5(3), 599–630 (2005)
4. Colombo, R.M., Corli, A.: Well posedness for multilane traffic models. Ann. Univ. Ferrara Sez. VII Sci. Mat. 52(2), 291–301 (2006)
5. Colombo, R.M., Corli, A., Rosini, M.D.: Non local balance laws in traffic models and crystal growth. ZAMM Z. Angew. Math. Mech. 87(6), 449–461 (2007)
6. Colombo, R.M., Goatin, P., Rosini, M.D.: On the modelling and management of traffic. ESAIM: Mathematical Modelling and Numerical Analysis 45(05), 853–872 (2011)
7. Dafermos, C.M., Hsiao, L.: Hyperbolic systems and balance laws with inhomogeneity and dissipation. Indiana Univ. Math. J. 31(4), 471–491 (1982)
8. Garavello, M., Piccoli, B.: Traffic flow on networks, vol. 1 of AIMS Series on Applied Mathematics. American Institute of Mathematical Sciences (AIMS), Springfield, MO, Conservation laws models
9. Gazis, D.C., Herman, R., Weiss, G.H.: Density oscillations between lanes of a multilane highway. Operations Research 10(5), 658–667 (1962)
10. Klar, A., Wegener, R.: A hierarchy of models for multilane vehicular traffic. I. Modeling. SIAM J. Appl. Math. 59(3), 983–1001 (1999) (electronic)
11. Klar, A., Wegener, R.: A hierarchy of models for multilane vehicular traffic. II. Numerical investigations. SIAM J. Appl. Math. 59(3), 1002–1011 (1999) (electronic)
12. Liu, T.P.: Quasilinear hyperbolic systems. Comm. Math. Phys. 68(2), 141–172 (1979)
13. Serre, D.: Systems of conservation laws, vol. 2. Cambridge University Press, Cambridge (2000)
14. Zhang, H.M., Jin, W.L.: A kinematic wave traffic flow model for mixed traffic. Presented at TRB 2002 Annual Meeting in Press Transportation Research Record 1802(1), 197–204 (2002)
15. Zhang, P., Liu, R.X., Wong, S.C., Dai, S.Q.: Hyperbolicity and kinematic waves of a class of multi-population partial differential equations. European Journal of Applied Mathematics 17(2), 171–200 (2006)

Chapter 12
Cost Functionals

Abstract. In this chapter we present some functionals related to traffic management and prove that they can be optimized thanks to the results proved in Sect. 6.5.

12.1 Introduction

Various reasonable goals to be pursued in the management of vehicular traffic can be described through suitable functionals, which have to be maximized or minimized along solutions to (11.2.1). In this chapter we propose some criteria to be optimized in a rational control of traffic [3]. We consider some reasonable cost functionals and show that, when computed along the solutions to (11.2.1), the well posedness results proved in Chap. 11 allow to prove that their regularity is sufficient to ensure the existence of optimal management strategies. As possible control parameters we consider below the inflow q_b or the maximal flow q_c at the constraint.

12.2 Queue Length

In this section we first rigorously define the functional assigning to a solution to (11.2.1) the length of the queue before the constraint at x_c. Then, we prove its lower semi–continuity and show through an example that upper semi–continuity does not hold. Therefore, this functional can be maximized but hardly minimized.

We consider the case in which the maximal through flow at $x_c \in \mathbb{R}_+$ is constant, i.e. $q_c(t) \equiv Q_c$ for a fixed constant $Q_c \in \,]0, f_{max}^{free}[$. To this aim, we introduce

$$A_c(\rho) = \left\{x \in [0, x_c[\, : \Psi\left(\rho(\xi+)\right) = f_{max}^{free} - Q_c \text{ for almost all } \xi \in [x, x_c[\right\}$$

the set of points immediately before x_c, where the traffic flows at the maximal possible level allowed by the constraint Q_c. Above Ψ is defined by (6.2.7)

M.D. Rosini: *Macroscopic Models for Vehicular Flows & Crowd Dynamics*, UCS, pp. 161–166.
DOI: 10.1007/978-3-319-00155-5_12

$$\Psi(\rho) = \text{sgn}\left(\rho - \rho_{max}^{free}\right) \left(f_{max}^{free} - f(\rho)\right) .$$

Note that $A_c(\rho)$ is well defined as soon as $\rho \in \mathscr{D}$, with $\mathscr{D}$ given as in (6.4.3) by

$$\mathscr{D} = \left\{\rho \in \mathbf{L^1}\left(\mathbb{R}_+;[0,R]\right) : \Psi(\rho) \in \mathbf{BV}\left(\mathbb{R}_+;\mathbb{R}\right)\right\} .$$

Define now the functional $\mathscr{L} : \mathscr{D} \to \mathbb{R}_+^*$ assigning to any traffic density $\rho \in \mathscr{D}$ the length of the queue before x_c by

$$\mathscr{L}(\rho) = \begin{cases} 0 & \text{if } A_c(\rho) = \emptyset \\ x_c - \inf A_c(\rho) & \text{otherwise .} \end{cases} \tag{12.2.1}$$

In the general case where $\rho = \rho(t)$ varies with time, also the queue length is time dependent: $\mathscr{L} = \mathscr{L}\left(\rho(t)\right)$. Refer to Fig. 12.1 for two visualizations of $\mathscr{L}(\rho)$ in the (x,t)–plane, on a solution to (11.2.1). Remark that $\mathscr{L}(\rho)$ measures the length of the queue due to the presence of the constraint at x_c. If further obstacles hinder the flow of traffic lowering the traffic flow below Q_c, then $\mathscr{L}$ may even vanish.

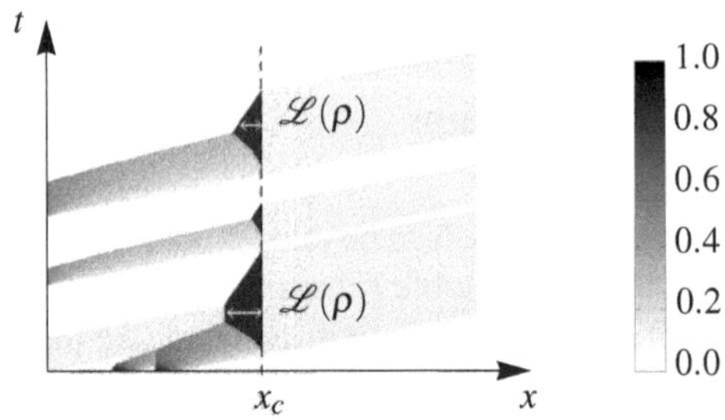

Fig. 12.1 The functional $\mathscr{L}(\rho)$ defined by (12.2.1) for a solution ρ to (11.2.1) is portrayed on the (x,t)–plane at two different times

Proposition 12.1. *The map* $\mathscr{L} : \mathscr{D} \to \mathbb{R}_+^*$ *defined by* (12.2.1) *is upper semi–continuous with respect to the* $\mathbf{L^1}$*–norm.*

Proof. Let us consider a sequence of functions ρ_n $\mathbf{L^1}$–converging to some $\tilde{\rho}$ on $[0,x_c]$:

$$\lim_{n\to+\infty} \|\rho_n - \tilde{\rho}\|_{\mathbf{L^1}([0,x_c];\mathbb{R})} = 0 .$$

Fix $\varepsilon \in \mathbb{R}_+$. If $A_c(\rho_n) = \emptyset$, then there exists some $N \in \mathbb{R}_+$ such that for all $n \geq N$

$$0 = \mathscr{L}(\rho_n) \leq \mathscr{L}(\tilde{\rho}) + \varepsilon .$$

Otherwise, there exists a subsequence (again labeled ρ_n) such that $A_c(\rho_n) \neq \emptyset$ for all $n \in \mathbb{N}$. Let $x_n = \inf A_c(\rho_n)$ and $\check{x} = \liminf_{n\to+\infty} x_n$. For any $\varepsilon \in \mathbb{R}_+$, there exists $N \in \mathbb{R}_+$ s.t. for all $n \geq N$

$$\mathscr{L}(\rho_n) \leq x_c - \check{x} + \varepsilon ,$$

and there exists a subsequence (again labeled x_n) such that $x_n \to \check{x}+$. Moreover, since ρ_n converges pointwise a.e. to $\tilde{\rho}$ on $[0,x_c]$ and $\rho_n(\xi) = \Psi^{-1}(f_{max}^{free} - Q_c)$ for almost all $\xi \in [x_n, x_c]$, we get that $\tilde{\rho}(\xi) = \Psi^{-1}(f_{max}^{free} - Q_c)$ for almost all $\xi \in [\check{x}, x_c]$. In fact, fix $\varepsilon \in \mathbb{R}_+$, then for any $\xi \in]\check{x}, x_c]$ such that $\lim_{n\to+\infty} \rho_n(\xi) = \tilde{\rho}(\xi)$, there exists $N \in \mathbb{R}_+$ sufficiently large such $x_n < \xi$ and $\left|\rho_n(\xi) - \tilde{\rho}(\xi)\right| \leq \varepsilon$ for all $n \geq N$. Since $\rho_n(\xi) = \Psi^{-1}(f_{max}^{free} - Q_c)$ a.e., we have $\left|\tilde{\rho}(\xi) - \Psi^{-1}(f_{max}^{free} - Q_c)\right| \leq \varepsilon$. This implies

$$\tilde{\rho}(\xi) = \Psi^{-1}\left(f_{max}^{free} - Q_c\right) \quad \text{for almost all } \xi \in]\check{x}, x_c] .$$

Hence for all $n \geq N$

$$\mathscr{L}(\tilde{\rho}) \geq x_c - \check{x} \geq \mathscr{L}(\rho_n) - \varepsilon ,$$

which concludes the proof. □

The above regularity is not in general sufficient to ensure the existence of minimizers for the queue length, as the next example points out.

Example 12.1. Fix $Q_c \in]0, f_{max}^{free}[$ and define the constant inflows $q_b^n = Q_c - n^{-1}$, $q_b^\infty = Q_c$ and the initial data $\overline{\rho}^n = f_*^{-1}(q_b^n)$, $\overline{\rho}^\infty = f_*^{-1}(Q_c)$. Call ρ^n, ρ^∞ the corresponding solutions to (11.2.1). Then, for any $n \in \mathbb{N}$, by (12.2.1) we have $\mathscr{L}\left(\rho^n(t)\right) = 0$. However, for all $t \in \mathbb{R}_+$, we have $\mathscr{L}\left(\rho^\infty(t)\right) = x_c \in \mathbb{R}_+$, coherently with Proposition 12.1 and showing that lower continuity may fail. □

12.3 Stop and Go Waves

A relevant criterion in the management of traffic dynamics is the minimization of **stop–and–go (start–stop)** phenomena and cluster formation, see for instance [4, 5, 8] or [6, Chap. 8] and the references therein. Note that a more regular traffic flow reasonably reduces both the probability of accidents as well as pollution.

Analytically, this amounts to minimize the total variation of traffic speed. This leads to consider the functional

$$\mathscr{J}(\rho) = \int_0^T \int_{\mathbb{R}_+} p(x) \, \mathrm{d}\left|\partial_x v(\rho)\right| \, \mathrm{d}t \tag{12.3.1}$$

measuring on the time interval $[0, T]$ the total variation $\left|\partial_x v(\rho)\right|$ of the traffic speed $v(\rho)$ weighted by a weight $p(x) \in [0, 1]$, higher in more dangerous road segments.

Proposition 12.2. *If $p \in \mathbf{C_c^0}(\mathbb{R}_+; [0, 1])$, then the functional $\mathscr{J} : \mathscr{D} \to \mathbb{R}_+^*$ defined in* (12.3.1) *is lower semi–continuous with respect to the $\mathbf{L^1}$–norm.*

Proof. Recall that by hypothesis **(F)** we have that v is a Lipschitz function. Hence, for any $t \in \mathbb{R}_+$, as soon as $\rho(t)$ is in **BV**, also the function $[x \to v\left(\rho(t, x)\right)]$ is in **BV** and its derivative $\left|\partial_x v(\rho)\right|$ is a Radon measure on $\mathbb{R}$. The inner integral in (12.3.1) is thus the integral of p with respect to the time dependent total variation measure $\left|\partial_x v(\rho)\right|$. Therefore, by [7, Theorem 6.19]

$$
\begin{aligned}
&\int_{\mathbb{R}_+} p(x)\ \mathrm{d}\left|\partial_x v(\rho)\right| = \\
&= \sup\left\{\int_{\mathbb{R}_+} \varphi(x)\ \mathrm{d}\left|\partial_x v(\rho)\right| \ :\ \varphi \in \mathbf{C_c^0}(\mathbb{R}_+;\mathbb{R}_+^*),\ |\varphi| \le p\right\} \\
&= \sup\left\{-\int_{\mathbb{R}_+} \partial_x\varphi(x)\ \left|\partial_x v(\rho)\right|\ \mathrm{d}x \ :\ \varphi \in \mathbf{C_c^0}(\mathbb{R}_+;\mathbb{R}_+^*),\ |\varphi| \le p\right\}.
\end{aligned}
$$

Let ρ_n be a sequence in $\mathbf{BV}(\mathbb{R};[0,\rho_m])$ $\mathbf{L^1}$–converging to $\tilde{\rho}$, then $\left|\partial_x v(\rho_n)\right|$ converges to $\left|\partial_x v(\tilde{\rho})\right|$ in $\mathbf{L^1}$. Therefore, for all $\varphi \in \mathbf{C_c^0}(\mathbb{R}_+;\mathbb{R}_+^*)$ and $t \in [0,T]$, the map

$$
\left[\rho \mapsto -\int_{\mathbb{R}_+} \partial_x\varphi(x)\ \left|\partial_x v(\rho)\right|\ \mathrm{d}x\right]
$$

is continuous in $\mathbf{L^1}$ and the integrand in (12.3.1) is lower semi–continuous. By Fatou's lemma and (12.3.1) we have

$$
\begin{aligned}
&\liminf_{n\to+\infty} \mathscr{J}(\rho_n) \ge \\
&\ge \int_0^T \liminf_{n\to+\infty} \sup\left\{-\int_{\mathbb{R}_+} \partial_x\varphi(x)\ \left|\partial_x v(\rho_n)\right|\ \mathrm{d}x \ :\ \varphi \in \mathbf{C_c^0}(\mathbb{R}_+;\mathbb{R}_+^*),\ |\varphi| \le p\right\}\ \mathrm{d}t \\
&\ge \int_0^T \sup\left\{-\int_{\mathbb{R}_+} \partial_x\varphi(x)\ \left|\partial_x v(\rho)\right|\ \mathrm{d}x \ :\ \varphi \in \mathbf{C_c^0}(\mathbb{R}_+;\mathbb{R}_+^*),\ |\varphi| \le p\right\}\ \mathrm{d}t \\
&= \mathscr{J}(\rho),
\end{aligned}
$$

proving the sequential lower semi–continuity of $\mathscr{J}$. □

We end the section recalling that in [2] it is proved the lower semi–continuity of the functional $\mathscr{J}$ for more general weight functions p.

12.4 Travel Times

From the point of view of drivers, key quantities determining the quality of traffic are related to the time necessary to reach the destination. With reference to (11.2.1), we neglect the initial datum, i.e. let $\overline{\rho} = 0$, and assume that the inflow q_b is non–zero and with support contained in, say, $[0, \tau_o]$. The total quantity of vehicles entering the road is thus $Q_{\text{in}} = \int_0^{\tau_o} q_b(t)\ \mathrm{d}t$, which is assumed to be finite and, obviously, strictly positive. Then, the **mean arrival time** at $x = \overline{x} \in \mathbb{R}_+$ is

$$
T_a(\overline{x}) = \frac{1}{Q_{\text{in}}} \int_{\mathbb{R}_+} t\ f\left(\rho(t,\overline{x})\right)\ \mathrm{d}t\ , \tag{12.4.1}
$$

see also [1, Eq. (5.3)], where $\rho = \rho(t,x)$ is the solution to (11.2.1). The **mean travel time** between the points 0 and $\bar{x}$ can then be easily computed:

$$T_t(\bar{x}) = \frac{1}{Q_{\text{in}}} \int_{\mathbb{R}_+} t \, \Big[f\big(\rho(t,\bar{x})\big) - f\big(\rho(t,0)\big) \Big] \, \mathrm{d}t \, . \tag{12.4.2}$$

To evaluate the regularity of the functionals (12.4.1) and (12.4.2), we consider below the case in which the flow at $x = 0$ is assigned as boundary datum, so that $f\big(\rho(t,0)\big) = q_b(t)$. The notations are the same as those of Corollary 6.1, in particular τ_c is a positive time satisfying (6.5.6).

Proposition 12.3. *Under the same assumptions and notations of Corollary 6.1, the mean arrival times $T_a(\bar{x})$, $T'_a(\bar{x})$ and the mean travel times $T_t(\bar{x})$, $T'_t(\bar{x})$ on the segment $[0,\bar{x}]$ satisfy the following Lipschitz estimates:*

$$\begin{aligned}
\left|T_a(\bar{x}) - T'_a(\bar{x})\right| \le{}& \frac{\tau_c}{\min\{Q_{\text{in}}, Q'_{\text{in}}\}} \left(1 + \frac{\tau_c\, f_{max}^{free}}{2 \, \min\{Q_{\text{in}}, Q'_{\text{in}}\}}\right) \left\|q_b - q'_b\right\|_{\mathbf{L}^1([0,\tau_o];\mathbb{R})} \\
&+ \frac{2\,\tau_c}{\min\{Q_{\text{in}}, Q'_{\text{in}}\}} \left\|q_c - q'_c\right\|_{\mathbf{L}^1([0,\tau_c];\mathbb{R})} \\
\left|T_t(\bar{x}) - T'_t(\bar{x})\right| \le{}& \frac{\tau_c}{\min\{Q_{\text{in}}, Q'_{\text{in}}\}} \left(2 + \frac{\tau_c\, f_{max}^{free}}{\min\{Q_{\text{in}}, Q'_{\text{in}}\}}\right) \left\|q_b - q'_b\right\|_{\mathbf{L}^1([0,\tau_o];\mathbb{R})} \\
&+ \frac{2\,\tau_c}{\min\{Q_{\text{in}}, Q'_{\text{in}}\}} \left\|q_c - q'_c\right\|_{\mathbf{L}^1([0,\tau_c];\mathbb{R})} \, .
\end{aligned}$$

Proof. Straightforward computations give:

$$\begin{aligned}
T_a - T'_a ={}& \frac{1}{Q_{\text{in}}} \int_{\mathbb{R}_+} t \, f\big(\rho(t,\bar{x})\big) \, \mathrm{d}t - \frac{1}{Q_{\text{in}}} \int_{\mathbb{R}_+} t \, f\big(\rho'(t,\bar{x})\big) \, \mathrm{d}t \\
&+ \frac{1}{Q_{\text{in}}} \int_{\mathbb{R}_+} t \, f\big(\rho'(t,\bar{x})\big) \, \mathrm{d}t - \frac{1}{Q'_{\text{in}}} \int_{\mathbb{R}_+} t \, f\big(\rho'(t,\bar{x})\big) \, \mathrm{d}t \\
={}& \frac{1}{Q_{\text{in}}} \left(\int_{\mathbb{R}_+} t \, f\big(\rho(t,\bar{x})\big) \, \mathrm{d}t - \int_{\mathbb{R}_+} t \, f\big(\rho'(t,\bar{x})\big) \, \mathrm{d}t \right) \\
&+ \left(\frac{1}{Q_{\text{in}}} - \frac{1}{Q'_{\text{in}}} \right) \int_{\mathbb{R}_+} t \, f\big(\rho'(t,\bar{x})\big) \, \mathrm{d}t \\
T_t - T'_t ={}& T_a - T'_a - \frac{1}{Q_{\text{in}}} \int_{\mathbb{R}_+} t \, q_b(t) \, \mathrm{d}t + \frac{1}{Q'_{\text{in}}} \int_{\mathbb{R}_+} t \, q'_b(t) \, \mathrm{d}t \\
={}& T_a - T'_a - \frac{1}{Q_{\text{in}}} \int_{\mathbb{R}_+} t \, q_b(t) \, \mathrm{d}t + \frac{1}{Q_{\text{in}}} \int_{\mathbb{R}_+} t \, q'_b(t) \, \mathrm{d}t \\
&- \frac{1}{Q_{\text{in}}} \int_{\mathbb{R}_+} t \, q'_b(t) \, \mathrm{d}t + \frac{1}{Q'_{\text{in}}} \int_{\mathbb{R}_+} t \, q'_b(t) \, \mathrm{d}t
\end{aligned}$$

$$= T_a - T_a' + \frac{1}{Q_{\text{in}}} \int_{\mathbb{R}_+} t \, \left(q_b'(t) - q_b(t)\right) \, dt + \left(\frac{1}{Q_{\text{in}}'} - \frac{1}{Q_{\text{in}}}\right) \int_{\mathbb{R}_+} t \, q_b'(t) \, dt \, .$$

The final estimates now easily follow from Corollary 6.1. □

Remark that evaluating the exit time T_e, i.e. the time at which no more vehicle are left in the segment $[0,x_c]$, yields a possibly discontinuous functional, see Fig. 13.6, right.

12.5 Density Dependent Functionals

A further class of integral functionals is that of integrals of functions φ of the density weighted by a weight w, say

$$\mathscr{F}(\rho) = \int_0^T \int_a^b \varphi\left(\rho(t,x)\right) w(t,x) \, dx \, dt$$

for a time $T \in \mathbb{R}_+$ and points $b > a > 0$. Reasonable examples of choices of the function φ are:

- To have all vehicles travel at a speed as near as possible to a desired optimal speed $\bar{v}$ along a given road segment $[a,b]$, choose for instance $\varphi(\rho) = \left(v(\rho) - \bar{v}\right)^2$.
- To maximize the traffic flow along $[a,b]$, choose $\varphi(\rho) = f(\rho)$.

As soon as the weight w and the cost function φ are continuous, it is immediate to prove that $\mathscr{F}$ is also continuous. This, together with Theorem 6.3, through an application of Weierstraß theorem, allows to prove the existence of choices of the initial/boundary data and/or of the constraint that optimize $\mathscr{F}$.

References

1. Ancona, F., Marson, A.: Scalar non-linear conservation laws with integrable boundary data. Nonlinear Anal. 35(6, Ser. A: Theory Methods), 687–710 (1999)
2. Colombo, R.M., Groli, A.: Minimising stop and go waves to optimise traffic flow. Appl. Math. Lett. 17(6), 697–701 (2004)
3. Colombo, R.M., Goatin, P., Rosini, M.D.: On the modelling and management of traffic. ESAIM: Mathematical Modelling and Numerical Analysis 45(05), 853–872 (2011)
4. Edie, L.C., Foote, R.S.: Traffic flow in tunnels, vol. 37, pp. 334–344 (1958)
5. Kerner, B.S., Konhäuser, P.: Cluster effect in initially homogenous traffic flow. Physical Review E 48(4) (1993)
6. Klar, A.: Kinetic and Macroscopic Traffic Flow Models. School of Computational Mathematics: Computational Aspects in Kinetic Models, XX edn. (2002)
7. Rudin, W.: Real and complex analysis. Mathematics Series. McGraw-Hill (1987)
8. Tomer, E., Safonov, L., Madar, N., Havlin, S.: Optimization of congested traffic by controlling stop-and-go waves. Phys. Rev. E (3) 65(6), 065101, 4 (2002)

Chapter 13
Numerical Applications

Abstract. Most of the examples presented in Chap. 11 lead to constrained initial–boundary value problems that are numerically integrated in the present chapter.

13.1 Introduction

In this chapter we point out the main features and advantages of the macroscopic models presented in Chap. 11 in terms of their applicability and ability to reproduce realistic vehicular traffic. To this aim we numerically integrate them. We mainly use as a numerical algorithm the wave front tracking method [6] described in Chap. 5. Our choice of using it as a numerical tool is motivated by our need to accurately compute the *position* of a shock in the solution to a conservation law as in the present framework it corresponds, for instance, to queues tails [4, 5] and it is therefore necessary to determine, for instance, the time necessary to a platoon of cars to pass through a prescribed point. As shown in Sect. 13.3, at least in the particular case considered therein, wave front tracking proves to be more precise and efficient of the classical Lax–Friedrichs method [8].

13.2 Passing through a Toll Gate

This section is devoted to some numerical integrations of (11.2.1). Our aim is only to show that this model features reasonable qualitative properties, hence we choose *normalized* parameters.

Let the real interval $[0,2]$ describe a segment of a highway with a toll gate at its center $x = 1$, as in Fig. 13.1. The evolution of traffic is described by (11.2.1) with, for instance, $f(\rho) = \rho\ (1-\rho)$. For simplicity, we assume that the initial density distribution is $\overline{\rho}(x) = 0.3$ for $x \in [0.2, 1]$ and $\overline{\rho}(x) = 0$ for $x \in [0,2] \setminus [0.2, 1]$. We also assume that no vehicles are entering the interval, namely $q_b(t) \equiv 0$. The threshold of

M.D. Rosini: *Macroscopic Models for Vehicular Flows & Crowd Dynamics*, UCS, pp. 167–173.
DOI: 10.1007/978-3-319-00155-5_13

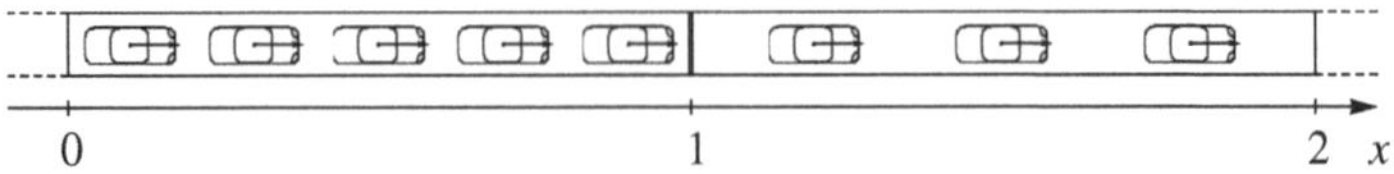

Fig. 13.1 A toll gate sited at $x = 1$

the through flow at the gate is $q_c(t) \equiv 0.1$. Then, the computed time necessary for all the vehicles to pass the toll gate is $t \approx 2.4$ and the evolution described by (11.2.1) is displayed in Fig. 13.2. As it has to be expected, the toll gate causes the rise of

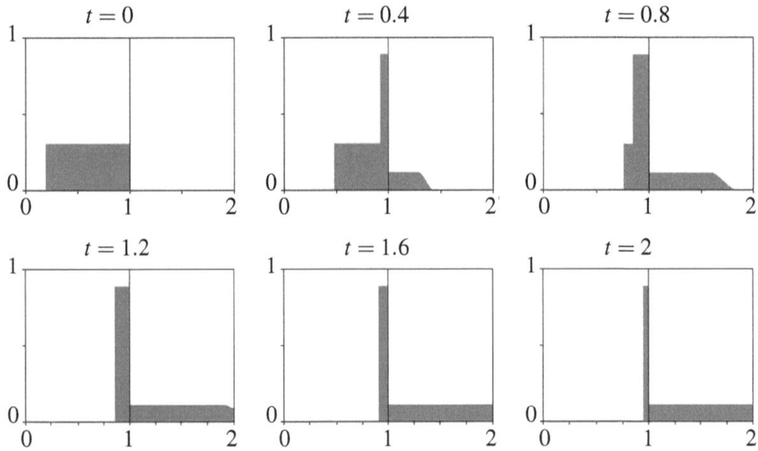

Fig. 13.2 Numerical integrations of (11.2.1) using Rusanov scheme, with $f(\rho) = \rho\ (1-\rho)$; $\overline{\rho}(x) = 0.3\ \chi_{[0.2,1]}$ and $q(t) \equiv 0.1$. The constraint at $x = 1$ is treated as suggested in [1].

a queue to the left of the gate. This queue first increases an then decreases, finally disappearing when all vehicles passed the gate.

We now let both the initial density $\overline{\rho}$ of vehicles and the efficiency of the gate q_c vary, while keeping the other parameters fixed as well as the segment $[0.2, 1]$ along which the cars are initially uniformly distributed. The time T that is necessary for all vehicles to pass the gate is then a function of $\overline{\rho}$ and q_c, that is $T = T(\overline{\rho}, q_c)$. As it has to be expected, this function is monotone in both variables, see Fig. 13.3. Note that as $q_c \searrow 0$, obviously, $T \to +\infty$. Hence, in Fig. 13.3, T is computed only for $q_c \geq 0.04$. Note the vertical segments in the level curves of T in Fig. 13.3, right. They realistically correspond to the gate being sufficiently efficient to avoid the rising of queues. On the contrary, as soon as the toll gate influences the traffic flow, T is well approximated by a function of the ratio $\overline{\rho}/q_c$, as also dimensional considerations suggest.

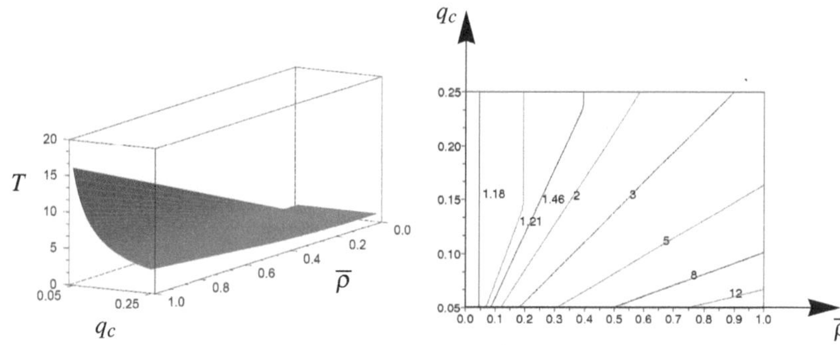

Fig. 13.3 A density of $\overline{\rho} \in [0.1, 1]$ vehicles is uniformly distributed on $[0.2, 1]$. A toll gate is sited at $x = 1$ and its through flow is $q_c \in [0.05, 0.25]$. T is the time necessary for all vehicles to pass the gate. Left, 3D diagram and, right, the level curves with $\overline{\rho}$ on the horizontal axis and q_c on the vertical one.

13.3 Lax–Friedrichs vs. Wave Front Tracking

Assume that along a road, at some point x_c, an obstacle hinders the flow of traffic. This situation can be described through the equations in (11.2.1). For simplicity, we

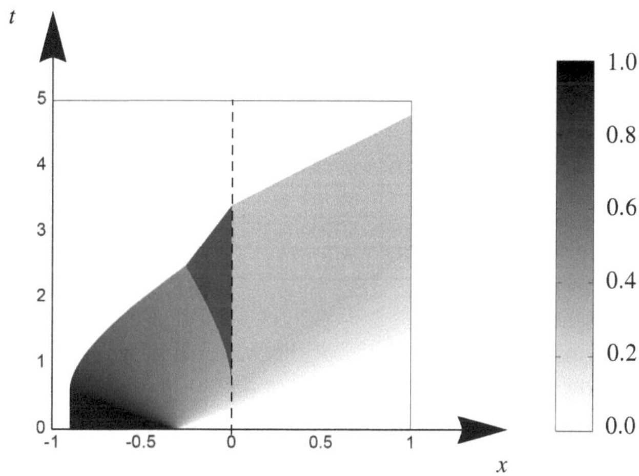

Fig. 13.4 The (x,t)–plane showing the numerical integration of (11.2.1) with $x_c = 0$, $\overline{\rho} = \chi_{[-0.9,-0.3]}$, $q_c(t) \equiv 0.2$, $q_b(t) \equiv 0$. The darker regions represent areas with higher density.

choose the normalized flux $f(\rho) = \rho\,(1-\rho)$ and we take $x_c = 0$, $\overline{\rho} = \chi_{[-0.9,-0.3]}$, $q_c(t) \equiv 0.2$, $q_b(t) \equiv 0$. Then, the exact solution can be explicitly computed via an elementary application of the wave front tracking method, see Chap. 5 and [2, 3, 6, 7]. In particular, by using the conservation of ρ, we obtain the following exact expression for the exit time T_e, i.e. the time at which the last vehicle passes through $x = 1$:

$$T_e = \frac{25}{4} - \frac{13}{4\sqrt{5}} \qquad (\approx 4.7965558146\ldots) . \tag{13.3.1}$$

For the numerical integration we use both the wave front tracking and the Lax–Friedrichs methods, the latter adapted to the present constrained situation as in [1], see [9, Sect. 12.5] as a general reference. The results are summarized in

Table 13.1 Results and CPU times of the numerical integration of (11.2.1) with $v(\rho) = 1 - \rho$, $x_c = 0$, $\overline{\rho} = \chi_{[-0.9,-0.3]}$, $q_c(t) \equiv 0.2$, $q_b(t) \equiv 0$ and different numerical parameters, with the wave front tracking algorithm and with the Lax–Friedrichs method (with CFL = 0.5). The solution is portrayed in Fig. 13.4, while the exact exit time is given in (13.3.1).

	$\Delta\rho$	Exit Time	CPU Time (s)	Relative Error
Wave Front Tracking	4.00e-03	4.79564272	0.32	-1.90e-02 %
	2.00e-03	4.79615273	0.59	-8.40e-03 %
	1.00e-03	4.79640870	1.18	-3.07e-03 %
	5.00e-04	4.79653693	2.36	-3.94e-04 %
	2.50e-04	4.79660132	4.95	9.49e-04 %
	1.25e-04	4.79656903	10.60	2.76e-04 %
	6.25e-05	4.79655291	24.48	-6.06e-05 %

	Δx	Exit Time	CPU Time (s)	Relative Error
Lax–Friedrichs	4.00e-03	4.94600000	1.69	3.12e-00 %
	2.00e-03	4.87000000	5.18	1.53e-00 %
	1.00e-03	4.83300000	18.90	7.60e-01 %
	5.00e-04	4.81475000	73.40	3.79e-01 %
	2.50e-04	4.80562500	295.99	1.89e-01 %
	1.25e-04	4.80100000	1213.41	9.27e-02 %
	6.25e-05	4.79878125	5264.29	4.64e-02 %

Table 13.1. Remark that, in the case of the wave front tracking scheme, we discretize the ρ variable, so that the mesh size has to be understood in ρ units. On the other hand, as it is more usual in the numerics for conservation laws, the Lax–Friedrichs scheme relies on the discretization of the space and time variables. It is important to observe that the choice of the initial datum may strongly affect the CPU time in the case of the wave front tracking algorithm. Indeed the number of operations carried out with this method depends on the number of waves, which in turn is determined by the initial datum. On the contrary, the number of operations in the Lax–Friedrichs method only depends on the mesh size.

We stress however that the use of the wave front tracking method allows for a much greater precision, in spite of the fact that the exit time is, in general, *not* a continuous function, see Fig. 13.6, right. Indeed, most numerical methods introduce viscosity or averaging of the ρ variable. The wave front tracking method, by its nature, computes piecewise constant solutions and keeps track of the location of the points of jump. Thus, in computing quantities such as the exit time, wave front tracking appears particularly suitable.

13.4 Synchronizing Traffic Lights

Consider the segment $[0,3]$ of a road $\mathbb{R}_+^*$ with two traffic lights, one at $x_{c1} = 1$ and one at $x_{c2} = 2$, respectively characterized by maximal flows q_{c1} and q_{c2}. We use the normalized speed law $v(\rho) = 1 - \rho$. Assume that the two traffic lights have the same fixed frequency of red/green light, say 1 time unit for each regime, so that $q_{c2}^{\tau}(t) = q_{c1}(t - \tau)$ for a delay $\tau \in [0,2[$. Moreover, we set $q_{c2}^{\tau}(t) = 0$ during the red interval and $q_{c2}^{\tau}(t) = f_{max}^{free} = 0.25$ during the green one. At $x = 0$ the inflow q_b is constant throughout the time interval $[0,4]$ and then vanishes. We choose the sample values $q_b = f(\rho_b)$, with $\rho_b = 0.01, 0.1, 0.2, 0.3, 0.4, 0.5$. Then the following refinement of (11.2.1) describes the evolution of traffic

$$
\begin{aligned}
&\partial_t \rho + \partial_x f(\rho) = 0 && (t,x) \in \mathbb{R}_+^2 && (13.4.1a)\\
&\rho(0,x) = 0 && x \in \mathbb{R}_+ && (13.4.1b)\\
&f\left(\rho(t,0)\right) = q_b(t) && t \in \mathbb{R}_+ && (13.4.1c)\\
&f\left(\rho(t,x_{c1})\right) \le q_{c1}(t) && t \in \mathbb{R}_+ && (13.4.1d)\\
&f\left(\rho(t,x_{c2})\right) \le q_{c2}^{\tau}(t) && t \in \mathbb{R}_+\,, && (13.4.1e)
\end{aligned}
$$

where

$$
\begin{aligned}
&q_b(t) = f(\rho_b)\, \chi_{[0,4]}(t) && \rho_b = 0.01, 0.1, 0.2, 0.3, 0.4, 0.5 && (13.4.1f)\\
&q_{c2}^{\tau}(t) = q_{c1}(t-\tau) && q_{c1}(t) = 0.25\, \chi_{[0,1] \cup [2,3] \cup [4,5] \cup [6,7]}(t)\,. && (13.4.1g)
\end{aligned}
$$

We seek the optimal synchronization of the two traffic lights. The existence of such an optimal delay τ is ensured by the analytical results in Sect. 6.5, through Weierstraß Theorem. Indeed, Theorem 6.3 can easily be extended to the case of (13.4.1). The map $[\tau \to q_{c2}^{\tau}]$ is continuous in $\mathbf{L}^1$ and by Proposition 12.3 the mean travel time T_t defined in (12.4.2) is an $\mathbf{L}^1$–continuous function of the constraint.

In order to get an estimate of the optimal delay, we integrate (13.4.1) with $\tau = 0, 0.05, 0.10, 0.15, \ldots, 0.50$, see Fig. 13.5. The graphs of the resulting mean arrival time T_a and exit time T_e are in Fig. 13.6.

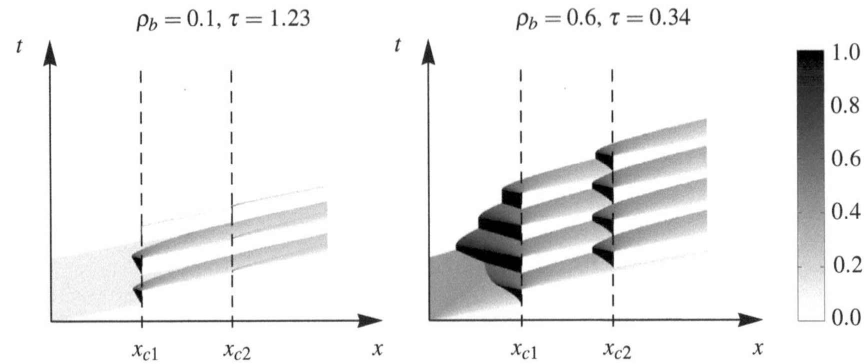

Fig. 13.5 Two solutions to (13.4.1), corresponding to $\rho_b = 0.1$, $\tau = 1.23$, left, and to $\rho_b = 0.5$, $\tau = 0.34$, right. The corresponding mean arrival times and exit times are part of Fig. 13.6.

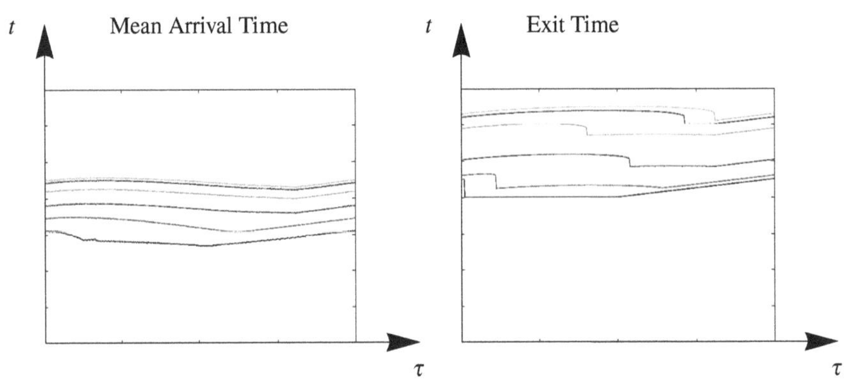

Fig. 13.6 Mean arrival times, left, and exit times, right, as functions of τ. Both graphs refer to (13.4.1), with the lower graphs corresponding to the lower inflows. Coherently with Proposition 12.3, the functions on the left are Lipschitz continuous, whereas the ones on the right display jump discontinuities. Two particular solutions are in Fig. 13.5.

References

1. Andreianov, B., Goatin, P., Seguin, N.: Finite volume schemes for locally constrained conservation laws. Numerische Mathematik 115, 609–645 (2010)
2. Bressan, A.: Hyperbolic systems of conservation laws. Oxford Lecture Series in Mathematics and its Applications, vol. 20. Oxford University Press, Oxford (2000)
3. Colombo, R.M.: Wave front tracking in systems of conservation laws. Appl. Math. 49(6), 501–537 (2004)
4. Colombo, R.M., Goatin, P., Rosini, M.D.: Conservation laws with unilateral constraints in traffic modeling. In: Mussone, L., Crisalli, U. (eds.) Transport Management and Land-Use Effects in Presence of Unusual Demand, Atti del convegno SIDT 2009 (June 2009)
5. Colombo, R.M., Goatin, P., Rosini, M.D.: On the modelling and management of traffic. ESAIM: Mathematical Modelling and Numerical Analysis 45(05), 853–872 (2011)

6. Dafermos, C.M.: Polygonal approximations of solutions of the initial value problem for a conservation law. J. Math. Anal. Appl. 38, 33–41 (1972)
7. Holden, H., Risebro, N.H.: *Front tracking for hyperbolic conservation laws*. Applied Mathematical Sciences, vol. 152. Springer, New York (2002)
8. Lax, P.D.: Weak solutions of nonlinear hyperbolic equations and their numerical computation. Communications on Pure and Applied Mathematics 7(1), 159–193 (1954)
9. LeVeque, R.J.: Finite volume methods for hyperbolic problems. Cambridge Texts in Applied Mathematics. Cambridge University Press, Cambridge (2002)

Chapter 14
Non–equilibrium Traffic Models

Abstract. This chapter deals with the PW model and the AR model, that represent the first two non–equilibrium models for vehicular traffic.

14.1 Introduction

Equilibrium models for traffic flows restrict the traffic dynamics to the curve of equilibrium states by imposing a constitutive equation that gives a speed–density relation. In reality, however, traffic is generally observed in non–equilibrium, see Fig. 10.1. Non–equilibrium theories attempt to relax such restriction through substituting the speed–density relation with a partial differential equation that describes the rate of change of travel speed, so that non–equilibrium sates are allowed. The most well know non–equilibrium models are the PW model [10, 11] and the AR model [1], that will be presented in this chapter.

14.2 Generalized PW Models

Lighthill and Whitham end their article [7] recognizing the limits of the LWR model in describing unstable light traffic flows. Their suggestion was to investigate the structure of shocks by introducing "higher–order" relations, akin to the conservation of momentum in fluid dynamics. The firsts to follow this suggestion were Payne [10] and later Whitham [11], who presented a non–equilibrium traffic model of the form

$$\partial_t \rho + \partial_x (\rho\, v) = 0 \tag{14.2.1a}$$

$$\partial_t v + v\, \partial_x v + \frac{1}{\rho}\, \partial_x p(\rho) = \frac{V(\rho) - v}{\tau}\,, \tag{14.2.1b}$$

M.D. Rosini: *Macroscopic Models for Vehicular Flows & Crowd Dynamics*, UCS, pp. 175–190.
DOI: 10.1007/978-3-319-00155-5_14

where $p(\rho)$ is the "pressure" with $p'(\rho) \geq 0$ and $\left(2\, p'(\rho) + \rho\, p''(\rho)\right) > 0$, $\tau \in \mathbb{R}_+$ is the relaxation time and $V(\rho)$ is the generalized equilibrium velocity, meant to mimic driver's maximum out–of–danger velocity and given by a steady–state relation such as, for instance, equations in (10.1.2) used in the case of equilibrium traffic models.

In the acceleration equation, namely (14.2.1b), $v\, \partial_x v$ is the convection term and describes changes in the mean velocity due to inflowing and outflowing vehicles; the term $\rho^{-1}\, \partial_x p(\rho)$ represents the anticipation and describes the driversŠ anticipation on spatially changing traffic conditions downstream; finally $\tau^{-1}(V(\rho) - v)$ describes the tendency of traffic flow to relax to an equilibrium velocity V. In fact, the material derivative $\partial_t v + v\, \partial_x v$ expresses the acceleration of the vehicles and responds negatively to the increase of concentration downstream, and positively (respectively, negatively) to travel speeds that are lower (respectively, higher) than the corresponding equilibrium speed for the same concentration. As a result, travel speed v usually differs from the equilibrium speed V, but this difference is reduced over time because of the relaxation effects. The parameter τ gives the strength of relaxation and can be interpreted as driver reaction time.

Various expressions for p have been proposed, see [8] for an overview. Here we recall

$$\begin{aligned} &\text{Payne [10]}: \quad && p'(\rho) = \frac{1}{2\,\tau}\, |V'(\rho)| \\ &\text{Whitham [11]}: \quad && p'(\rho) = D \qquad \text{for a suitable constant } D \in \mathbb{R}_+\,. \end{aligned}$$

Once we assume a constant decrease–rate of the equilibrium velocity V, then the so called **PW model** can be derived by system (14.2.1), which is referred to as the **generalized PW model**. In this case the above two definitions of p coincide and

$$p(\rho) = C_o^2\, \rho\,,$$

where $C_o \in \mathbb{R}_+$ is the **traffic sound speed**.

Aiming to write (14.2.1) in the conservative form, we introduce the new variable

$$q = \rho\, v\,,$$

that has a clear physical meaning: it is the traffic flow.

Remark 14.1. As pointed out in Example 3.3, the notion of weak solution is not invariant under nonlinear transformations of the independent variables. More precisely, two conservation laws resulting by a nonlinear transformation of the independent variables are equivalent only in the class of smooth solutions, but not in the class of weak solutions. In the present case, the choice of the "correct" form of the acceleration equation do not come from physical arguments, since it is not derived from any physical principle but from a constitutive model. As such, the choice of a second independent variable beside ρ is somewhat arbitrary and has only an analytical meaning. □

By choosing ρ, q as the new independent variables, system (14.2.1) writes [5]

$$\partial_t u + \partial_x f(u) = G(u) , \tag{14.2.2a}$$

where

$$u = \begin{pmatrix} \rho \\ q \end{pmatrix}, \quad f(u) = \begin{pmatrix} q \\ \dfrac{q^2}{\rho} + p(\rho) \end{pmatrix}, \quad G(u) = \begin{pmatrix} 0 \\ \dfrac{\rho\, V(\rho) - q}{\tau} \end{pmatrix}. \tag{14.2.2b}$$

According to condition (D.2) of Sect. 9.1, Eq. (14.2.2) is a balance law rather than a conservation law. The quasilinear form of system (14.2.2) is

$$\partial_t u + a(u)\, \partial_x u = G(u) , \tag{14.2.3a}$$

where

$$a(u) = \begin{pmatrix} 0 & 1 \\ -\dfrac{q^2}{\rho^2} + p'(\rho) & \dfrac{2\,q}{\rho} \end{pmatrix}. \tag{14.2.3b}$$

The eigenvalues and the corresponding right eigenvectors of the matrix $a(u)$ are

$$\lambda_1(u) = \frac{q}{\rho} - \sqrt{p'(\rho)} \qquad \lambda_2(u) = \frac{q}{\rho} + \sqrt{p'(\rho)} \tag{14.2.4a}$$

$$r_1(u) = \begin{pmatrix} 1 \\ \lambda_1(u) \end{pmatrix} \qquad r_2(u) = \begin{pmatrix} 1 \\ \lambda_2(u) \end{pmatrix}. \tag{14.2.4b}$$

By the assumptions on p, system (14.2.3) is strictly hyperbolic and condition (R.1) of Sect. 9.1 is satisfied. Moreover, the characteristic fields are genuinely nonlinear because

$$-\nabla\lambda_1(u) \cdot r_1(u) = \nabla\lambda_2(u) \cdot r_2(u) = \frac{\sqrt{p'(\rho)}}{\rho} + \frac{p''(\rho)}{2\sqrt{p'(\rho)}} .$$

Although both generalized PW model (14.2.1) and equilibrium model (10.1.1) are hyperbolic, they differ in a number of significant ways:

(1) an equilibrium model is a conservation law, while a generalized PW model is a balance law;
(2) differently from any equilibrium model, a generalized PW model can be unstable [11];
(3) an equilibrium model has a unique characteristic speed and therefore it has one kind of shocks and one kind of rarefaction waves, while a generalized PW model has two characteristic speeds and consequently it has two kinds of shocks and two kinds of rarefaction waves;
(4) the characteristic speed $\lambda(\rho) = v(\rho) + \rho\, v'(\rho)$ of an equilibrium model is always slower than traffic, while a generalized PW model has a characteristic

speed $\lambda_1(u)$ always slower than traffic and a characteristic speed $\lambda_2(u)$ always *faster* than traffic;

(5) condition (R.3) of Sect. 9.1 is satisfied by any equilibrium model, but not by a generalized PW model.

The above mentioned differences have profound consequences on the behavior of the corresponding solutions. First, differently from an equilibrium model, a generalized PW model can simulate vehicle clusters effect in traffic when it is unstable [4]. Furthermore, only in the case of generalized PW models the waves may grow in magnitude and eventually become shocks in the form of **roll waves**, namely a series of smooth, monotonic profiles separated by jumps, as it is proved in [6, 11] for the case of the PW model, $p'(\rho) \equiv C_o^2$. Recall that roll wave solutions are particularly interesting because of their similarity to experimentally observed stop–and–go waves. However, a generalized PW model allows that part of the information always travels *faster* than the velocity $v = q/\rho$ of vehicles! This is due to the fact that any disturbance propagates at both characteristic speeds reaching drivers from front and *behind*. This imply, that the vehicles behind have effect on the vehicles ahead. The consequence is an unrealistic behavior of the vehicles in certain circumstances, as it is more clear, for instance, in the next example.

Example 14.1 (see [1, Appendix]). Assume that in front of a driver there is a more dense traffic traveling with higher speed, see Fig. 14.1. In this case, the density $\rho = \rho(t,x)$ is increasing with respect to x, but decreasing with respect to $(x - v\,t)$. Therefore, a generalized PW model (14.2.1) predicts that this driver would slow down and even reach negative velocities! On the contrary, any reasonable real driver would accelerate, since the denser traffic travels faster than him. □

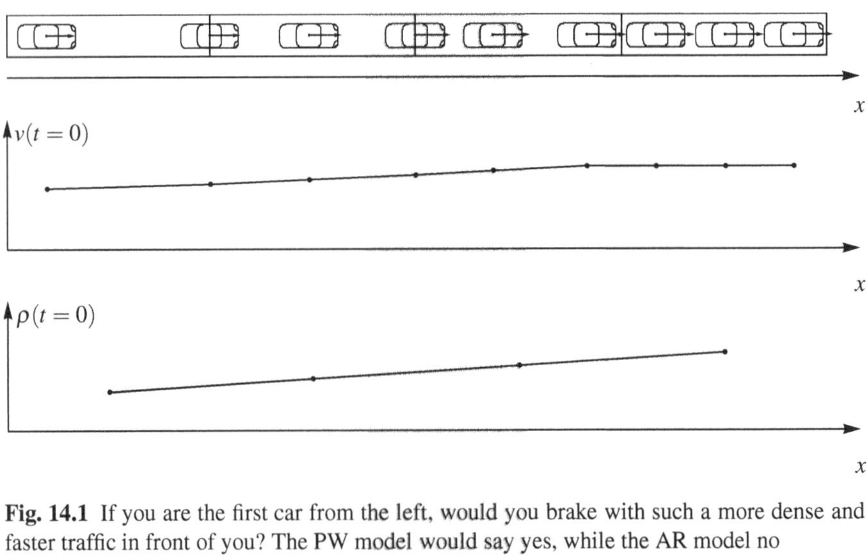

Fig. 14.1 If you are the first car from the left, would you brake with such a more dense and faster traffic in front of you? The PW model would say yes, while the AR model no

Example 14.2. Consider the case when there is a traffic jam ahead $x = 0$ as in the Example 10.1, but now assume that no car moves towards it, namely consider the initial datum

$$\rho(0,x) = \begin{cases} 0 & \text{if } x < 0 \\ \rho_m & \text{if } x \geq 0 \end{cases} \qquad v(0,x) = 0 \qquad x \in \mathbb{R}\,.$$

The correct solution

$$\rho(t,x) = \begin{cases} 0 & \text{if } x < 0 \\ \rho_m & \text{if } x \geq 0 \end{cases} \tag{14.2.5}$$

is then selected by the LWR model, see (10.2.4) with $\rho_l = 0$, but not by the PW model. In fact, (14.2.5) does not satisfy (14.2.1). □

Daganzo [2] presented a neat study of the severe drawbacks of the generalized PW models, and essentially concluded by rejecting those models for good reasons which are recalled below. The drawbacks of the generalized PW models are that they do not remove all the shocks. Furthermore, they are derived as approximation to the microscopic car–following model neglecting terms of order higher than one. This argument might be successful if speeds and spacings for a line of vehicles vary slowly with position and time during a reaction time. Indeed, in this case speeds and spacings can be closely approximated by smooth functions of space and time. However, this is not the case and higher–order term should be included. Also, they include an unrealistic relaxation mechanism for speed adjustments, since they imply that the desired speed distribution is a property of the road and not of the drivers [9]. One of the characteristic speed is greater than the macroscopic fluid velocity. Roughly speaking, this means that the future conditions of a traffic platoon are, in part, determined by what is happening behind it, and this is highly undesirable, especially because it happens also under heavy traffic, when passing is not allowed. This is in contrast with the condition (D.1) of Sect. 9.1.

14.3 AR Model

To repair the PW model, Aw and Rascle [1] and later (independently) Zhang [12] replace the space derivative of the "pressure" in (14.2.1b) by a convective derivative, $\partial_t + v\,\partial_x$. The resulting model is called the **AR model** and writes

$$\partial_t \rho + \partial_x(\rho\, v) = 0 \tag{14.3.1a}$$

$$\partial_t \left(v + p(\rho)\right) + v\,\partial_x \left(v + p(\rho)\right) = 0\,. \tag{14.3.1b}$$

Here, the function p is a smooth increasing function satisfying the following assumptions:

$$\begin{aligned} &p(\rho) \sim \rho^\gamma \text{ for } \rho \text{ small, } \gamma \in \mathbb{R}_+ \\ &2\,p'(\rho) + \rho\, p''(\rho) > 0 \text{ for all } \rho\,. \end{aligned} \tag{14.3.1c}$$

As prototype of functions p we can take

$$p(\rho) = \rho^\gamma , \qquad \gamma \in \mathbb{R}_+ .$$

By subtracting the Eq. (14.3.1a) multiplied by p' to the Eq. (14.3.1b), we obtain the quasilinear equation

$$\partial_t u + a(u)\, \partial_x u = 0_2 , \tag{14.3.2a}$$

where

$$u = \begin{pmatrix} \rho \\ v \end{pmatrix} , \qquad a(u) = \begin{pmatrix} v & \rho \\ 0 & v - \rho\, p'(\rho) \end{pmatrix} . \tag{14.3.2b}$$

The eigenvalues and the corresponding right eigenvectors of the matrix $a(u)$ are

$$\lambda_1(u) = v - \rho\, p'(\rho) \qquad \lambda_2(u) = v \tag{14.3.3a}$$

$$r_1(u) = \frac{1}{2p'(\rho) + \rho\, p''(\rho)} \begin{pmatrix} -1 \\ p'(\rho) \end{pmatrix} \qquad r_2(u) = \begin{pmatrix} 1 \\ 0 \end{pmatrix} . \tag{14.3.3b}$$

Therefore condition (R.1) of Sect. 9.1 is satisfied for $\rho \in \mathbb{R}_+$. More precisely, the system (14.3.2) is strictly hyperbolic, except for $\rho = 0$, where the two eigenvalues coalesce.

The first characteristic field is genuinely nonlinear and the second characteristic field is linearly degenerate because

$$\nabla\lambda_1(u) \cdot r_1(u) = 1 \qquad , \qquad \nabla\lambda_2(u) \cdot r_2(u) = 0 .$$

Therefore, depending on the data, the waves of the first family are either rarefaction or shock waves, while the waves of the second family are always contact discontinuities. The corresponding Riemann invariants are

$$w_1(u) = v + p(\rho) \qquad , \qquad w_2(u) = \lambda_2(u) . \tag{14.3.4}$$

For any fixed $u^l = (\rho^l, v^l)^T$, $u^r = (\rho^r, v^r)^T \in \mathbb{R}_+^* \times \mathbb{R}_+^*$, with $u^l \neq u^r$, we want to study the solution of the Riemann problem associated to (14.3.2) with initial datum of the form

$$u(0,x) = \begin{cases} u^l & \text{if } x < 0 \\ u^r & \text{if } x \geq 0 . \end{cases} \tag{14.3.5}$$

Proposition 14.1. *The solution to the Riemann problem* (14.3.2), (14.3.5) *is an entropy 1–rarefaction iff*

$$w_1(u^l) = w_1(u^r) \qquad \textit{and} \qquad \rho^r < \rho^l . \tag{14.3.6}$$

In this case, the solution is

$$u(t,x) = \begin{cases} u^l & \text{if } x < \lambda_1(u^l)\,t \\ \phi\,(x/t) & \text{if } \lambda_1(u^l)\,t \leq x < \lambda_1(u^r)\,t \\ u^r & \text{if } x \geq \lambda_1(u^r)\,t\,, \end{cases} \tag{14.3.7a}$$

where

$$\phi(\xi) = \begin{pmatrix} \left(p(\rho^l) + \dfrac{\lambda_1(u^l) - \xi}{\gamma+1}\right)^{1/\gamma} \\ v^l + \dfrac{\xi - \lambda_1(u^l)}{\gamma+1} \end{pmatrix}. \tag{14.3.7b}$$

Proof. By Theorem 7.1 the solution to the Riemann problem (14.3.2), (14.3.5) is an entropy 1–rarefaction iff

$$w_1(u^l) = w_1(u^r) \qquad \text{and} \qquad \lambda_1(u^r) > \lambda_1(u^l)\,,$$

that is equivalent to (14.3.6). In order to explicitly compute the solution, we have to solve the system

$$\begin{aligned} &\phi'(\xi) = r_1\left(\phi(\xi)\right) & \xi > \lambda_1(u^l) \\ &\phi\left(\lambda_1(u^l)\right) = u^l\,. \end{aligned}$$

By the assumptions (14.3.1c) on p and (14.3.3), the above system can be easily solved and (14.3.7) is proved to be its unique solution. □

Proposition 14.2. *The solution to the Riemann problem* (14.3.2), (14.3.5) *is an entropy 1–shock iff*

$$w_1(u^l) = w_1(u^r) \qquad \textit{and} \qquad \rho^r > \rho^l\,. \tag{14.3.8}$$

In this case, the solution is

$$u(t,x) = \begin{cases} u^l & \text{if } x < \sigma\,t \\ u^r & \text{if } x \geq \sigma\,t\,, \end{cases} \tag{14.3.9a}$$

where

$$\sigma = \frac{\rho^r\,v^r - \rho^l\,v^l}{\rho^r - \rho^l}\,. \tag{14.3.9b}$$

Proof. In order to study the 1–shock solutions, the system (14.3.1) has to be written in a conservative form. By adding the Eq. (14.3.1a) multiplied by $\left(v + p(\rho)\right)$ to the Eq. (14.3.1b) multiplied by ρ, we obtain the conservation law

$$\partial_t y + \partial_x f(y) = 0_2\,, \tag{14.3.10a}$$

where

$$y = \begin{pmatrix} \rho \\ \rho\ (v + p(\rho)) \end{pmatrix}, \qquad f(y) = v(y)\, y\,. \tag{14.3.10b}$$

Let us underline that the choice of the conserved quantities is far from being indifferent, see Example 3.3. Here, as in [1], we choose $y_1 = \rho$ and the "momentum" $y_2 = \rho\ (v + p(\rho))$ as independent variables and deduce their conservation. As a consequence, the velocity v has to be expressed as a function of y, namely $v(y) = -p(y_1) + y_2/y_1$. The corresponding Rankine–Hugoniot conditions (7.1.8) can be written as

$$\begin{aligned} \left[y_2^r - y_1^r\, p(y_1^r)\right] - \left[y_2^l - y_1^l\, p(y_1^l)\right] &= \sigma\, \left[y_1^r - y_1^l\right] \\ \left[\left(\frac{y_2^r}{y_1^r} - p(y_1^r)\right)\, y_2^r\right] - \left[\left(\frac{y_2^l}{y_1^l} - p(y_1^l)\right)\, y_2^l\right] &= \sigma\, \left[y_2^r - y_2^l\right], \end{aligned} \tag{14.3.11}$$

and the Lax conditions (7.3.16) are given by

$$\begin{aligned} \sigma &< \frac{y_2^l}{y_1^l} - p(y_1^l) - y_1^l p'(y_1^l) \\ \frac{y_2^r}{y_1^r} - p(y_1^r) - y_1^r p'(y_1^r) < \sigma &< \frac{y_2^r}{y_1^r} - p(y_1^r)\,. \end{aligned} \tag{14.3.12}$$

Going back to the original variables, (14.3.11) becomes

$$\left[\rho\,(v-\sigma)\right] = 0 \tag{14.3.13a}$$

$$\rho^l\left(v^l - \sigma\right)\left[v + p(\rho)\right] = 0\,, \tag{14.3.13b}$$

and (14.3.12)

$$\sigma < \lambda_1(u^l) \tag{14.3.14a}$$

$$\lambda_1(u^r) < \sigma < \lambda_2(u^r)\,. \tag{14.3.14b}$$

By (14.3.14a) it follows that $(v^l - \sigma)$ is positive, and therefore (14.3.13b) is equivalent to

$$w_1(u^r) = w_1(u^l)\,,$$

and together with (14.3.12), we obtain that the solution to the Riemann problem (14.3.2), (14.3.5) is an entropy 1–shock iff (14.3.8) holds. □

We underline that system (14.3.2) is a **Temple system**, that is the shock and rarefaction curves coincide.

Proposition 14.3. *The solution to the Riemann problem* (14.3.2), (14.3.5) *is a 2–contact discontinuity iff*

$$v^l = v^r\,. \tag{14.3.15}$$

In this case, the solution is

$$u(t,x) = \begin{cases} u^l & \text{if } x < v^l\, t \\ u^r & \text{if } x \geq v^l\, t\,. \end{cases} \tag{14.3.16}$$

Proof. By Theorem 7.3 the solution to the Riemann problem (14.3.2), (14.3.5) is a 2–contact discontinuity iff $w_2(u^l) = w_2(u^r)$ or, equivalently, iff (14.3.15) holds. □

Observe that a 2–contact discontinuity is discontinuous only in the ρ–coordinate.

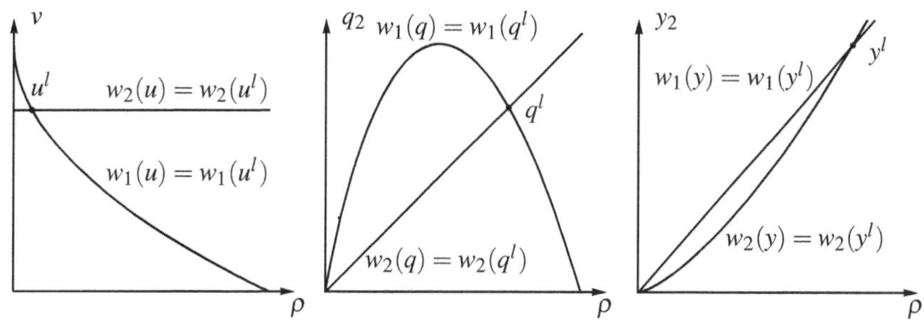

Fig. 14.2 Riemann invariants w_1, w_2 defined in (14.3.4) and represented in different coordinates, in the case $\gamma < 1$

We draw in Fig. 14.2 curves like $w_1(u) = w_1(u^l)$ in the u, $q = (\rho, \rho\, v)^T$, and y planes. We note that the correspondence between u and q or y are no longer one to one at the vacuum $\rho = 0$. It is sometimes more convenient to treat this case in the u–plane, taking in mind that the velocity is not defined when $\rho = 0$, and that all the vacuum states $u = (0, v)$ have to be identified with the origin $(0,0)$.

Proposition 14.4. *For any given positive maximal speed* v_m*,*

$$\mathcal{U}(v_m) = \left\{ (\rho, v)^T \in \mathbb{R}_+^* \times \mathbb{R}_+^* \;:\; v + p(\rho) \leq v_m \right\} \tag{14.3.17}$$

is an invariant set for the Riemann problem (14.3.2), (14.3.5).

Proof. Fix $v_m \in \mathbb{R}_+$. We have to check that starting with an arbitrary initial datum in $\mathcal{U}(v_m)$, the solution remains in $\mathcal{U}(v_m)$. Therefore, the goal is to solve (14.3.2) for any Riemann initial datum (14.3.5) in $\mathcal{U}(v_m)$, see Sect. 7.3.3. In general, we first connect the left state u^l to a middle state u_o by a wave of the first family, and then we connect this intermediate state u_o to the right state u^r by a contact discontinuity of the second family, which is always the faster wave. This is to avoid problems of inconsistency of the solutions.

Case 1: $\rho^l, \rho^r \in \mathbb{R}_+$ and $v^r \leq v^l$. In this case, the unique solution is a 1–shock from u^l to the intersection u_o between the strictly decreasing curve $w_1(u) = w_1(u^l)$ with the straight line $w_2(u) = w_2(u^r)$, followed by a 2–contact discontinuity between u_o and u^r, see Fig. 14.4. We can easily check that the alternative

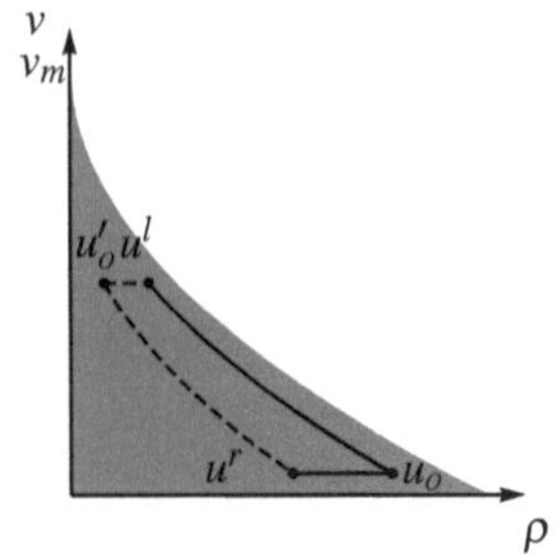

Fig. 14.3 Solid line: The unique solution to the Riemann problem (14.3.2), (14.3.5) in the **Case 1**. Dashed line: An inconsistent solution. Shaded area: The set $\mathscr{U}(v_m)$ defined in (14.3.17).

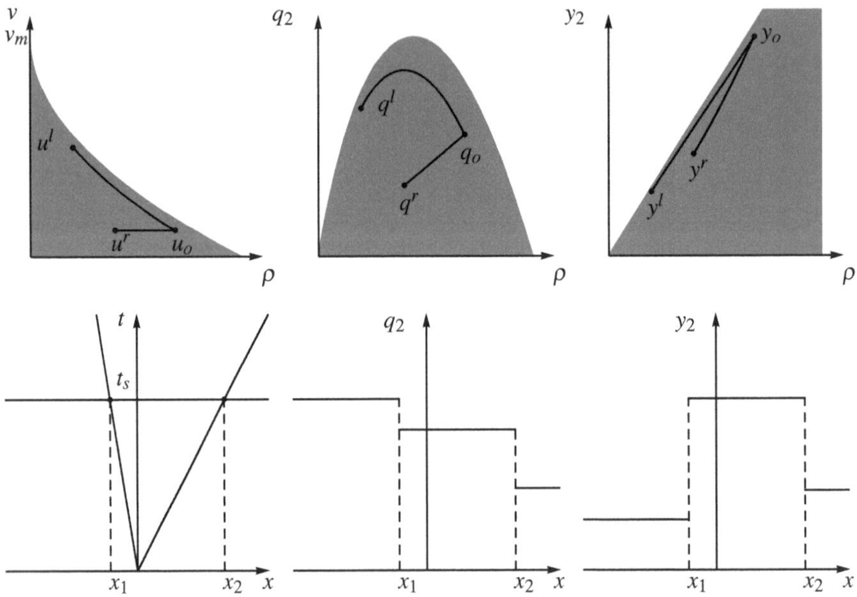

Fig. 14.4 **Case 1**: $u^l, u^r \in \mathscr{U}(v_m)$, with $\rho^r \neq 0 \neq \rho^l$ and $v^r \leq v^l$. In this case the solution is a 1–shock connecting u^l and u_o, followed by a 2–contact discontinuity connecting u_o and u^r.

solution would start with a 2–contact discontinuity, which would be faster than the subsequent 1–shock, which is of course inconsistent, see Fig. 14.3.

Case 2: $\rho^l, \rho^r \in \mathbb{R}_+$ and $v^l \leq v^r \leq w_1(u^l)$. In this case, the unique solution is a 1–rarefaction from u^l to the intersection u_o between the strictly decreasing curve $w_1(u) = w_1(u^l)$ with the straight line $w_2(u) = w_2(u^r)$, followed by a 2–contact discontinuity between u_o and u^r, see Fig. 14.5.

Let us move to the case where the solution to the Riemann problem (14.3.2), (14.3.5) involves the vacuum $\rho = 0$. We recall that at this point the system (14.3.2) is not strictly hyperbolic, and the velocity is neither physically, nor mathematically defined in the q or y planes.

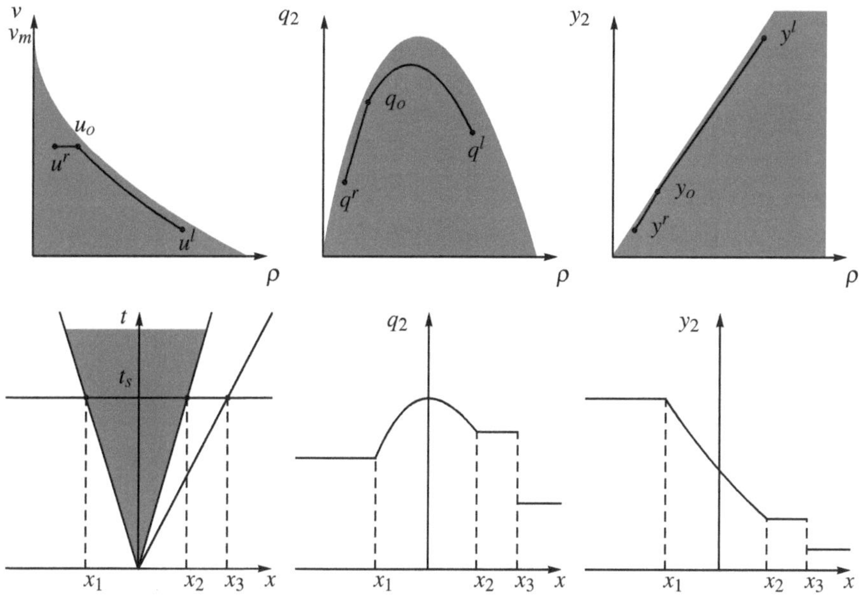

Fig. 14.5 Case 2: $u^l, u^r \in \mathscr{U}(v_m)$, with $\rho^r \neq 0 \neq \rho^l$ and $v^l \leq v^r \leq w_1(u^l)$. In this case the solution is a 1–rarefaction connecting u^l and u_o, followed by a 2–contact discontinuity connecting u_o and u^r.

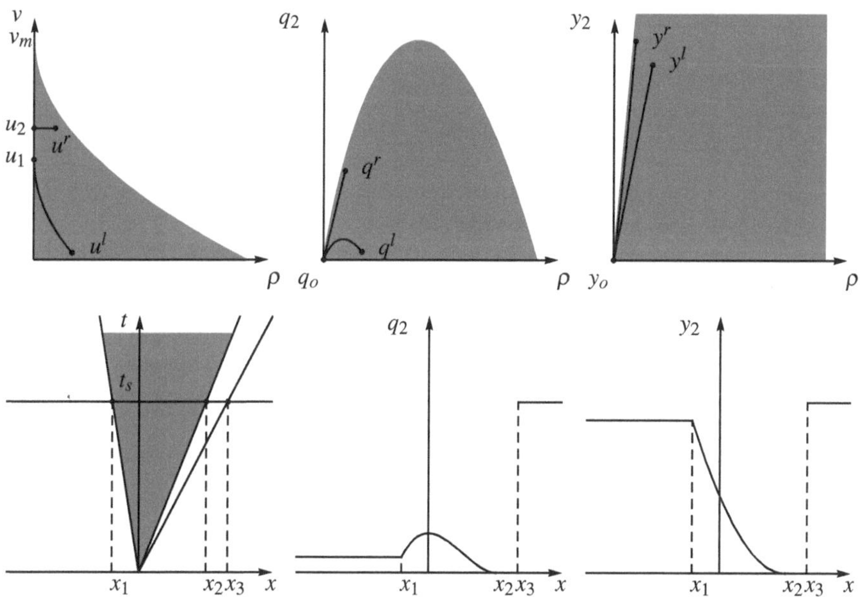

Fig. 14.6 Case 3: $u^l, u^r \in \mathscr{U}(v_m)$, with $\rho^r \neq 0 \neq \rho^l$ and $w_1(u^l) < v^r$. In this case the solution is a 1–rarefaction connecting u^l and u_1, followed by a fake *vacuum wave* which connects two vacuum states u_1 and u_2, followed by a 2–contact discontinuity connecting u_2 and u^r.

Case 3: $\rho^l, \rho^r \in \mathbb{R}_+$ and $w_1(u^l) < v^r$. In this case, the unique solution is a 1–rarefaction from u^l to the intersection u_1 between the strictly decreasing curve $w_1(u) = w_1(u^l)$ with the axis $\rho = 0$, followed by a 2–contact discontinuity connecting $u_2 = (0, v^r)^T$ to u^r, see Fig. 14.6.

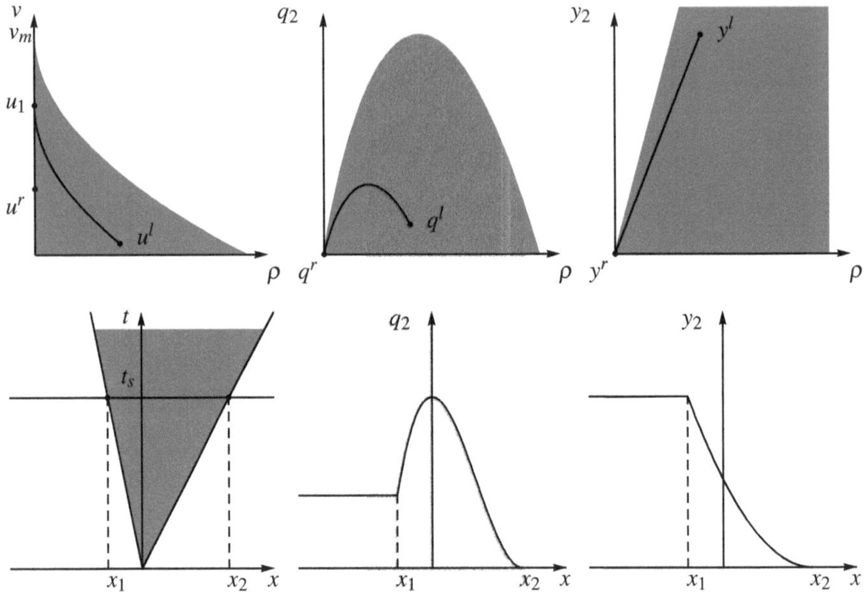

Fig. 14.7 **Case 4**: $q^l, q^r \in \mathscr{U}(v_m)$ and $\rho^r = 0 \neq \rho^l$. In this case the solution is a 1–rarefaction connecting q^l and q^r.

Case 4: $\rho^l \in \mathbb{R}_+$ and $\rho^r = 0$. In this case, the unique solution is a 1–rarefaction from u^l to the intersection u_1 between the strictly decreasing curve $w_1(u) = w_1(u^l)$ with the axis $\rho = 0$, see Fig. 14.7.

Case 5: $\rho^l = 0$ and $\rho^r \in \mathbb{R}_+$. In this case, the unique solution is a 2–contact discontinuity connecting u^l with to $u_1 = (0, v^l)$, see Fig. 14.8.

The above analysis concludes the proof and proves that the solutions of any Riemann problem with initial datum in $\mathscr{U}(v_m)$ takes values in $\mathscr{U}(v_m)$. □

Proposition 14.5. *There is continuous dependence of the solution with respect Riemann data only away from the vacuum.*

Proof. Clearly the solution depends continuously on the data, except near by the vacuum. Indeed, when the Riemann datum involves the vacuum, then such continuous dependence can be lost, as the next examples show.

1. Consider the **Case 5** described in the proof of Proposition 14.4. Assume that the velocity v^l is so big that $v^r < v^l$, $w_1(u^r) < w_1(u^l)$, see Fig. 14.8. Let us slightly perturb the left state and substitute it with $u^l_\varepsilon = (\varepsilon, v^l)$, being $0 < \varepsilon \ll 1$, see Fig. 14.9. Then, the solution consists of a 1–shock wave connecting u^l_ε with

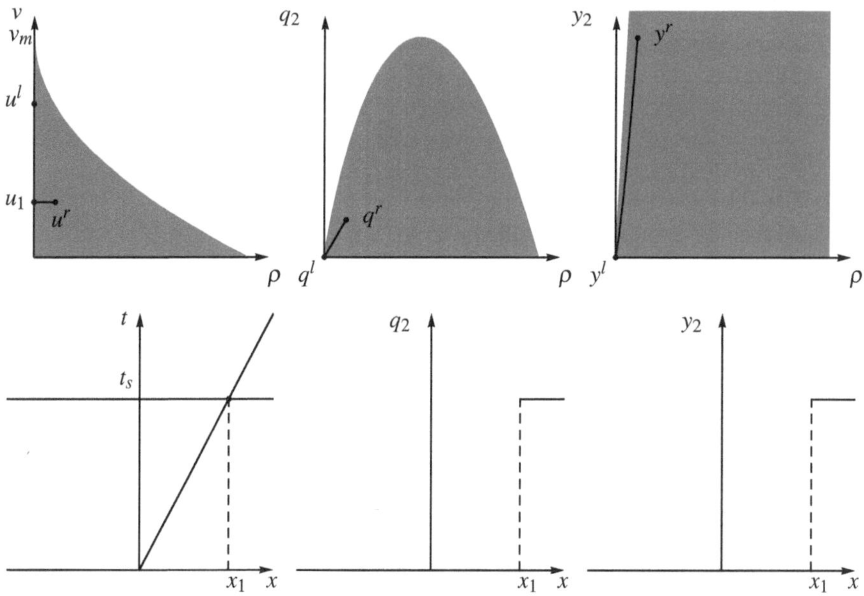

Fig. 14.8 **Case 5**: $q^l, q^r \in \mathscr{U}(v_m)$ and $\rho^l = 0 \neq \rho^r$. In this case the solution is a 2–contact discontinuity connecting q^l with to q^r.

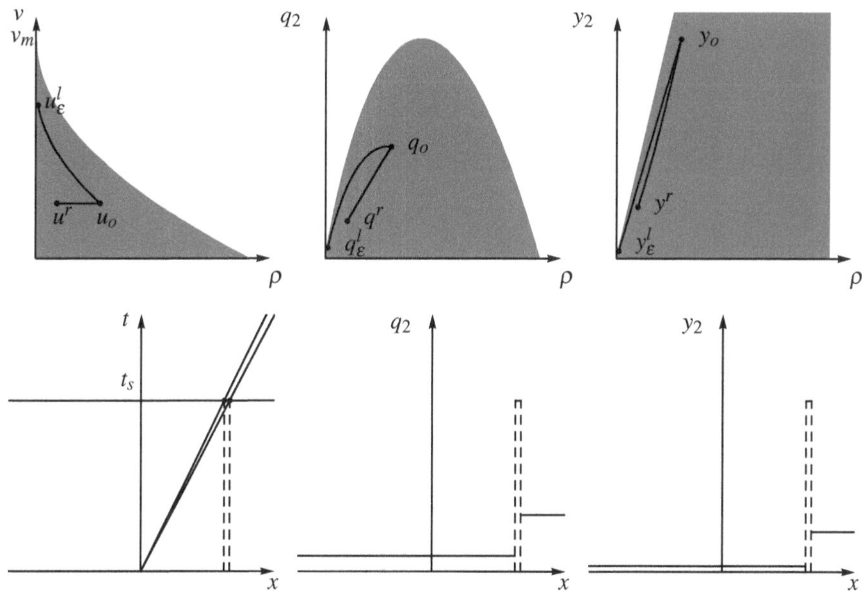

Fig. 14.9 **Case 5 perturbed**: $v^r < v^l$, $w_1(u^r) < w_1(u^l)$. The solution between $q^l_\varepsilon = (\varepsilon, q^l)$ and q^r is a 1–shock connecting q^l_ε and q_o followed by a 2–contact discontinuity connecting q_o and q^r. This is an example of non–continuous dependence of the solution from the initial datum.

the intersection u_o of the decreasing curve $w_1(u) = w_1(u^l)$ and the line $v = v^r$, almost immediately followed by a 2–contact discontinuity connecting u_o to u^r. Therefore, the shock between u^l_ε and u_o has much larger amplitude than the original contact discontinuity between u^l and u^r since by hypothesis $w_1(u^l) = w_1(u_o) > w_1(u^r)$. Therefore, under a small perturbation of the Riemann datum, the solution has dramatically changed and a big oscillation appeared.

2. Consider the **Case 4** described in the proof of Proposition 14.4 and slightly perturb the right state introducing $u^r_\varepsilon = (\varepsilon, v^r)$, being $0 < \varepsilon \ll 1$.

   a. Assume that $w_1(u^l) \geq v^r \geq v^l$. Then, the solution of the perturbed problem consists of a possibly null 1–rarefaction wave between u^l and the intersection u_o of the decreasing curve $w_1(u) = w_1(u^l)$ with $w_2(u) = w_2(u^r)$, followed by a 2–contact discontinuity connecting u_o to u^r_ε, see Fig. 14.10. Therefore, under a small perturbation of the Riemann datum, the solution has dramatically changed and this is the second example of discontinuous dependence with respect to the initial datum.
   b. Assume that $v^l \geq v^r$. Then, the solution of the perturbed problem is still more dramatically different from the original one. Indeed in this case the solution consists of a (possibly large) shock between u^l and the intersection u_o of the decreasing curve $w_1(u) = w_1(u^l)$ and $w_2(u) = w_2(u^r)$, followed by a large 2–contact discontinuity connecting u_o to u^r_ε, see Fig. 14.11.

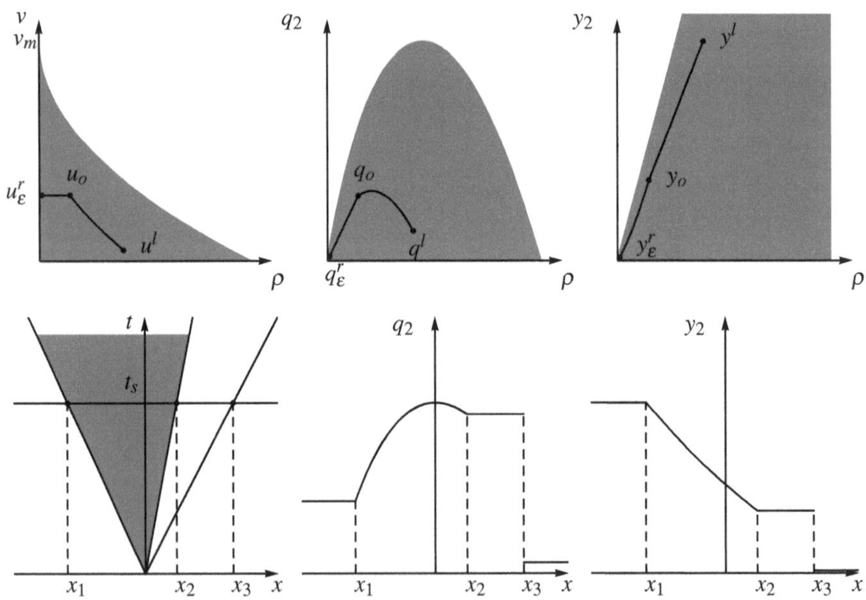

Fig. 14.10 Case 4 perturbed: If $w_1(u^l) \geq v^r \geq v^l$, the solution between $q^r_\varepsilon = (\varepsilon, q^r_2)$ and q^r is a 1–shock connecting q^l_ε and q_o followed by a 2–contact discontinuity connecting q_o and q^r. This is an example of non–continuous dependence of the solution from the initial datum.

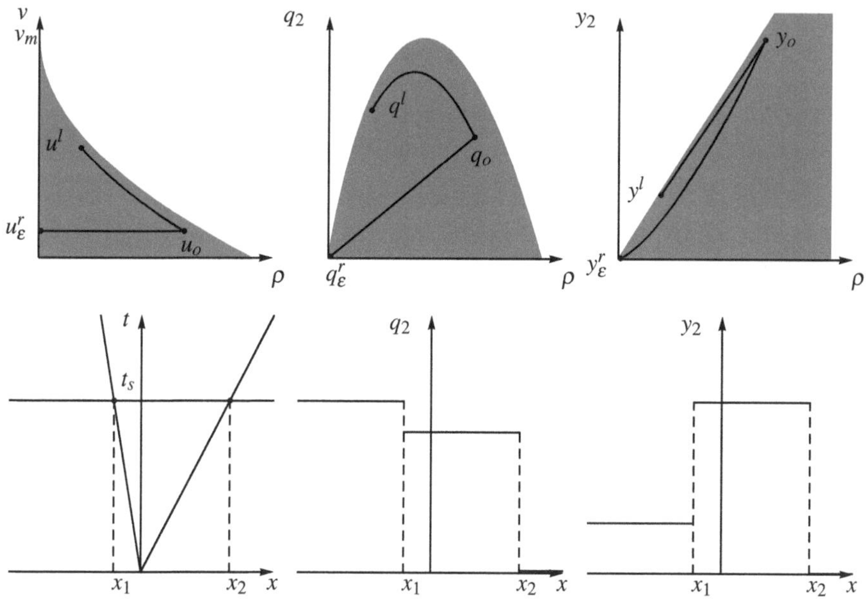

Fig. 14.11 Case 4 perturbed: If $v^l \geq v^r$, the solution between $q^r_\varepsilon = (\varepsilon, q^r_2)$ and q^r is a 1–shock connecting q^l_ε and q_o followed by a 2–contact discontinuity connecting q_o and q^r. This is an example of non–continuous dependence of the solution from the initial datum.

These three examples show that the solution, in general, does not continuously depend on the initial datum. □

We collect the achieved results in the following theorem.

Theorem 14.1. *Consider the three formally equivalent systems* (14.3.1), (14.3.2), (14.3.10).

1. *Any of these systems is strictly hyperbolic, except at the origin, and condition (R.1) of Sect. 9.1 is satisfied.*
2. *For any fixed $v_m \in \mathbb{R}_+$ and function p satisfying* (14.3.1c), *there exists a unique solution to any Riemann problem with datum $q^{lr} \in \mathscr{U}(v_m)$, set defined in* (14.3.17), *and satisfies the conditions (R.2)–(R.5) given in Sect. 9.1:*

   - *The solution remains in $\mathscr{U}(v_m)$ for all the times. As a consequence, the velocity and the density remain non–negative and bounded from above. In particular condition (R.2) of Sect. 9.1 is satisfied.*
   - *The propagation speed of any wave involving a state $u = (\rho, v)^T$ is at most equal to its velocity v. Thus, no information travels faster than the velocity of vehicles and condition (R.3) of Sect. 9.1 is satisfied.*
   - *Condition (R.4) of Sect. 9.1 is satisfied: braking corresponds to a shock, accelerating to a rarefaction.*
   - *Near the vacuum, the solution presents instabilities and also condition (R.5) of Sect. 9.1 is satisfied.*

A drawback of the AR model appears evident in the Case 4 studied in the proof of Proposition 14.4. Indeed, as already observed in [1], in this case the maximal velocity reached is $v^l + p(\rho^l)$, which is greater than the velocity of the cars, v^l.

The AR model has also another drawback [3]: the maximal speed reached by vehicles on an *empty* road depends on the initial data!

References

1. Aw, A., Rascle, M.: Resurrection of "second order" models of traffic flow. SIAM J. Appl. Math. 60(3), 916–938 (2000) (electronic)
2. Daganzo, C.F.: Requiem for high-order fluid approximations of traffic flow. Trans. Res. 29B(4), 277–287 (1995)
3. Goatin, P.: The Aw-Rascle vehicular traffic flow model with phase transitions. Mathematical and Computer Modelling 44(3-4), 287–303 (2006)
4. Jin, W.L., Zhang, H.M.: The formation and structure of vehicle clusters in the Payne-Whitham traffic flow model. Transp. Res. B 37, 207–223 (2003)
5. Jin, W.L., Zhang, H.M.: The formation and structure of vehicle clusters in the Payne-Whitham traffic flow model. Transportation Research Part B: Methodological 37(3), 207–223 (2003)
6. Kuhne, R.D.: Macroscopic Freeway Model for Dense Traffic: Stop-Start Waves and Incident Detection, pp. 20–42 (1984)
7. Lighthill, M.J., Whitham, G.B.: On kinematic waves. II. A theory of traffic flow on long crowded roads. Proc. Roy. Soc. London. Ser. A. 229, 317–345 (1955)
8. Liu, G., Lyrintzis, A.S., Michalopoulos, P.G.: Improved High-Order Model for Freeway Traffic Flow. Transportation Research Record 1644(1), 37–46 (1998)
9. Paveri-Fontana, S.L.: On Boltzmann-like treatments for traffic flow: A critical review of the basic model and an alternative proposal for dilute traffic analysis. Transportation Research 9(4), 225–235 (1975)
10. Payne, H.J.: Models of freeway traffic and control. Math. Models Publ. Sys. Simul. Council Proc. (28), 51–61 (1971)
11. Whitham, G.B.: Linear and nonlinear waves. Pure and Applied Mathematics. Wiley-Interscience, John Wiley & Sons, New York (1974)
12. Zhang, H.M.: A non-equilibrium traffic model devoid of gas-like behavior. Transportation Research Part B: Methodological 36(3), 275–290 (2002)

Part III
Models for Pedestrian Traffic

Chapter 15
General Concepts

Abstract. In this chapter we introduce the general concepts dealing with the description of crowd dynamics characterized by a *large* number of individuals. We also define the so called panic and highlight its dynamic effects, such as the Braess' paradox for pedestrian flows. We finally explain from the modeling point of view the reasons of the fail of the classical theory for conservation laws to attempt at the description of the arise of panic and, consequently, justify the introduction of a non–classical theory.

15.1 Introduction

In the sociological literature, see for instance [29], it is often stated that the irrationality of crowd behavior is a misconception originated in the aristocracy. Only after the French Revolution scientists started to doubt about this theory. For this reason, the study of pedestrian flows has not received particular attention for a long time. Indeed, crowds must behave rationally to conceive the possible existence of equations governing pedestrian flow. This explains why the knowledge of pedestrian flow is at present inadequate and behind that of other transport modes.

A typical phenomenon that characterizes crowd dynamics is the so called "panic". In general, **panic** is considered as a sudden terror which dominates or replaces thinking, see for instance [5, 18, 29, 36, 46]. Both animals and humans are vulnerable to panic. Panic is infectious, in the sense that one individual's panic may easily spread to other members of the group nearby and soon to the entire group. It typically occurs in disaster or violent situations. It is believed to originate from biological responses in the brains and endocrine systems, for instance in the case of herd animals as the response to predators. Often, a large stampede eliminates everything along its path, possibly including some individuals of the escaping group, too. Deaths from stampedes occur primarily from compressive asphyxiation, and usually not from trampling. These accidents are referred to as **crowd crushes**. At the individual level, warning signs of an incumbent crowd crush include the rise of density to roughly

M.D. Rosini: *Macroscopic Models for Vehicular Flows & Crowd Dynamics*, UCS, pp. 193–201.
DOI: 10.1007/978-3-319-00155-5_15

more than four people per square meter. Having this in mind, we "mathematically" individuate the panic by associating it with very high densities, as it will be more clear in the next chapter.

Since decades, the needs of emergency services have been posing several requests to mathematicians, engineers, architects, sociologists, ... aiming at the prevention or, at least, at a rational management of panic. At the design and planning levels, engineers and architects try to prevent the possible rise of panic, usually by preventing congestion locations and determining the most efficient escape routes. In fact, pedestrians evacuating a closed space accumulate near to door exits. The rise of panic may create a dramatic fall in the overall people outflow. The most effective methods adopted to speed up the evacuation of a large room are often non–intuitive. For instance, a tall column placed in front of the door exit may be helpful, as the obstacle reduces the inter–pedestrian pressure in front of the door, decreasing the magnitude of clogging and making the overall outflow higher and more regular. This is known as the **Braess' paradox for pedestrian flows**. Optimal management problems about the shape and the position of such obstacle are crucial issues, but still not completely clear and still under investigation.

Hopefully, modern developments may help prevent some of the approximately two thousand deaths that annually occur in accidents owing to, or related to, crowding. In Table 15.1 we give a non–exhaustive survey of severe crowd accidents. Often, lower evacuation times could have helped preventing or at last diminishing the effects of these accidents.

Human stampedes most often occur during religious pilgrimages, professional sporting and music events. They also often occur in times of mass panic, as a result of a fire or explosion, as people try to get away.

Models for pedestrian flows are currently under intense investigation and are receiving an increasing attention in the specialized literature, which is testified by the growing number of papers published in international journals. In Fig. 15.1 is represented a non–exhaustive survey of the number of papers having in the title the word "pedestrian" and published in Elsevier or Springer journals versus year. Often, these models are of a microscopic nature, i.e. they postulate some rules for the behavior of each individual and then consider many individuals, see [19, 21–24, 27, 35, 38–40, 44] and the references therein. Fewer articles develop continuum, or macroscopic, models, where pedestrians are treated in an aggregate way and detailed interactions are overlooked, as for instance in [1, 3, 4, 15, 17, 25, 26, 30, 42]. The use of continuum models in the context of pedestrian flows is not justified, *a priori*, by the number of individuals, obviously far lower than the typical number of molecules in fluid dynamics. However, only the availability of reliable continuum models allows, *a posteriori*, to state and possibly solve optimal management problems. In fact, the aim of a good macroscopic model is to capture the essence of various features of real pedestrian flows and to reproduce them within an analytically treatable framework. Furthermore, microscopic approaches are computationally expensive, as each individual is represented by an ordinary differential equation to be solved at each time step, and as the number of individuals increases, so does the size of the system to be solved. On the other hand,

Table 15.1 A list with the main crowd accidents occurred in the recent years in the world. (Source: http://en.wikipedia.org/wiki/Stampede)

YEAR	DEAD	CITY	NATION	YEAR	DEAD	CITY	NATION
1711	245	Lyon	France	1998	70	Kathmandu	Nepal
1872	19	Ostrów	Poland	1998	118	Mecca	Saudi Arabia
1876	278	Brooklyn	USA	1999	53	Minsk	Belarus
1883	12	Brooklyn	USA	2001	43	Henderson	USA
1883	180	Sunderland	England	2001	126	Accra	Ghana
1896	1,389	Moscow	Russia	2003	21	Chicago	USA
1903	602	Chicago	USA	2003	100	West Warwick	USA
1908	16	Barnsley	England	2004	194	Buenos Aires	Argentina
1913	73	Michigan	USA	2004	251	Mecca	Saudi Arabia
1941	4,000	Chongqing	China	2005	300	Wai	India
1942	354	Genoa	Italy	2005	265	Maharashtra	India
1943	173	London	England	2005	1,000	Baghdad	Iraq
1946	33	Bolton	England	2006	345	Mecca	Saudi Arabia
1956	124	Yahiko	Japan	2006	74	Pasig City	Philippines
1971	66	Glasgow	England	2006	51	Ibb	Yemen
1979	11	Cincinnati	USA	2007	12	Chililabombwe	Zambia
1982	66	Moscow	Russia	2008	12	Mexico City	Mexico
1985	39	Brussels	Belgium	2008	23	Omdurman	Sudan
1988	93	Tripureswhor	Nepal	2008	147	Jodhpur	India
1989	96	Sheffield	England	2008	162	Himachal Pradesh	India
1990	1,426	Al-Mu'aysam	Saudi Arabia	2008	147	Jodhpur	India
1991	40	Orkney	South Africa	2009	19	Abidjan	Côte d'Ivoire
1991	42	Chalma	Mexico	2010	71	Kunda	India
1993	21	Hong Kong	Cina	2010	63	Amsterdam	Netherlands
1993	73	Madison	USA	2010	21	Duisburg	Germany
1994	270	Mecca	Saudi Arabia	2010	347	Phnom Penh	Cambodia
1994	113	Nagpur	India	2011	102	Kerala	India
1996	82	Guatemala City	Guatemala	2011	16	Haridwar	India

the macroscopic models are computationally less expensive because they have fewer design details in terms of interaction among the pedestrians and between the pedestrians and their environment. As a consequence, macroscopic models minimize the number of equations and parameters. Thus, it is desirable to use macroscopic models if a good model can be found satisfactorily to describe the pedestrian flows.

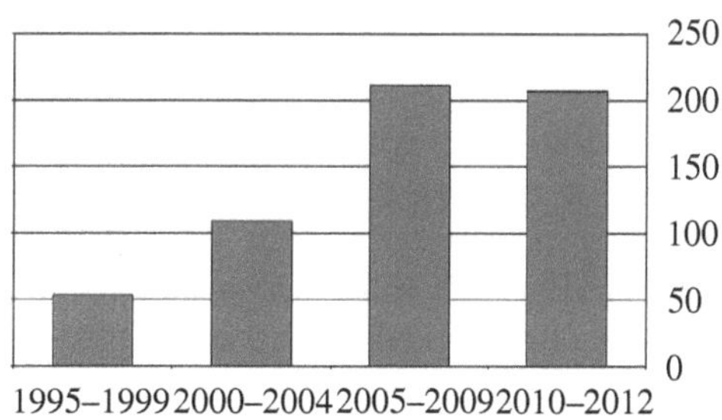

Fig. 15.1 Number of papers having in the title the word "pedestrian" and published in Elsevier or Springer journals versus year

Remark 15.1. The modeling and analysis of pedestrian flows can be performed at different scales. Beyond the microscopic and continuum models, in the literature are also proposed kinetic and cellular automata models, see for instance [2, 6, 32, 37, 43, 45]. □

The understanding and modeling of the multi–scale and multi–physic phenomena involved by crowd–structure interaction make a contribution from different research field necessary to achieve general and conclusive results. For instance, the pedestrian walking behavior has been extensively studied in the field of biomechanics, while the crowd modeling belongs to transportation, physics and applied mathematics research fields and the structural design of efficient facilities for pedestrians to civil engineering. The convergence of these multidisciplinary knowledges would represent a significant advance in the comprehension of the phenomena involved in pedestrian flows.

Applications related to civil engineering and architecture have been among the main practical motivations and final goals of these studies. A recent issue arises by the onset of panic conditions, which substantially modify the crowd dynamics. Therefore, the design of structures, such as stadia grandstands or public buildings, can not be simply based on normal crowd conditions. Indeed, crowd behavior in panic have to be taken into account both for evacuation purpose and to prevent the structural collapse due to congestion phenomena.

Another difficulty in modeling pedestrians movements derive from their behavior of living systems. It is well understood, in the case of crowd, that human and animal behavior follow specific strategies that modify laws of classical mechanics. This is a specific characteristic of all living systems even in the case of low scales such as insects or cells. Therefore, the coupling of living and mechanical systems have to be taken into account in a comprehensive modeling approach. It is worth to recall the two main different aspects which characterize these two type of systems. First, mechanical systems follow rules of continuum mechanics according to conservation laws and are constant in time; while living systems follow rules generated by their self–organized ability in responding the contingent situations. Second, a mechanical system is represented by continuum models, namely by a system with an infinite number of freedom; while a living system is a discrete system, that is, a system with finite degrees of freedom.

Aim of this last part of the book is to describe the **CR model**, proposed by Colombo and Rosini in [14] and further investigated in [7–13, 16, 17, 20, 30, 31, 34, 41, 47], together with its applications. Actually, the CR model is the unique macroscopic model capable to predict the crowd behavior in panic situations. In particular, this model describes the possible overcompressions in a crowd and the fall in the outflow through a door of a panicking crowd jam. The next Sect. 15.2 is addressed to describe the real situations that we want to model and to explain why the classical theory can not be useful. Chapter 16 is devoted to define a proper non–entropic Riemann solver and to analyze the corresponding Cauchy problem. Applications of the model are discussed in Chap. 17, pointing out cases in which the Braess' paradox occurs.

15.2 The Need of a Non–classical Theory

The situation that we want to describe is the evacuation of pedestrians from a narrow corridor or a bridge, mathematically represented by the interval $[0,D]$. It is assumed that the escaping pedestrians have to pass through an exit *"door"* sited at $x = D$. Before reaching it, they have to go through an *"obstacle"* at, say, $x = d$ whose role is to regulate the evacuation process.

A possible benchmarking of crowd dynamics models may be based on the answers that different models give to the following questions:

Panic: When, where, how and why does the panic arise?
Clog doors: When, how and why does the efficiency of the exit fall down?
Braess' paradox: When, how and why is the obstacle helpful in the evacuation?

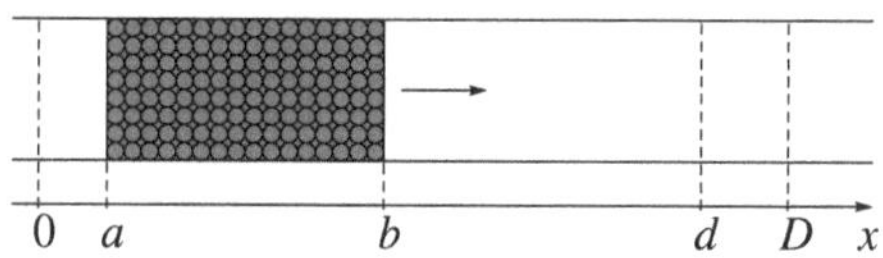

Fig. 15.2 Evacuation of a corridor $[0,D]$ through the exit door at $x = D$. At the initial time pedestrians are uniformly distributed in $[a,b]$. The flow is regulated by an obstacle at $x = d$.

Assume that no crowd crash occurs and that the total number of pedestrians is conserved. We also assume that the average velocity v of the pedestrians at time t and location x is a function of the crowd density $\rho(t,x)$, namely $v = v(\rho)$, so that the crowd flow is $f(\rho) = \rho\, v(\rho)$. Then, we are led to the conservation law

$$\partial_t \rho + \partial_x f(\rho) = 0\,, \qquad (15.2.1)$$

analogous to the classical LWR model [28, 33] for vehicular traffic, see Chap. 10.

One might be now lead to force pedestrian flow to follow the same description provided in the case of vehicular traffic by the classical LWR model. This would amount first to introduce also for pedestrians a speed law and a fundamental diagram, roughly speaking, such as those in Fig. 15.3. Then, the standard classical definitions of entropy weak solutions could be applied. However, the resulting model

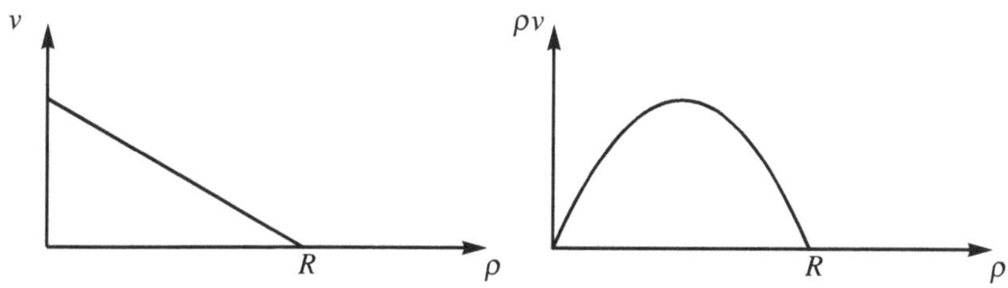

Fig. 15.3 The typical speed–density diagram, left, and the fundamental diagram, right, for the classical LWR model [28, 33] for vehicle traffic, see Chap. 10

would not be able to capture relevant patterns that are typical of crowd dynamics and that are not present in vehicular traffic. In particular, the resulting description of the behavior of pedestrians in panic situations would be hardly acceptable. More than that, the very definition of *panic* would be difficult.

From the analytical point of view, we stress that classical solutions to (15.2.1) satisfy the maximum principle. This elementary analytical result prevents any increase in the maximal density, in contrast with a realistic description of panic, when a sort of *overcompression* arises and is often a cause of major accidents.

The CR model [14] relies on an extension of the interval of the possible crowd densities: beyond the interval $[0,R]$ of the standard densities, the *panic states* $\rho \in]R,R^*]$ are introduced. Therefore, the speed law and the fundamental diagram

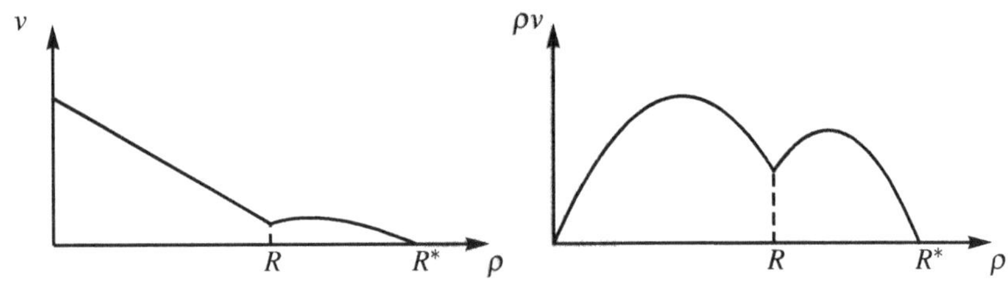

Fig. 15.4 The speed–density diagram, left, and the fundamental diagram, right, used in [14] to model pedestrian flows.

proposed in [14] are of the form as those here displayed in Fig. 15.4. However, to avoid the implications of the maximum principle, also the very definition of solution needs to be suitably modified, as described in detail in the next chapter. From the physical point of view, the main assumption of the CR model was recently experimentally confirmed in [20]. From the analytical point of view, this model is an example of a conservation law in which non–entropy solutions have a physical motivation and a global existence result for the Cauchy problem with large datum is available [16]. Furthermore, the availability of efficient numerical schemes [8, 9] developed ad–hoc to approximate the solutions of the CR model makes it practically usable for simulations of real world starting from real data.

References

1. Bellomo, N.: Modeling complex living systems. In: Modeling and Simulation in Science, Engineering and Technology.Birkhäuser Boston Inc., Boston (2008); A kinetic theory and stochastic game approach
2. Bellomo, N., Bellouquid, A.: On the modelling of vehicular traffic and crowds by kinetic theory of active particles. In: Naldi, G., Pareschi, L., Toscani, G., Bellomo, N. (eds.) Mathematical Modeling of Collective Behavior in Socio-Economic and Life Sciences, Modeling and Simulation in Science, Engineering and Technology, pp. 273–296. Birkhäuser, Boston (2010)

3. Bellomo, N., Dogbé, C.: On the modelling crowd dynamics from scaling to hyperbolic macroscopic models. Math. Models Methods Appl. Sci. 18(suppl.), 1317–1345 (2008)
4. Bellomo, N., Dogbe, C.: On the Modeling of Traffic and Crowds: A Survey of Models, Speculations, and Perspectives. SIAM Rev. 53(3), 409–463 (2011)
5. Berk, R.: Collective Behavior. W. C. Brown Co. (1974)
6. Burlatsky, S., Atrazhev, V., Erikhman, N., Narayanan, S.: A Novel Kinetic Model to Simulate Evacuation Dynamics. In: Klingsch, W.W.F., Rogsch, C., Schadschneider, A., Schreckenberg, M. (eds.) Pedestrian and Evacuation Dynamics 2008, pp. 611–618. Springer, Heidelberg (2010)
7. Chalons, C.: Numerical Approximation of a Macroscopic Model of Pedestrian Flows. SIAM J. Sci. Comput. 29, 539–555 (2007)
8. Chalons, C.: Transport-Equilibrium Schemes for Pedestrian Flows with Nonclassical Shocks. In: Schadschneider, A., Pöschel, T., Kühne, R., Schreckenberg, M., Wolf, D.E. (eds.) Traffic and Granular Flow 2005, pp. 347–356. Springer, Heidelberg (2007)
9. Chalons, C., Goatin, P., Seguin, N.: General constrained conservation laws. Application to pedestrian flow modeling (to appear)
10. Colombo, R.M., Facchi, G., Maternini, G., Rosini, M.D.: On the continuum modeling of crowds. American Mathematical Society (AMS), Providence (2009)
11. Colombo, R.M., Goatin, P., Maternini, G., Rosini, M.D.: Macroscopic Models for Pedestrian Flows. In: Big Events and Transport: The Transportation Requirements for the Management of Large Scale Events, pp. 11–22. IUAV – TTL Research Unit (2010)
12. Colombo, R.M., Goatin, P., Rosini, M.D.: A macroscopic model for pedestrian flows in panic situations. In: Proceedings of the 4th Polish-Japanese Days. GAKUTO International Series. Mathematical Sciences and Applications, vol. 32, pp. 255–272 (2010)
13. Colombo, R.M., Goatin, P., Rosini, M.D.: Conservation laws with unilateral constraints in traffic modeling. In: Mussone, L., Crisalli, U. (eds.) Transport Management and Land-Use Effects in Presence of Unusual Demand, Atti del Convegno SIDT 2009 (June 2009)
14. Colombo, R.M., Rosini, M.D.: Pedestrian flows and non-classical shocks. Math. Methods Appl. Sci. 28(13), 1553–1567 (2005)
15. Colombo, R.M., Rosini, M.D.: Well posedness of balance laws with boundary. J. Math. Anal. Appl. 311(2), 683–702 (2005)
16. Colombo, R.M., Rosini, M.D.: Existence of nonclassical solutions in a Pedestrian flow model. Nonlinear Analysis: Real World Applications 10(5), 2716–2728 (2009)
17. Coscia, V., Canavesio, C.: First-order macroscopic modelling of human crowd dynamics. Math. Models Methods Appl. Sci. 18(suppl.), 1217–1247 (2008)
18. Diethelm, O.: The Nosological Position of Panic Reactions. Am. J. Psychiatry 90(6), 1295–1316 (1934)
19. Farkas, I., Helbing, D., Vicsek, T.: Human waves in stadiums. Phys. A 330(1-2), 18–24 (2003); Randomness and complexity (Eilat, 2003)
20. Helbing, D., Johansson, A., Al-Abideen, H.Z.: Dynamics of crowd disasters: An empirical study. Physical Review E (Statistical, Nonlinear, and Soft Matter Physics) 75(4), 046109 (2007)
21. Helbing, D., Schönhof, M., Stark, H.U., Hoℓyst, J.A.: How individuals learn to take turns: emergence of alternating cooperation in a congestion game and the prisoner's dilemma. Adv. Complex Syst. 8(1), 87–116 (2005)
22. Helbing, D., Siegmeier, J., Lämmer, S.: Self-organized network flows. Netw. Heterog. Media 2(2), 193–210 (2007) (electronic)

23. Hoogendoorn, S., Bovy, P.H.L.: Simulation of pedestrian flows by optimal control and differential games. Optimal Control Appl. Methods 24(3), 153–172 (2003)
24. Hoogendoorn, S., Bovy, P.H.L.: Pedestrian Route-Choice and Activity Scheduling Theory and Models. Transp. Res. B (38), 169–190 (2004)
25. Hughes, R.L.: A continuum theory for the flow of pedestrians. Transportation Research Part B 36, 507–535 (2002)
26. Hughes, R.L.: The flow of human crowds. Annual Review of Fluid Mechanics 35, 169–182 (2003)
27. Johansson, A., Helbing, D., Shukla, P.K.: Specification of the social force pedestrian model by evolutionary adjustment to video tracking data. Adv. Complex Syst. 10(suppl. 2), 271–288 (2007)
28. Lighthill, M.J., Whitham, G.B.: On kinematic waves. II. A theory of traffic flow on long crowded roads. Proc. Roy. Soc. London. Ser. A. 229, 317–345 (1955)
29. McPhail, C.: The Myth of the Madding Crowd. Walter de Gruyter, New York (1991)
30. Piccoli, B., Tosin, A.: Time-evolving measures and macroscopic modeling of pedestrian flow. Preprint, arXiv:0811.3383v1 (submitted)
31. Piccoli, B., Tosin, A.: Pedestrian flows in bounded domains with obstacles. Continuum Mechanics and Thermodynamics (21), 85–107 (2009)
32. Rangel-Huerta, A., Munoz-Melendez, A.: Kinetic theory of situated agents applied to pedestrian flow in a corridor. Physica A: Statistical Mechanics and its Applications 389(5), 1077–1089 (2010)
33. Richards, P.I.: Shock waves on the highway. Operations Res. 4, 42–51 (1956)
34. Rosini, M.D.: Nonclassical interactions portrait in a macroscopic pedestrian flow model. J. Differential Equations 246(1), 408–427 (2009)
35. Schadschneider, A., Kirchner, A., Nishinari, K.: Cellular automata simulation of collective phenomena in pedestrian dynamics. In: Interface and Transport Dynamics. Lect. Notes Comput. Sci. Eng., vol. 32, pp. 390–405. Springer, Berlin (2003)
36. Smelser, N.: Theory of Collective Behavior. Free Pres, New York (1962)
37. Spencer, D.: Cities and Complexity: Understanding Cities with Cellular Automata, Agent-Based Models, and Fractals. The Journal of Architecture 14(3), 446–450 (2009)
38. Sugiyama, Y., Nakayama, A.: Modeling, simulation and observations for freeway traffic and pedestrian. In: Interface and Transport Dynamics. Lect. Notes Comput. Sci. Eng., vol. 32, pp. 406–421. Springer, Berlin (2003)
39. Tajima, Y., Nagatani, T.: Scaling behavior of crowd flow outside a hall. Physica A 292, 545–554 (2001)
40. Takimoto, K., Nagatani, T.: Spatio-temporal distribution of escape time in evacuation process. Physica A 320, 611–621 (2003)
41. Venuti, F., Bruno, L.: Crowd-structure interaction in lively footbridges under synchronous lateral excitation: A literature review. Physics of Life Reviews (6), 176–206 (2009) (in Press, corrected proof)
42. Venuti, F., Bruno, L., Bellomo, N.: Crowd dynamics on a moving platform: mathematical modelling and application to lively footbridges. Math. Comput. Modelling 45(3-4), 252–269 (2007)
43. Yamamoto, K., Kokubo, S., Nishinari, K.: Simulation for pedestrian dynamics by real-coded cellular automata (RCA). Physica A: Statistical Mechanics and its Applications 379(2), 654–660 (2007)
44. Yu, W., Johansson, A.: Modeling crowd turbulence by many-particle simulations. Physical Review E (Statistical, Nonlinear, and Soft Matter Physics) 76(4), 046105 (2007)

45. Yue, H., Guan, H., Zhang, J., Shao, C.: Study on bi-direction pedestrian flow using cellular automata simulation. Physica A: Statistical Mechanics and its Applications 389(3), 527–539 (2010)
46. Zhen, W., Mao, L., Yuan, Z.: Analysis of trample disaster and a case study – mihong bridge fatality in china in 2004. Safety Science 46(8), 1255–1270 (2008)
47. Zheng, X., Zhong, T., Liu, M.: Modeling crowd evacuation of a building based on seven methodological approaches. Building and Environment 44(3), 437–445 (2009)

Chapter 16
The CR Model

Abstract. In this chapter we describe the main characteristics of the CR model. After a brief introduction and a detailed study of all the possible wave interactions, we prove the existence of a weighted total variation that does not increase after any interaction. This is the main ingredient used in the last section to tackle the Cauchy problem through wave front tracking.

16.1 Introduction

In this section we introduce the CR model proposed by Colombo and Rosini [5]. It is expressed by a conservation law of the form

$$\partial_t \rho + \partial_x f(\rho) = 0 ,$$

where ρ is the (mean) density of pedestrians and f is the flow that satisfies the following properties, see Fig. 16.1:

(F.1) The Lipschitzianity of f is a minimal regularity requirement to ensure the finite speed of propagation of the waves (see Remark 3.5):

$$f \in \mathbf{W}^{1,\infty}\left([0,R^*];\mathbb{R}_+^*\right) .$$

(F.2) The flow vanishes iff the density is either zero or maximal:

$$f(\rho) = 0 \text{ iff } \rho \in \left\{0,R^*\right\} .$$

(F.3) Concavity is a standard technical assumption that avoids mixed waves:

the restrictions $f|_{[0,R]}$ and $f|_{[R,R^*]}$ are strictly concave.

(F.4) The maximal flow in standard situations exceeds that in panic:

M.D. Rosini: *Macroscopic Models for Vehicular Flows & Crowd Dynamics*, UCS, pp. 203–226.
DOI: 10.1007/978-3-319-00155-5_16

$$\max\{f(\rho) : \rho \in]0,R[\} > \max\left\{f(\rho) : \rho \in \left]R,R^*\right[\right\} .$$

(F.5) When entering the panic states, there is a small increase in the flow:

$$f \text{ has a local minimum at } \rho = R .$$

(F.6) The flow $f(R)$, i.e. the flow at the standard maximal density, is very small:

$$f(R) < \min\left\{f'(R+)\,R,\ -f'(R-)\,\left(R^* - R\right)\right\} .$$

Remark 16.1. Conditions **(F.4)–(F.6)** are not assumed in the original paper [5] and are not strictly necessary. Nevertheless, their introduction here allows for a sensible reduction of the amount of necessary technicalities and is justified by their explained meanings, see [1–4, 6, 10]. □

For notational convenience, for all $\rho_1, \rho_2 \in [0,R^*]$ with $\rho_1 \neq \rho_2$, we denote the point $(\rho_1, f(\rho_1))$ on the graph $f = f(\rho)$ as $P_f(\rho_1)$ and segment with endpoints $P_f(\rho_1)$ and $P_f(\rho_2)$ as

$$s(\rho_1,\rho_2) = \left\{ (\rho,f) \in \mathbb{R}^2 : \begin{array}{c} f = \dfrac{f(\rho_1)-f(\rho_2)}{\rho_1-\rho_2}\,(\rho-\rho_1)+f(\rho_1) \\ \rho \text{ between } \rho_1 \text{ and } \rho_2 \end{array} \right\} .$$

By the assumptions **(F.3)–(F.5)**, see Fig. 16.1, left, there exist a unique $R_M \in]0,R[$ and a unique $R_M^* \in]R,R^*[$ such that

$$f(R_M) = \max\{f(\rho) : \rho \in]0,R[\} > f(R_M^*) = \max\left\{f(\rho) : \rho \in \left]R,R^*\right[\right\} .$$

Furthermore, by **(F.6)**, there exists a unique $R_4 \in]R,R^*[$ (respectively, $R_1 \in]0,R[$)

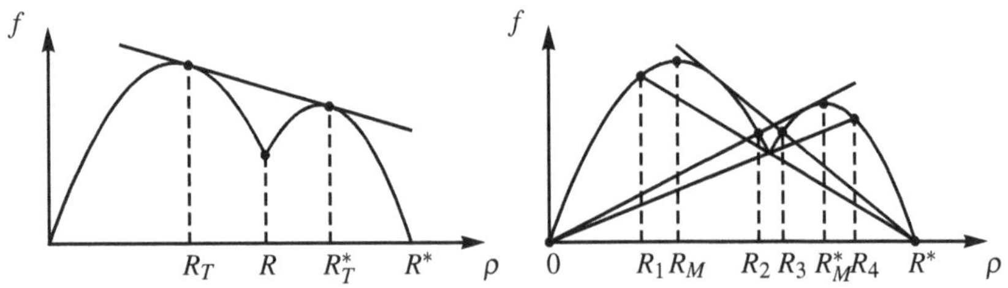

Fig. 16.1 A typical flow function f satisfying conditions **(F.1)–(F.9)** and notations

such that the segment $s(0,R_4)$ (respectively, $s(R_1,R^*)$) intersects $f = f(\rho)$ at the point $P_f(R)$, see Fig. 16.1, right.

Let us further introduce the auxiliary functions ψ, ϕ_ψ and ϕ_R. First, see Fig. 16.2, let $\psi(R) = R$ and, for $\rho \neq R$, let $\psi(\rho)$ be such that the segment $s(\rho, \psi(\rho))$ is

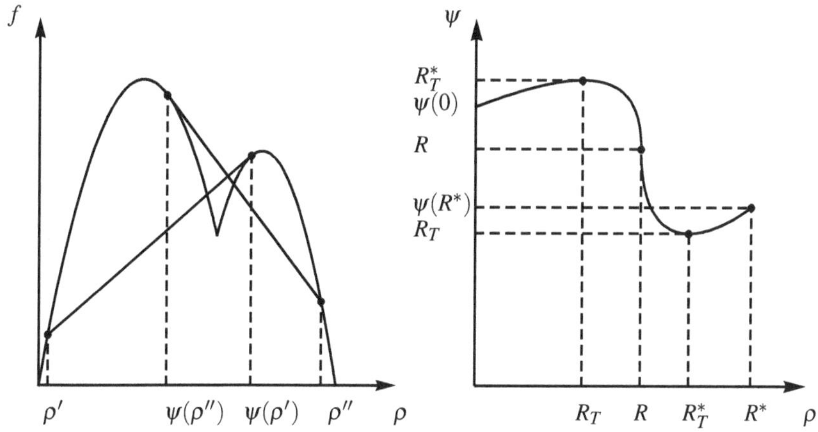

Fig. 16.2 The function ψ: its geometrical meaning, left, and its graph, right

tangent to the graph of f at $P_f\left(\psi(\rho)\right)$. By **(F.6)**, ψ is well defined and $\psi(\rho) \neq R$ for all $\rho \in [0,R^*] \setminus \{R\}$. We assume also that there exists only one couple $(R_T, R_T^*) \in \left]R_M, R\right[\times \left]R_M^*, R^*\right[$ such that $\psi(R_T) = R_T^*$ and $\psi(R_T^*) = R_T$, see Fig. 16.1, left, i.e.:

(F.7) $f\left(\psi(R^*)\right) < -f'(R^*)\left(R^* - \psi(R^*)\right)$.

Then, the segment $s(R_T, R_T^*)$ is the unique segment (of the form $s(\rho_1,\rho_2)$) tangent to $f = f(\rho)$ in two (distinct) points. These assumptions imply that ψ is increasing in $\left[0,R_T\right[\cup \left]R_T^*, R^*\right]$ and decreasing in $\left]R_T, R_T^*\right[$, while $\psi'(R_T) = \psi'(R_T^*) = 0$. Moreover, $R_T < \psi(R^*) < R < \psi(0) < R_T^*$. We now introduce $\phi_\psi : [0,R^*] \to [0,R^*]$. First, we define $\phi_\psi(R_T) = R_T$ and $\phi_\psi(R_T^*) = R_T^*$. If $\rho \in \left[0,R_T\right[\cup \left]R_T^*, R^*\right]$, then by **(F.6)** the segment $s\left(\rho,\psi(\rho)\right)$ intersects the graph of f, say in $P_f\left(\phi_\psi(\rho)\right)$. Let $R_2 = \phi_\psi(0)$ and $R_3 = \phi_\psi(R^*)$, see Fig. 16.1. If $\rho \in \left]R_T, R_2\right] \cup \left[R_3, R_T^*\right[$, then there exists a segment $s\left(\psi(\rho), \phi_\psi(\rho)\right)$ that intersects $f = f(\rho)$ in $P_f(\rho)$. Finally, we define $\phi_\psi(\rho) = 0$ for all $\rho \in \left]R_2, R\right[$ and $\phi_\psi(\rho) = R^*$ for all $\rho \in \left[R, R_3\right[$. By definition, $[\rho \mapsto \phi_\psi(\rho)]$ is a continuous decreasing function in $\left[0,R\right[$ and in $[R,R^*]$. Whether it exists, let $s\left(\rho,\phi_R(\rho)\right)$ be such that its intersection with $f = f(\rho)$ is $P_f(R)$. Let $R_1 = \phi_R(R^*)$ and $R_4 = \phi_R(0)$, see Fig. 16.1, then $\phi_R : [0,R_1] \cup [R_4,R^*] \to [0,R_1] \cup [R_4,R^*]$ is a continuous increasing function.

Finally, we concentrate our attention on the cases in which:

(F.8) $0 < R_1 < R_T$, $2R_2 > \psi(0)$ and $R_T^* < R_4 < R^*$.

(F.9) $[\rho \mapsto (\psi(\rho) - \rho)]$ is strictly decreasing in $[0,R]$.

We now define a non–entropic Riemann solver $\mathscr{R}_{CR}$ suited to the description of crowd dynamics, yielding physically reasonable solutions to all Riemann problems of the form

$$\partial_t \rho + \partial_x f(\rho) = 0 \quad , \qquad \rho(0,x) = \begin{cases} \rho^l & \text{if } x < 0 \\ \rho^r & \text{if } x \geq 0\,. \end{cases} \tag{16.1.1}$$

Recall that for any pair $(\rho^l, \rho^r) \in [0, R^*]^2$, according with Definition 4.5, $\mathscr{R}_{CR}[\rho^l, \rho^r]$ is the self similar weak solution to the Riemann problem (16.1.1) computed at time $t = 1$. Therefore, also the non–entropic shocks introduced by $\mathscr{R}_{CR}$ have to satisfy the Rankine–Hugoniot condition (3.4.4). Introduce two thresholds s and Δs such that

$$0 < \mathrm{s} < R_M\,, \quad \Delta\mathrm{s} > \mathrm{s} \quad \text{and} \quad R > \mathrm{s} + \Delta\mathrm{s} \geq \phi_\psi(\mathrm{s}) > R_T > R - \Delta\mathrm{s}\,. \quad (16.1.2)$$

Sufficient conditions for (16.1.2) to be satisfied are in Lemma 16.1. The solution to the Riemann problem (16.1.1) with $\rho^l, \rho^r \in [0, R^*]$ is selected through the following conditions, see Fig. 16.3, left:

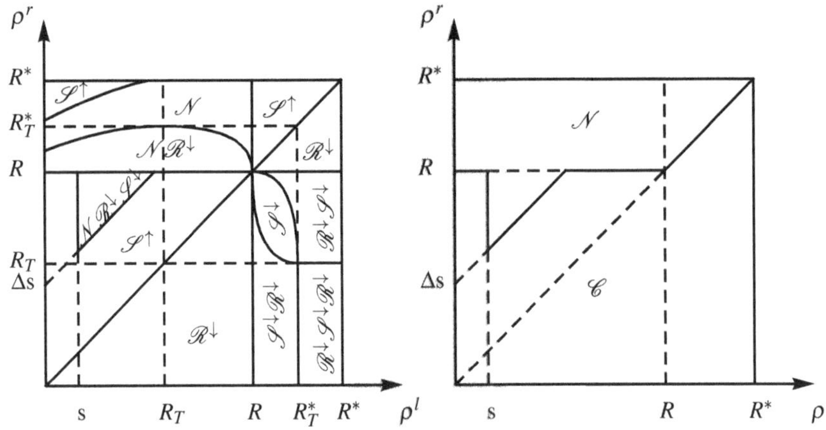

Fig. 16.3 Left: The Riemann solver $\mathscr{R}_{CR}$ selected by **(R.1)**–**(R.4)**. Here, for instance, $\mathscr{N}\mathscr{R}^\downarrow$ indicates that $\mathscr{R}_{CR}[\rho^l, \rho^r]$ is a non–entropic shock followed by a decreasing rarefaction. Right: The notation introduced in (16.1.3).

(R.1) If $\rho^l, \rho^r \in [0, R]$, then $\mathscr{R}_{CR}[\rho^l, \rho^r] = \mathscr{R}_e[\rho^l, \rho^r]$ unless

$$\rho^l > \mathrm{s} \quad \text{and} \quad \rho^r - \rho^l > \Delta\mathrm{s}\,.$$

In this case, $\mathscr{R}_{CR}[\rho^l, \rho^r]$ consists of a non–entropic shock between ρ^l and $\psi(\rho^l)$, followed by the entropy solution between $\psi(\rho^l)$ and ρ^r.

(R.2) If $\rho^r \leq \rho^l$, then $\mathscr{R}_{CR}[\rho^l, \rho^r] = \mathscr{R}_e[\rho^l, \rho^r]$.

(R.3) If $R \leq \rho^l < \rho^r$ or $\rho^l < R < \rho^r$ and the segment $s(\rho^l, \rho^r)$ does not intersect $f = f(\rho)$ for $\rho \neq R$, then the solution is a shock between ρ^l and ρ^r.

(R.4) If $\rho^l < R < \rho^r$ and the segment $s(\rho^l, \rho^r)$ intersects $f = f(\rho)$ for $\rho \neq R$, then $\mathscr{R}_{CR}[\rho^l, \rho^r]$ consists of a non–entropic shock between ρ^l and a panic state, followed by a possibly null entropic wave. More precisely:

(R.4.1) If $\rho^r \in \,]R, \psi(\rho^l)[$, then $\mathscr{R}_{CR}[\rho^l, \rho^r]$ consists of a non–entropic shock between ρ^l and $\psi(\rho^l)$, followed by a decreasing rarefaction between $\psi(\rho^l)$ and ρ^r.

(R.4.2) If $\rho^r \in [\psi(\rho^l), R^*]$, then $\mathscr{R}_{CR}[\rho^l, \rho^r]$ consists of a single non–entropic shock.

Recall that $\mathscr{R}_e$ denotes the entropic Riemann solver that selects the self similar entropy weak solutions, see Sect. 4.6 and Definition 4.5.

Remark 16.2. Conditions **(R.1)** and **(R.4)** imply that a non–entropic shock can not be adjacent to an entropic shock. In fact, any non–entropic shock is left–subsonic, i.e. its speed is smaller than the characteristic speed of the state to its left. Therefore, by the Lax condition (3.6.1), no wave can be on its left. Furthermore, any non–entropic shock defined by **(R.1)** or **(R.4.1)** is right–sonic, i.e. its speed is equal to the characteristic speed of the state to its right. Hence, by the Lax condition (3.6.1), the only wave that can be adjacent to such a non–entropic shock is a rarefaction on its right. Moreover, a non–entropic shock defined by **(R.4.2)** is right–supersonic, i.e. it has a speed higher than the characteristic speed of the state to its right. Thus, also in this case, by the Lax condition (3.6.1), no wave can follow it. □

To state the main properties concerning the Riemann solver $\mathscr{R}_{CR}$, the following subsets of the square $[0,R^*]^2$ are of use, see Fig. 16.3, right:

$$\mathscr{C}_{\mathscr{N}} = \left\{(\rho^l,\rho^r) \in [0,R^*]^2 \colon \rho^l \geq \rho^r \geq R\right\} \tag{16.1.3a}$$

$$\mathscr{N}_{\mathscr{C}} = \left\{(\rho^l,\rho^r) \in [0,R]^2 \colon \rho^l > \mathrm{s} \text{ and } \rho^r - \rho^l > \Delta\mathrm{s}\right\} \tag{16.1.3b}$$

$$\mathscr{C} = \Big([0,R^*] \times [0,R]\Big) \cup \mathscr{C}_{\mathscr{N}} \setminus \mathscr{N}_{\mathscr{C}} \tag{16.1.3c}$$

$$\mathscr{N} = \Big([0,R^*[\times]R,R^*]\Big) \cup \mathscr{N}_{\mathscr{C}} \setminus \mathscr{C}_{\mathscr{N}} \,. \tag{16.1.3d}$$

Theorem 16.1. *Let* $f \,:\, [0,R^*] \to [0,+\infty[$ *satisfy assumptions* ***(F.1)–(F.8)****. Choose thresholds* s *and* Δs *such that* (16.1.2) *holds. Then, there exists a unique Riemann solver* $\mathscr{R}_{CR} \,:\, [0,R^*]^2 \to \mathbf{BV}(\mathbb{R})$ *satisfying* ***(R.1)–(R.4)*** *and such that* $\rho(t,x) = \mathscr{R}_{CR}[\rho^l,\rho^r](x/t)$ *is a weak solution to* (16.1.1)*. Moreover,*

(P1) $\mathscr{R}_{CR}$ *is* $\mathbf{L}^1_{\mathbf{loc}}$*–continuous in* $\mathscr{C}$*, in* $\mathscr{N}$ *and also along the segment* $\rho^l = \rho^r$ *for* $\rho^l \in]R,R^*]$*;*
(P2) $\mathscr{R}_{CR}$ *is consistent in* $\mathscr{C}$ *and in* $\mathscr{N}$*.*

Proof. As a first step, we prove that **(R.1)–(R.4)** uniquely select a weak solution to any Riemann problem (16.1.1). By **(R.1)** and **(R.2)**, the Riemann solver $\mathscr{R}_{CR}$ coincides with $\mathscr{R}_e$ in $\mathscr{C}$. If $(\rho^l,\rho^r) \in \mathscr{N}_{\mathscr{C}}$, then, by **(R.1)**, $\mathscr{R}_{CR}[\rho^l,\rho^r]$ consists of a non–entropic shock from ρ^l to $\psi(\rho^l)$ followed by the entropy solution of the Riemann problem with states $\psi(\rho^l)$ and ρ^r, that is given by a rarefaction between $\psi(\rho^l)$ and $\psi(\rho^r)$, followed by an entropic shock between $\psi(\rho^r)$ and ρ^r. In fact, this juxtaposition is possible because by (16.1.2) we have

$$R > \rho^r > \rho^l + \Delta\mathrm{s} > \mathrm{s} + \Delta\mathrm{s} \geq \phi_\psi(\mathrm{s}) > \phi_\psi(\rho^l) \,.$$

If $R \leq \rho^l < \rho^r \leq R^*$, then $\mathscr{R}_{CR}[\rho^l,\rho^r]$ is an entropic shock by **(R.3)**. The remaining cases to be analyzed are those with $0 \leq \rho^l < R < \rho^r \leq R^*$ and are described by **(R.3)** and **(R.4)**. If $\rho^r < \psi(\rho^l)$, then the solution prescribed by **(R.4.1)** starts with a non–entropic shock which is right–sonic and, consequently, its juxtaposition with a rarefaction starting from its ending state is allowed. If $\psi(\rho^l) \leq \rho^r$, then the solution is given by a single shock, that is non-entropic iff the segment $s(\rho^l, \rho^r)$ intersects $f = f(\rho)$ for a $\rho \neq R$, as described by **(R.4.1)**; otherwise it is entropic, as described by **(R.3)**. Each step in the above construction is uniquely determined by **(R.1)**–**(R.4)**, yielding the uniqueness of the whole Riemann solver $\mathscr{R}_{CR}$. Furthermore, $\mathscr{R}_{CR}[\rho^l, \rho^r]$ is a weak solution to (16.1.1) because the Rankine–Hugoniot condition (3.4.4) are satisfied also by the introduced non–entropic shocks.

The entropic Riemann solver $\mathscr{R}_e$ is known to be $\mathbf{L^1_{loc}}$–continuous and consistent. Therefore it is sufficient to prove **(P1)** and **(P2)** in $\mathscr{N}$.

About **(P1)**, we have to show that all intermediate states and wave speeds in $\mathscr{R}_{CR}[\rho^l, \rho^r]$ are continuous functions of ρ^l and ρ^r for $(\rho^l, \rho^r) \in \mathscr{N}$. Indeed, in this case the various wave speeds are either difference quotients or derivatives of the flow computed at ρ^l, $\psi(\rho^l)$, $\psi(\rho^r)$ or ρ^r, which are continuous functions of (ρ^l, ρ^r) except at most at the point R, where f' is discontinuous. Hence, it is sufficient to study the cases $\rho^l = R$ and $\rho^r = R$. If $\rho^l = R$, then for all $(\rho^l_\varepsilon, \rho^r_\varepsilon) \in \mathscr{N}$ in a sufficiently small neighborhood of (R, ρ^r), $\mathscr{R}_{CR}[\rho^l_\varepsilon, \rho^r_\varepsilon]$ consists of a single shock and its speed approaches that of $\mathscr{R}_{CR}[R, \rho^r]$ as $(\rho^l_\varepsilon, \rho^r_\varepsilon) \to (R, \rho^r)$. If $\rho^r = R$, then, for all $(\rho^l_\varepsilon, \rho^r_\varepsilon) \in \mathscr{N}$ in a neighborhood sufficiently small of (ρ^l, R), $\mathscr{R}_{CR}[\rho^l_\varepsilon, \rho^r_\varepsilon]$ consists of a non–entropic shock between ρ^l_ε and $\psi(\rho^l_\varepsilon)$, followed by either a mixed wave ending with a shock having speed $f'\left(\psi(\rho^r_\varepsilon)\right)$, or by a rarefaction with fastest speed $f'(\rho^r_\varepsilon)$. In the former case, as $\rho^r_\varepsilon \nearrow R$, we have $\psi(\rho^r_\varepsilon) \searrow R$ and therefore $f'\left(\psi(\rho^r_\varepsilon)\right) \nearrow f'\left(\psi(R)+\right) = f'(R+)$. In the latter case, as $\rho^r_\varepsilon \searrow R$ we have $f'(\rho^r_\varepsilon) \nearrow f'(R+)$. Therefore the proof of **(P1)** is concluded.

We now prove **(P2)**, namely that the conditions **(C1)** and **(C2)** of Definition 4.6 hold in $\mathscr{N}$. Fix $(\rho^l, \rho^r) \in \mathscr{N}$. Then by Remark 16.2 in $\mathscr{R}_{CR}[\rho^l, \rho^r]$ there can not be a non–entropic shock adjacent to an entropic shock. Therefore, **(C2)** needs a verification only when the solution to $\mathscr{R}_{CR}[\rho^l, \rho^r]$ contains a rarefaction with $\bar{x}$ in its support. In this case, if $\rho^m = \mathscr{R}_{CR}[\rho^l, \rho^r](\bar{x})$, then $\mathscr{R}_{CR}[\rho^m, \rho^r] = \mathscr{R}_e[\rho^l, \rho^m]$ in $[\bar{x}, +\infty[$, while an inspection of the above construction of $\mathscr{R}_{CR}$ proves that $\mathscr{R}_{CR}[\rho^l, \rho^m]$ coincides with the restriction of $\mathscr{R}_{CR}[\rho^l, \rho^r]$ to $]-\infty, \bar{x}[$. Fix now also ρ^m such that $(\rho^l, \rho^m), (\rho^m, \rho^r) \in \mathscr{N}$. If $(\rho^l, \rho^m) \in \mathscr{N}_{\mathscr{C}}$, then by (16.1.2) we have $R > \rho^m > \mathrm{s} + \Delta\mathrm{s} > R_T$ and therefore ρ^r has to be greater than R. We have to distinguish two cases: $\rho^r > \psi(\rho^m)$ and $\rho^r \leq \psi(\rho^m)$. In both cases does not exist any $\bar{x}$ such that $\mathscr{R}_{CR}[\rho^l, \rho^m](\bar{x}) = \mathscr{R}_{CR}[\rho^m, \rho^r](\bar{x})$. In the first case, the reason is because the two solutions can not be juxtaposed. In the latter case, this is a consequence of the right continuity assumption for the functions in $\mathbf{BV} \cap \mathbf{L}^\infty$, see Remark 2.2, even if the two solutions can be placed side by side. Finally, if $(\rho^l, \rho^m) \in \mathscr{N} \setminus \mathscr{N}_{\mathscr{C}}$, then it has to be $R < \rho^m < \rho^r$ and the corresponding two solutions can not be juxtaposed. This ends the proof of **(C1)** in $\mathscr{N}$. □

Proposition 16.1. *The Riemann solver $\mathscr{R}_{CR}$ is not consistent in $[0, R^*]^2$. More precisely, with reference to Definition 4.6, $\mathscr{R}_{CR}$ satisfies the condition* **(C2)** *but not* **(C1)**.

Proof. The condition **(C2)** is already checked in the proof of Theorem 16.1. Let us consider the following examples:

- $\rho^l, \rho^m \in [0, R[$, $\mathscr{R}_{CR}[\rho^l, \rho^m]$ is an increasing shock, $\rho^r > \psi(\rho^l)$ and the segment $s(\rho^l, \rho^m)$ has slope less than the segment $s(\rho^l, \rho^r)$;
- $\rho^l, \rho^m \in [0, R[$, $\mathscr{R}_{CR}[\rho^l, \rho^m]$ is an increasing shock, $R < \rho^r < \psi(\rho^l)$ and $s(\rho^l, \rho^m)$ has slope less than the segment $s\left(\rho^l, \psi(\rho^l)\right)$;
- $\rho^l, \rho^r \in [0, R[$, $\mathscr{R}_{CR}[\rho^l, \rho^r]$ is an increasing shock and $\psi(\rho^r) < \rho^m = \psi(\rho^l)$.

The above examples show that $\mathscr{R}_{CR}$ does not satisfy the condition **(C1)**. □

Proposition 16.2. *$\mathscr{R}_{CR}$ is not $\mathbf{L}^1_{\mathbf{loc}}$–continuous in $[0, R^*]^2$.*

Proof. Fix $\varepsilon \in \mathbb{R}_+$ sufficiently small. With reference to the following two examples

- $\rho^l \in [0, \mathrm{s}] \cup [\mathrm{s} + \Delta\mathrm{s}, R[$, $\rho^r = R$ and $\rho^r_\varepsilon = R + \varepsilon$;
- $\rho^l \in]\mathrm{s}, \mathrm{s} + \Delta\mathrm{s}[$, $\rho^r = \rho^l + \Delta\mathrm{s}$ and $\rho^r_\varepsilon = \rho^l + \Delta\mathrm{s} + \varepsilon$;

we have that in both cases $\lim_{\varepsilon \searrow 0} \left\| \mathscr{R}_{CR}[\rho^l, \rho^r_\varepsilon] - \mathscr{R}_{CR}[\rho^l, \rho^r] \right\|_{\mathbf{L}^1(\mathbb{R};\mathbb{R})} \neq 0$. Analogously, with reference to the following two examples

- $\rho^r \in]\mathrm{s} + \Delta\mathrm{s}, R]$, $\rho^l = \mathrm{s}$ and $\rho^l_\varepsilon = \mathrm{s} + \varepsilon$;
- $\rho^r \in]\mathrm{s} + \Delta\mathrm{s}, R]$, $\rho^l = \rho^r - \Delta\mathrm{s}$ and $\rho^l_\varepsilon = \rho^r - \Delta\mathrm{s} - \varepsilon$;

we have that in both cases $\lim_{\varepsilon \searrow 0} \left\| \mathscr{R}_{CR}[\rho^l_\varepsilon, \rho^r] - \mathscr{R}_{CR}[\rho^l, \rho^r] \right\|_{\mathbf{L}^1(\mathbb{R};\mathbb{R})} \neq 0$. □

16.2 Study of the Interactions

In this section we study all the possible wave interactions when a *single* left wave connecting two states ρ^l and ρ^m interacts with a *single* right wave connecting two states ρ^m and ρ^r. We use, for instance, the usual notation $\mathscr{R}_{CR}[\rho^l, \rho^m] = \mathscr{N}$ when the left incoming wave is a non–entropic shock. We indicate whether the wave is increasing or decreasing by adding, respectively, an up arrow or a down arrow.

(A) $\mathscr{R}_{CR}[\rho^l, \rho^m] = \mathscr{R}^{\downarrow}$ with $\rho^l \leq R$.

(A.1) $\mathscr{R}_{CR}(\rho^m, \rho^r) = \mathscr{R}^{\downarrow}$. In this case the two waves do not interact.

(A.2) $\mathscr{R}_{CR}(\rho^m, \rho^r) = \mathscr{S}^{\uparrow}$ with $\rho^r \leq R$.

(A.2.a) If $\rho^r \leq \rho^l$, then the outgoing wave is a possible null rarefaction from ρ^l to ρ^r.

(A.2.b) If $\rho^l < \rho^r$ and $\mathscr{R}_{CR}[\rho^l, \rho^r] = \mathscr{S}^{\uparrow}$, then the outgoing wave is an entropic shock from ρ^l to ρ^r.

(A.2.c) If $\rho^m \leq \mathrm{s} < \rho^l$ and $\rho^r - \rho^l > \Delta\mathrm{s}$, see Fig. 16.4.a, then the outgoing wave is a non–entropic shock from ρ^l to $\psi(\rho^l)$, followed by a rarefaction from $\psi(\rho^l)$ to $\psi(\rho^r)$ and by a possible null shock from $\psi(\rho^r)$ to ρ^r.

(A.3) $\mathscr{R}_{CR}(\rho^m, \rho^r) = \mathscr{N}$.

(A.3.a) If $\rho^l \leq R_T$ and $\rho^r < \psi(\rho^l)$, see Fig. 16.4.b, then the outgoing wave is a non–entropic shock from ρ^l to $\psi(\rho^l)$, followed by a rarefaction from $\psi(\rho^l)$ to ρ^r.

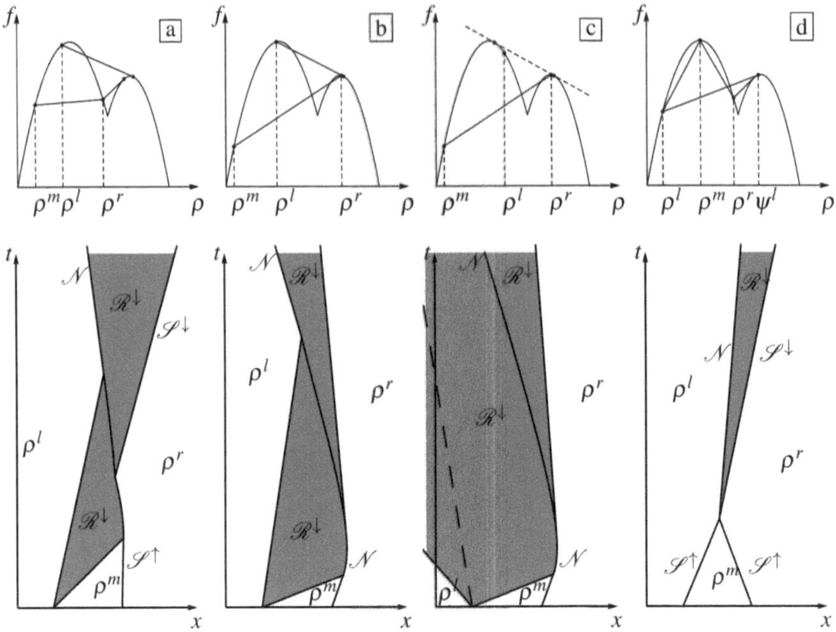

Fig. 16.4 Examples of interactions involving non–entropic shocks. Above $\psi^l = \psi(\rho^l)$.

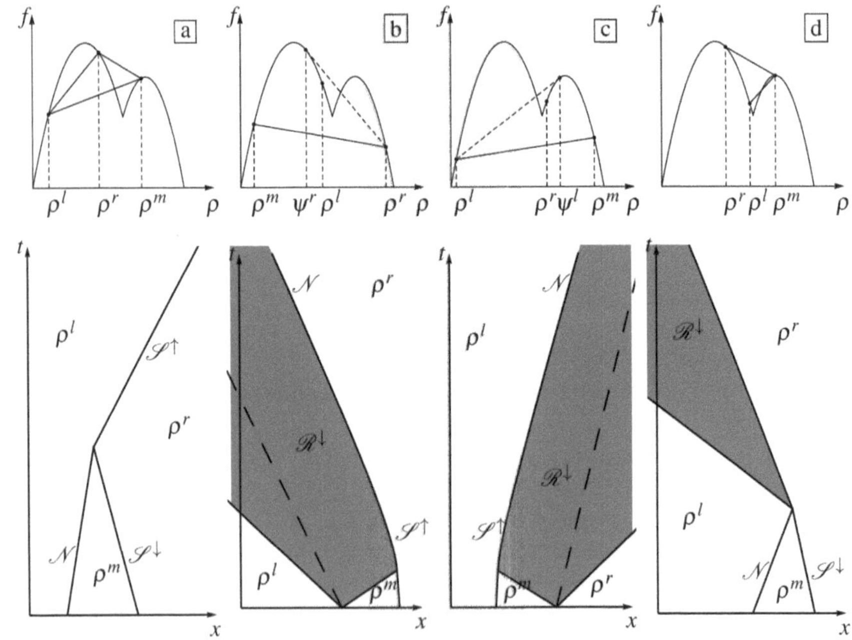

Fig. 16.5 Examples of interactions involving non–entropic shocks. Above $\psi^* = \psi(\rho^*)$.

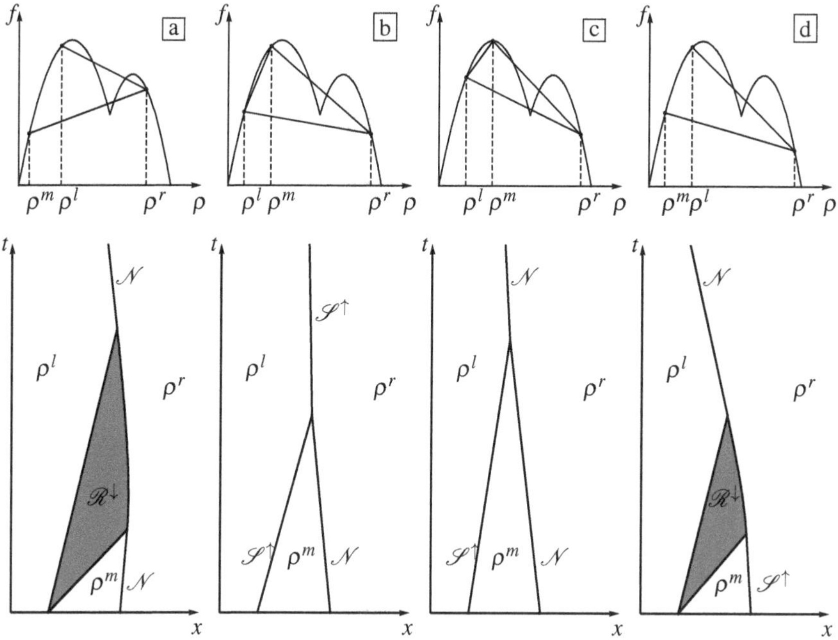

Fig. 16.6 Examples of interactions involving non–entropic shocks

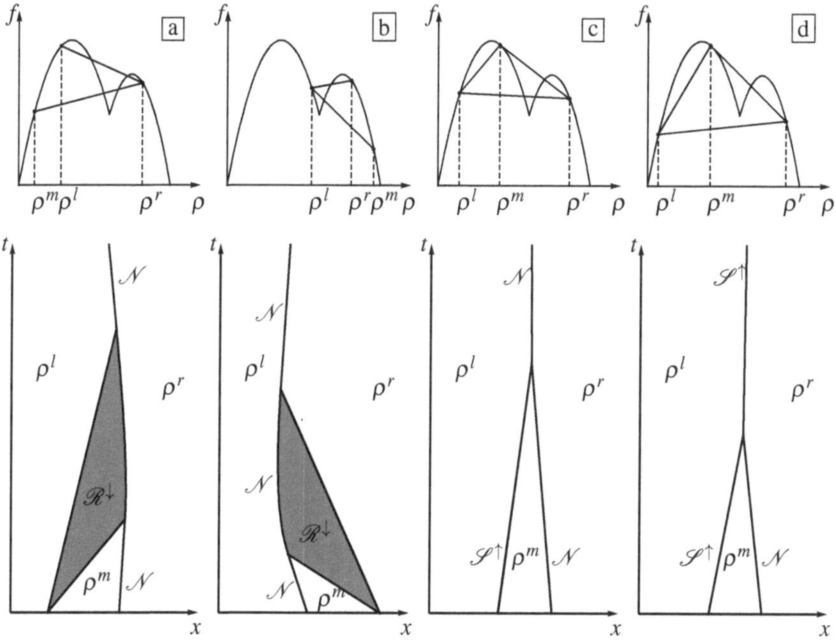

Fig. 16.7 Examples of interactions involving non–entropic shocks

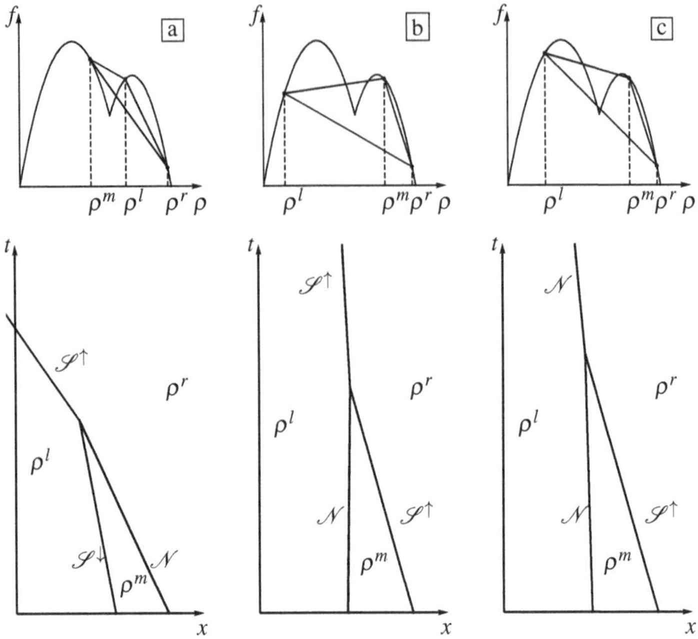

Fig. 16.8 Examples of interactions involving non–entropic shocks

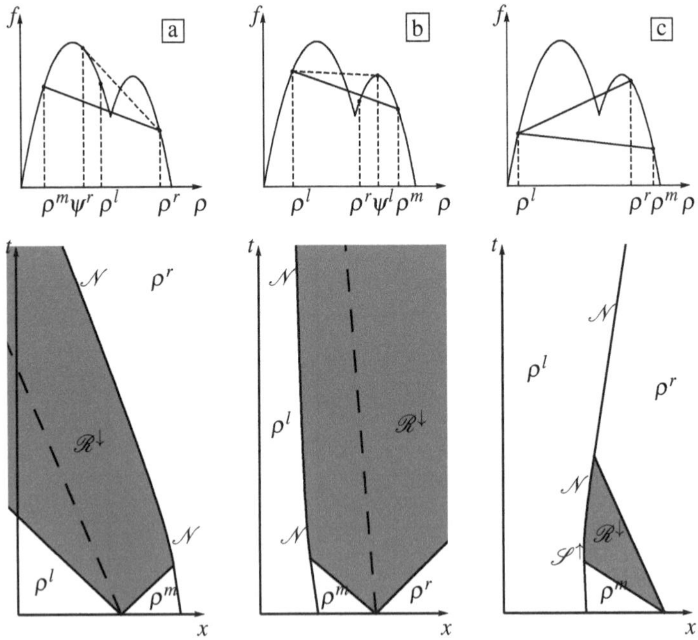

Fig. 16.9 Examples of interactions involving non–entropic shocks. Above $\psi^* = \psi(\rho^*)$.

(A.3.b) If $\rho^l \le R_T$ and $\psi(\rho^l) \le \rho^r$, see Fig. 16.6.a, then the outgoing wave is a non–entropic shock from ρ^l to ρ^r.

(A.3.c) If $\rho^m < R_T < \rho^l$ and $\rho^r \le R_T^*$, see Fig. 16.4.c, then the outgoing wave is a rarefaction from ρ^l to R_T, followed by a non–entropic shock from R_T to R_T^* and by a possible null rarefaction from R_T^* to ρ^r.

(A.3.d) If $R_T < \rho^l \le \psi(\rho^r)$ and $R_T^* < \rho^r$, see Fig. 16.7.a, then the outgoing wave is a non–entropic shock from ρ^l to ρ^r.

(A.3.e) If $\rho^m < \psi(\rho^r) < \rho^l$ and $R_T^* < \rho^r$, see Fig. 16.9.a, then the outgoing wave is a rarefaction from ρ^l to $\psi(\rho^r)$ followed by a non–entropic shock from $\psi(\rho^r)$ to ρ^r.

(A.3.f) If $\psi(\rho^r) \le \rho^m$, then the two waves do not interact.

(A.4) $\mathscr{R}_{CR}(\rho^m, \rho^r) = \mathscr{S}^\uparrow$ with $\rho^r \ge R_4$.

(A.4.a) If $\rho^l \le \psi(\rho^r)$ and $s(\rho^l, \rho^r)$ intersects $f = f(\rho)$, see Fig. 16.6.d, then the outgoing wave is non–entropic shock from ρ^l to ρ^r.

(A.4.b) If $\rho^l \le \psi(\rho^r)$ and $s(\rho^l, \rho^r)$ does not intersect $f = f(\rho)$, then the outgoing wave is an entropic shock from ρ^l to ρ^r.

(A.4.c) If $\rho^l > \psi(\rho^r)$, see Fig. 16.5.b, then the outgoing wave is a rarefaction from ρ^l to $\psi(\rho^r)$ followed by a non–entropic shock from $\psi(\rho^r)$ to ρ^r.

(B) $\mathscr{R}_{CR}[\rho^l, \rho^m] = \mathscr{S}^\uparrow$ with $\rho^m \le R$.

(B.1) $\mathscr{R}_{CR}(\rho^m, \rho^r) = \mathscr{R}^\downarrow$.

(B.1.a) If $\rho^r \le \rho^l$, then the outgoing wave is a possible null rarefaction from ρ^l to ρ^r.

(B.1.b) If $\rho^l < \rho^r$, then the outgoing wave is an entropic shock from ρ^l to ρ^r.

(B.2) $\mathscr{R}_{CR}(\rho^m, \rho^r) = \mathscr{S}^\uparrow$ with $\rho^r \le R$.

(B.2.a) If $\mathscr{R}_{CR}[\rho^l, \rho^r] = \mathscr{S}^\uparrow$, then the outgoing wave is an entropic shock from ρ^l to ρ^r.

(B.2.b) If $s < \rho^l < \rho^r - \Delta s \le \rho^m$, see Fig. 16.4.d, then the outgoing wave is a non–entropic shock from ρ^l to $\psi(\rho^l)$ followed by a rarefaction from $\psi(\rho^l)$ to $\psi(\rho^r)$ and by a possible null shock from $\psi(\rho^r)$ to ρ^r.

(B.3) $\mathscr{R}_{CR}(\rho^m, \rho^r) = \mathscr{S}^\uparrow$ with $\rho^m = R$.

(B.3.a) If $\rho^l < R_1$ and $\rho^r > \phi_R(\rho^l)$, then the outgoing wave is an entropic shock from ρ^l to ρ^r.

(B.3.b) If $\rho^l < R_1$ and $\rho^r \le \phi_R(\rho^l)$, then the two waves do not interact.

(B.3.c) If $\rho^l \ge R_1$, then the two waves do not interact.

(B.4) $\mathscr{R}_{CR}(\rho^m, \rho^r) = \mathscr{N}$. For any $\rho^m > R_T$, if $P_f(\rho^m)$ lies below $s(0, \rho^r)$ let $\Phi(\rho^m, \rho^r) = 0$, otherwise let $\Phi(\rho^m, \rho^r) = \rho^*$ be such that $s(\rho^*, \rho^r)$ intersects $f = f(\rho)$ in $P_f(\rho^m)$.

(B.4.a) If $\rho^m \le R_T$ and $s(\rho^l, \rho^r)$ intersects $f = f(\rho)$, see Fig. 16.6.c, then the outgoing wave is a non–entropic shock from ρ^l to ρ^r.

(B.4.b) If $\rho^m \le R_T$ and $s(\rho^l, \rho^r)$ does not intersect $f = f(\rho)$, see Fig. 16.6.b, then the outgoing wave is an entropic shock from ρ^l to ρ^r.

(B.4.c) If $\rho^m > R_T$, $\rho^l < \Phi(\rho^m, \rho^r)$ and $s(\rho^l, \rho^r)$ intersects the graph of f, see Fig. 16.7.c, then the outgoing wave is a non–entropic shock from ρ^l to ρ^r.

(B.4.d) If $\rho^m > R_T$, $\rho^l < \Phi(\rho^m, \rho^r)$ and $s(\rho^l, \rho^r)$ does not intersect $f = f(\rho)$, see Fig. 16.7.d, then the outgoing wave is an entropic shock from ρ^l to ρ^r.

(B.4.e) If $\rho^m > R_T$ and $\rho^l \geq \Phi(\rho^m, \rho^r)$, then the two waves do not interact.

(B.5) $\mathscr{R}_{CR}(\rho^m, \rho^r) = \mathscr{S}^{\uparrow}$ with $\rho^r \geq R_4$. The outgoing wave is an entropic shock from ρ^l to ρ^r.

(C) $\mathscr{R}_{CR}[\rho^l, \rho^m] = \mathscr{N}$.

(C.1) $\mathscr{R}_{CR}(\rho^m, \rho^r) = \mathscr{S}^{\downarrow}$.

(C.1.a) If $\rho^l < R_T$, $\rho^m = \psi(\rho^l)$ and $P_f(\rho^r)$ belongs to $s(\rho^l, \rho^m)$, then the two waves do not interact.

(C.1.b) If $\rho^l < R_T$, $\rho^m \geq \psi(\rho^l)$ and $P_f(\rho^r)$ does not belong to $s(\rho^l, \rho^m)$, see Fig. 16.5.a, then the outgoing wave is an entropic shock from ρ^l to ρ^r.

(C.1.c) If $\rho^l \geq R_T$, $\rho^m = \psi(\rho^l)$ and $\rho^r = \rho^l$, then the two waves do not interact.

(C.1.d) If $\rho^l > R_T$, $\rho^m \geq \psi(\rho^l)$ and $\rho^r < \rho^l$, see Fig. 16.5.d, then the outgoing wave is a rarefaction from ρ^l to ρ^r.

(C.2) $\mathscr{R}_{CR}(\rho^m, \rho^r) = \mathscr{R}^{\downarrow}$.

(C.2.a) If $\rho^r < \psi(\rho^l)$, see Fig. 16.9.b, then the outgoing wave is a non–entropic shock from ρ^l to $\psi(\rho^l)$ followed by a rarefaction from $\psi(\rho^l)$ to ρ^r.

(C.2.b) If $\psi(\rho^l) \leq \rho^r$, see Fig. 16.7.b, then the outgoing wave is a non–entropic shock from ρ^l to ρ^r.

(C.3) $\mathscr{R}_{CR}(\rho^m, \rho^r) = \mathscr{S}^{\uparrow}$.

(C.3.a) If $s(\rho^l, \rho^r)$ intersects $f = f(\rho)$, see Fig. 16.8.c, then the outgoing wave is a non–entropic shock from ρ^l to ρ^r.

(C.3.b) If $s(\rho^l, \rho^r)$ does not intersect $f = f(\rho)$, see Fig. 16.8.b, then the outgoing wave is an entropic shock from ρ^l to ρ^r.

(D) $\mathscr{R}_{CR}[\rho^l, \rho^m] = \mathscr{S}^{\uparrow}$ with $\rho^l \leq R_1$ and $\rho^m \geq R_4$.

(D.1) $\mathscr{R}_{CR}(\rho^m, \rho^r) = \mathscr{R}^{\downarrow}$.

(D.1.a) If $\rho^r < \psi(\rho^l)$, see Fig. 16.5.c, then the outgoing wave is a non–entropic shock from ρ^l to $\psi(\rho^l)$ followed by a rarefaction from $\psi(\rho^l)$ to ρ^l.

(D.1.b) If $\psi(\rho^l) \leq \rho^r < \phi_R(\rho^l)$, see Fig. 16.9.c, then the outgoing wave is a non–entropic shock from ρ^l to ρ^r.

(D.1.c) If $\rho^r \geq \phi_R(\rho^l)$, then the outgoing wave is an entropic shock from ρ^l to ρ^r.

(D.2) $\mathscr{R}_{CR}(\rho^m, \rho^r) = \mathscr{S}^{\uparrow}$. The outgoing wave is an entropic shock from ρ^l to ρ^r.

(E) $\mathscr{R}_{CR}[\rho^l, \rho^m] = \mathscr{S}^{\downarrow}$.

(E.1) $\mathscr{R}_{CR}(\rho^m, \rho^r) = \mathscr{R}^{\downarrow}$.

(E.1.a) If $\rho^r < \psi(\rho^l)$, then the outgoing wave is an entropic shock from ρ^l to $\psi(\rho^l)$ followed by a rarefaction from $\psi(\rho^l)$ to ρ^r.

(E.1.b) If $\psi(\rho^l) \leq \rho^r$, then the outgoing wave is an entropic shock from ρ^l to ρ^r.

(E.2) $\mathscr{R}_{CR}(\rho^m, \rho^r) = \mathscr{S}^{\uparrow}$.

(E.2.a) If $\rho^l \leq \psi(\rho^r)$, then the outgoing wave is an entropic shock from ρ^l to ρ^r.

(E.2.b) If $\psi(\rho^r) < \rho^l$, then the outgoing wave is a rarefaction from ρ^l to $\psi(\rho^r)$ followed by a possible null shock $\psi(\rho^r)$ to ρ^r.

(E.3) $\mathscr{R}_{CR}(\rho^m, \rho^r) = \mathscr{N}$. If $\rho^l = \psi(\rho^m)$, let $\Phi(\rho^l, \rho^m) = \rho^l$, while if $\rho^l < \psi(\rho^m)$, let $\Phi(\rho^l, \rho^m) = \rho^*$ be such that $s(\rho^l, \rho^*)$ intersects $f = f(\rho)$ in $P_f(\rho^l)$.

(E.3.a) If $\rho^r \leq \Phi(\rho^l, \rho^m)$, then the two waves do not interact.

(E.3.b) If $\Phi(\rho^l,\rho^m) < \rho^r$, see Fig. 16.8.a, then the outgoing wave is an entropic shock from ρ^l to ρ^r.

(F) $\mathscr{R}_{CR}[\rho^l,\rho^m] = \mathscr{R}^{\downarrow}$ with $\rho^m \geq R$.

(F.1) $\mathscr{R}_{CR}(\rho^m,\rho^r) = \mathscr{S}^{\downarrow}$.

(F.1.a) If $\rho^l \leq \psi(\rho^r)$, then the outgoing wave is an entropic shock from ρ^l to ρ^r.

(F.1.b) If $\psi(\rho^r) < \rho^l$, then the outgoing wave is a rarefaction from ρ^l to $\psi(\rho^r)$ followed by an entropic shock from $\psi(\rho^r)$ to ρ^r.

(F.2) $\mathscr{R}_{CR}(\rho^m,\rho^r) = \mathscr{R}^{\downarrow}$.

(F.2.a) If $\rho^r \geq R$ then the two waves do not interact.

(F.2.b) If $\rho^r < R_T$, $\rho^m = R$ and $\rho^l > R_T^*$, then the outgoing wave is a rarefaction from ρ^l to R_T^*, followed by an entropic shock from R_T^* to R_T and by a rarefaction from R_T to ρ^r.

(F.2.c) If $\rho^r < \psi(\rho^l)$, $\rho^m = R$ and $\rho^l \leq R_T^*$, then the outgoing wave is an entropic shock from ρ^l to $\psi(\rho^l)$ followed by a rarefaction from $\psi(\rho^l)$ to ρ^r.

(F.2.d) If $R_T \leq \rho^r$, $\rho^m = R$ and $\rho^l > R_T^*$, then the outgoing wave is a rarefaction from ρ^l to $\psi(\rho^r)$ followed by an entropic shock from $\psi(\rho^r)$ to ρ^r.

(F.2.e) If $\psi(\rho^l) \leq \rho^r$, $\rho^m = R$ and $\rho^l \leq R_T^*$, then the outgoing wave is an entropic shock from ρ^l to ρ^r.

(F.3) $\mathscr{R}_{CR}(\rho^m,\rho^r) = \mathscr{S}^{\uparrow}$.

(F.3.a) If $\rho^r \leq \rho^l$, then the outgoing wave is a possible null rarefaction from ρ^l to ρ^r.

(F.3.b) If $\rho^l < \rho^r$, then the outgoing wave is an entropic shock from ρ^l to ρ^r.

(G) $\mathscr{R}_{CR}[\rho^l,\rho^m] = \mathscr{S}^{\uparrow}$ with $R \leq \rho^l$.

(G.1) $\mathscr{R}_{CR}[\rho^m,\rho^r] = \mathscr{S}^{\downarrow}$.

(G.1.a) If $\rho^r < \psi(\rho^l)$, then the outgoing wave is an entropic shock from ρ^l to $\psi(\rho^l)$ followed by a rarefaction from $\psi(\rho^l)$ to ρ^r.

(G.1.b) If $\psi(\rho^l) \leq \rho^r$, then the outgoing wave is an entropic shock from ρ^l to ρ^r.

(G.2) $\mathscr{R}_{CR}[\rho^m,\rho^r] = \mathscr{R}^{\downarrow}$.

(G.2.a) If $\rho^r \leq \rho^l$, then the outgoing wave is a possible null rarefaction from ρ^l to ρ^r.

(G.2.b) If $\rho^l < \rho^r$, then the outgoing wave is an entropic shock from ρ^l to ρ^r.

(G.3) $\mathscr{R}_{CR}[\rho^m,\rho^r] = \mathscr{S}^{\uparrow}$. The outgoing wave is an entropic shock from ρ^l to ρ^r.

16.3 A Weighted Total Variation

In this section we study the total variation of the solution to the Riemann problem (16.1.1) with f satisfying assumptions **(F.1)–(F.9)**. For any fixed constant $W \in \mathbb{R}_+$, introduce the **weighted total variation** $\mathrm{TV}_\mathrm{w} : \mathbf{BV}(\mathbb{R};\mathbb{R}) \to [0,+\infty[$, see Fig. 16.10, defined as it follows. Let $\mathscr{R}_{CR}(\rho^l,\rho^r)$ be a single wave, then

- $\mathrm{TV}_\mathrm{w}(\mathscr{R}_{CR}[\rho^l,\rho^r]) = \left|\rho^r - \rho^l\right|$ if $\mathscr{R}_{CR}[\rho^l,\rho^r]$ is an entropic or non–entropic shock with $\rho^r \in \,]R,R^*]$;
- $\mathrm{TV}_\mathrm{w}(\mathscr{R}_{CR}[\rho^l,\rho^r]) = W\left|\rho^r - \rho^l\right|$ otherwise, i.e. if $\rho^r \leq R$ or $\rho^r < \rho^l$.

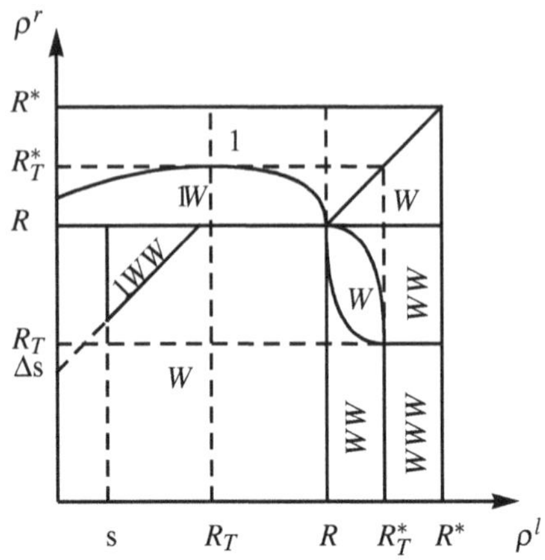

Fig. 16.10 The weighted total variation $\mathrm{TV_w}$ introduced in Sect. 16.3. The notation $1W$ means that the first wave has weight 1 and the second wave has weight W.

Proposition 16.3. *Assume that*

$$\frac{\phi_\psi(0)}{\psi(0)} \leq \frac{\Delta \mathrm{s}}{\psi(\mathrm{s}) - \mathrm{s}}, \tag{16.3.1}$$

and consider a constant W *such that*

$$W > 1 \quad \textit{and} \quad \frac{\phi_\psi(0)}{\psi(0)} \leq \frac{W+1}{2W} \leq \frac{\Delta \mathrm{s}}{\psi(s) - s}. \tag{16.3.2}$$

Then the map $[t \mapsto \mathrm{TV_w}(\rho_n(t))]$ *is a non–increasing function.*

Proof. The theory for entropic Riemann solver ensures that the total variation does not increase after an interaction of entropic waves, see Chap. 4. Thus, it is not limitative to consider only the interactions which involves waves with different weights or non–entropic shocks.

1. In the case described by Fig. 16.4.a, it results that $\mathrm{TV_w}$ decreases iff

$$(W+1)(\psi(\rho^l) - \rho^l) + 2W\rho^m < 2W\rho^r.$$

   Since $\rho^r \in]\mathrm{s}+\Delta\mathrm{s}, R[$ and $\rho^m \leq s < \rho^l$, by **(F.9)** we have

$$\begin{aligned} &2W\rho^r > 2W(\mathrm{s}+\Delta\mathrm{s}) \\ &(W+1)\left(\psi(\rho^l) - \rho^l\right) + 2W\rho^m < (W+1)\psi(s) + (W-1)s. \end{aligned}$$

   Therefore (16.3.2) ensures that $\mathrm{TV_w}$ decreases.
2. In the case described by Fig. 16.4.b, it results that $\mathrm{TV_w}$ decreases iff

$$\psi(\rho^l) - \rho^l < \rho^r - \rho^m.$$

   Indeed, being $\rho^r \geq \psi(\rho^m)$ and $\rho^m < \rho^l$, by **(F.9)** we have

$$\rho^r - \rho^m \geq \psi(\rho^m) - \rho^m > \psi(\rho^l) - \rho^l.$$

3. In the case described by Fig. 16.4.c, it results that $\mathrm{TV_w}$ decreases iff
$$R_T^* - R_T < \rho^r - \rho^m .$$
Being $\rho^r \geq \psi(\rho^m)$ and $\rho^m < R_T$, by **(F.9)** we have
$$\rho^r - \rho^m \geq \psi(\rho^m) - \rho^m > R_T^* - R_T .$$
Therefore $\mathrm{TV_w}$ decreases.
4. In the case described by Fig. 16.4.d, it results that $\mathrm{TV_w}$ decreases iff
$$(W+1)\psi(\rho^l) + (W-1)\rho^l < 2W\rho^r .$$
Since $\rho^r > \rho^l + \Delta\mathrm{s}$ and $s < \rho^l$, by **(F.9)** we have
$$\begin{aligned} &2W\rho^r > 2W(\rho^l + \Delta\mathrm{s}) \\ &(W+1)\psi(\rho^l) + (W-1)\rho^l < (W+1)\left(\psi(\mathrm{s}) - \mathrm{s}\right) + 2W\rho^l . \end{aligned}$$
Therefore (16.3.2) ensures that $\mathrm{TV_w}$ decreases.
5. In the case described by Fig. 16.5.a, it results that $\mathrm{TV_w}$ iff
$$2W\rho^r < (W+1)\rho^m + (W-1)\rho^l .$$
Since $\rho^l \geq 0$, $\rho^m \geq \psi(0)$ and $\rho^r < R_2$, by condition (16.3.2), $\mathrm{TV_w}$ decreases.

The reader can easily check that the remanning cases follow directly from the condition (16.3.2) and complete the proof. □

Sufficient conditions for (16.1.2), (16.3.1) and (16.3.2) to be satisfied are in the following lemma.

Lemma 16.1. *Assume that*
$$\psi(0) \leq 2\,\phi_\psi(0) \quad \textit{and} \quad \left(\psi'(0) - 1\right)\ \phi_\psi(0) > \left(\phi_\psi{}'(0) - 1\right)\ \psi(0) . \qquad (16.3.3)$$
Then, there exists $\rho^* \in \,]0, R_T[$ *such that for any* $\mathrm{s} \in \,]0, \rho^*[$ *there exist* $\Delta\mathrm{s}$ *and* W *satisfying* (16.1.2), (16.3.1) *and* (16.3.2).

Proof. Introduce the function $\Upsilon\colon\ [0, R_T] \to \mathbb{R}_+^*$ by
$$\Upsilon(\rho) = \frac{\phi_\psi(\rho) - \rho}{\psi(\rho) - \rho} .$$
By definition, $\Upsilon(0) = \phi_\psi(0)/\psi(0) \in \,]0,1[$ and $\Upsilon(R_T) = 0$. Furthermore, by (16.3.3), we have $\Upsilon'(0) > 0$. Therefore, there exists $\rho^* \in \,]0, R_T[$ such that $\Upsilon(\rho) > \phi_\psi(0)/\psi(0)$ for all $\rho \in \,]0, \rho^*[$ and $\Upsilon(\rho^*) = \phi_\psi(0)/\psi(0)$. Fix $s \in \mathbb{R}_+$ such that $\mathrm{s} < \min\{\rho^*, R_M\}$. Then, it is possible to choose $\Delta\mathrm{s} \in \mathbb{R}_+$ such that $\max\{R - R_T, \phi_\psi(\mathrm{s}) - \mathrm{s}\} < \Delta\mathrm{s} < R - \mathrm{s}$. Clearly (16.1.2) holds true. Concerning (16.3.2), observe that

$$1 > \frac{\Delta s}{\psi(s) - s} > \Upsilon(s) > \frac{\phi_\psi(0)}{\psi(0)} \geq \frac{1}{2}.$$

To complete the proof, observe that $\tau(W) = (W+1)/(2W)$ is a decreasing continuous function for $W \in \mathbb{R}_+$, $\tau(1) = 1$ and $\lim_{W \to +\infty} \tau(W) = 1/2$. □

16.4 Numerical Example

In this section we consider a numerical example characterized by a flow function f with the same shape of that one performed by the first Fig. **(b)** of [8]. In fact, see Fig. 16.11, we consider the flow function

$$f(\rho) = \max\left\{2\rho(35-\rho),\, 3(50-\rho)(\rho-25)\right\} \quad , \qquad \rho \in [0,50]\,. \tag{16.4.1}$$

By using the same notations of Sect. 16.1, with the above choice for f, we have

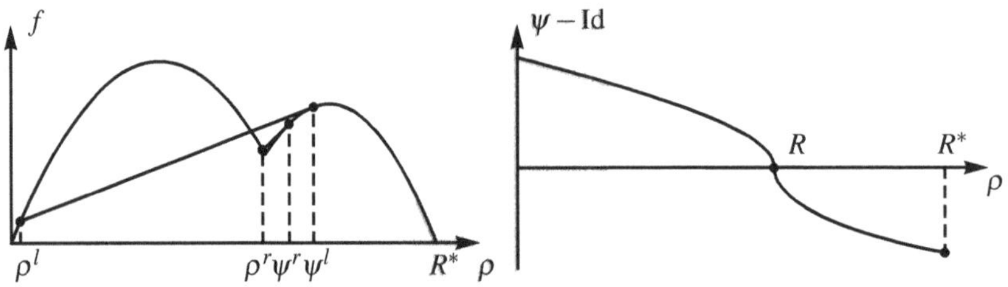

Fig. 16.11 Left: Construction of the solution to the Riemann problem (16.1.1) with f given by (16.4.1), $\rho^l = 1.1$ and $\rho^r = 29.7$. Above $\psi^* = \psi(\rho^*)$. Right: The function $[\rho \mapsto \psi(\rho) - \rho]$.

$$\begin{aligned} R &= 30 & R^* &= 50 \\ R_M &= 35/2 & R_M^* &= 75/2 \\ R_1 &= 25/2 & R_2 &= 5\left(30\sqrt{2}-31\right)/2 \\ R_3 &= 5\left(41-4\sqrt{30}\right)/3 & R_4 &= 125/3 \\ R_T &= 5\left(62-19\sqrt{6}\right)/4 & R_T^* &= 5\left(93-19\sqrt{6}\right)/6 \end{aligned}$$

and $\psi(\rho) = \rho + \operatorname{sgn}(30-\rho)\sqrt{\dfrac{2(125-\rho)|30-\rho|}{5+\operatorname{sgn}(30-\rho)}}$, $\rho \in [0,50]$.

Proposition 16.4. *The flow* $f\,:\,[0,50] \to [0,35/2]$ *given by* (16.4.1) *satisfies the conditions* ***(F.1)–(F.9)****. Furthermore, there exist* s, Δs *and* W *such that the conditions* (16.1.2) *and* (16.3.2) *are also satisfied.*

Proof. Clearly, the conditions **(F.1)–(F.5)** are satisfied, see Fig. 16.11. Observe that the conditions **(F.6)–(F.9)** are satisfied because

$$\frac{f(R)}{R} = 10 < 45 = f'(R+) \quad \text{and} \quad \frac{f(R)}{R^* - R} = 15 < 50 = -f'(R-)$$

$$-f'(R^*) = 75 > 10\left(13 - 2\sqrt{30}\right) = \frac{f\left(\psi(R^*)\right)}{R^* - \psi(R^*)}$$

$$\left[\rho \mapsto \psi(\rho) - \rho\right] \text{ is decreasing on } [0,50], \text{ see Fig. 16.11, right.}$$

By definition of ϕ_ψ we have $\phi_\psi(\rho) = 2\rho + \sqrt{3(\rho^2 - 155\rho + 3750)} - \frac{155}{2}$. Therefore, the conditions (16.1.2) and (16.3.2) hold with

$$\mathrm{s} = 1\,, \quad \phi_\psi(\mathrm{s}) = 2\sqrt{2697} - \frac{151}{2}\,, \quad \Delta\mathrm{s} = \frac{5}{2}\left(30\sqrt{2} - 31\right)\,, \quad W = \frac{31}{20}\,.$$

□

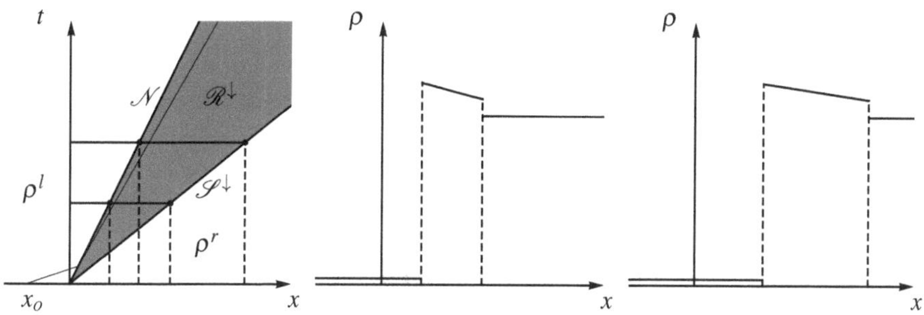

Fig. 16.12 Left: Representation in the (x,t)–plane of the solution to the Riemann problem (16.1.1) with f given by (16.4.1), $\rho^l = 1.1$ and $\rho^r = 29.7$. The entropy solution would consist of a single shock, while the non–entropy solution selected through **(R.1)**–**(R.4)** is given by a non–entropic shock, $\mathscr{N}$, followed by a decreasing rarefaction, $\mathscr{R}^\downarrow$, and a decreasing shock, $\mathscr{S}^\downarrow$. The thin line represents the path of a pedestrian staring from $x = x_o \in \mathbb{R}_-$. Center: Solution at time $t = 4$. Right: Solution at time $t = 7$. See also Fig. 16.11. Note that the maximum principle is violated.

As a corollary, we can apply to the case here considered the techniques developed in the previous sections, build a Riemann solver that satisfies the conditions **(R.1)**–**(R.4)** and prove that the total variation of the solution is uniformly bounded with respect to time. Finally, we observe that for any $\rho^l, \rho^r \in \left]1,30\right[$ with $\rho^r - \rho^l > 5(30\sqrt{2} - 31)/2$, see Fig. 16.11, the solution to the Riemann problem (16.1.1) with f given by (16.4.1) is

$$\rho(t,x) = \begin{cases} \rho^l & \text{if } x < \sigma(\rho^l)\, t \\ \frac{1}{6}\left(225 - \frac{x}{t}\right) & \text{if } \sigma(\rho^l)\, t \le x < \sigma(\rho^r)\, t \\ \rho^r & \text{if } x \ge \sigma(\rho^r)\, t\,, \end{cases}$$

where $\sigma(\rho) = 225 - 6\rho - 2\sqrt{3(125-\rho)(30-\rho)}$. In particular, if we consider $\rho^l = 1.1$ and $\rho^r = 29.7$, then the solution is that one depicted in the Fig. 16.12.

16.5 The Cauchy Problem

This section is concerned with the Cauchy problem for the Eq. (15.2.1). The availability proved in Proposition 16.3 of a weighted total variation $\mathrm{TV_w}$, that does not increase after an interaction, allows to tackle the Cauchy problem through the wave front tracking method [7] introduced in Chap. 5.

Let $\overline{\rho} \in \mathbf{L^1} \cap \mathbf{BV}\left(\mathbb{R};[0,R^*]\right)$ and consider the Cauchy problem associated to the Eq. (15.2.1)

$$\partial_t \rho + \partial_x f(\rho) = 0 \qquad (16.5.1a)$$

$$\rho(t=0) = \overline{\rho}\,. \qquad (16.5.1b)$$

Recall that a function $\rho \in \mathbf{C^0}\left(\mathbb{R}_+^*;\mathbf{L^1_{loc}}(\mathbb{R};[0,R^*])\right)$ is a weak solution to the Cauchy problem (16.5.1) if the initial condition (16.5.1b) holds a.e. and

$$\int_{\mathbb{R}_+}\int_{-\infty}^{+\infty} \left(\rho\;\partial_t\varphi + f(\rho)\;\partial_x\varphi\right)\mathrm{d}x\;\;\mathrm{d}t = 0 \qquad (16.5.2)$$

for every $\varphi \in \mathbf{C_c^\infty}\left(\mathbb{R}_+ \times \mathbb{R};\mathbb{R}\right)$, see Sect. 3.4.

Now we start the standard wave front tracking procedure to construct an approximate solution to (16.5.1). First, fix $n \in \mathbb{N}$ and define the mesh

$$\mathcal{M}_n = \left\{i\,2^{-n}R^*\;:\;i=0,1,\ldots,2^n\right\} \cup \left\{R_M\,,R_T\,,R\,,R_M^*,R_T^*\right\}\,.$$

Approximate the flow f by means of a piecewise linear and continuous flow f_n such that $f_n(\rho) = f(\rho)$ for all $\rho \in \mathcal{M}_n$. Introduce the functions $f_n^+ \;:\; [0,R^*] \times \mathcal{M}_n \to \mathbb{R}_+^*$, $\psi_n \;:\; [0,R^*] \to \left[R_T\,,R_T^*\right]$ and $\phi_{\psi_n} \;:\; [0,R_T[\,\cap\,\mathcal{M}_n \to\,]R_T\,,R] \cap \mathcal{M}_n$ letting

$$f_n^+(\rho,\overline{\rho}) = \begin{cases} \inf\left\{\begin{array}{l} F(\rho)\;:\;F \in \mathbf{PLC}\text{ is concave, and} \\ F(\rho) \geq f_n(\rho)\;\forall\rho \in [R,R^*] \cup \{0,\overline{\rho}\} \end{array}\right\} & \text{if } \overline{\rho} \in [0,R] \\ \\ \inf\left\{\begin{array}{l} F(\rho)\;:\;F \in \mathbf{PLC}\text{ is concave, and} \\ F(\rho) \geq f_n(\rho)\;\forall\rho \in [0,R] \cup \{\overline{\rho},R^*\} \end{array}\right\} & \text{if } \overline{\rho} \in [R,R^*] \end{cases}$$

$$\psi_n(\rho) = \begin{cases} \inf\left\{r \in\,]R,R^*]\;:\;f_n(r) = f_n^+(r,\rho)\right\} & \text{if } \rho \in [0,R[\\ R & \text{if } \rho = R \\ \sup\left\{r \in [0,R[\;:\;f_n(r) = f_n^+(r,\rho)\right\} & \text{if } \rho \in\,]R,R^*] \end{cases}$$

$$\phi_{\psi_n}(\rho) = \sup\left\{r \in\,]R_T\,,R] \cap \mathcal{M}_n\;:\;\frac{f_n(r) - f_n(\rho)}{r-\rho} \geq \frac{f_n\left(\psi_n(\rho)\right) - f_n(\rho)}{\psi_n(\rho) - \rho}\right\}\,.$$

Observe that f_n induces through **(R.1)–(R.4)** a Riemann solver $\mathscr{R}^n_{RC}$ that assigns to any Riemann datum $(\rho^l, \rho^r) \in (\mathscr{M}_n)^2$ a self similar, piecewise constant, weak solution $\mathscr{R}^n_{RC}(\rho^l, \rho^r)$ with range in $\mathscr{M}_n$ for all time $t \in \mathbb{R}^*_+$. More precisely, the following discretized version of Theorem 16.1 holds true:

Lemma 16.2. *Under the same assumptions of Theorem 16.1, for all $n \in \mathbb{N}$ there exists a unique Riemann solver $\mathscr{R}^n_{RC} : (\mathscr{M}_n)^2 \to \mathbf{BV}(\mathbb{R}; \mathscr{M}_n)$ that to any pair $(\rho^l, \rho^r) \in (\mathscr{M}_n)^2$ associates a weak solution to the Riemann problem*

$$\partial_t \rho + \partial_x f_n(\rho) = 0$$
$$\rho(0,x) = \begin{cases} \rho^l & \text{if } x < 0 \\ \rho^r & \text{if } x \geq 0 \end{cases}$$

*that satisfies **(R.1)–(R.4)** with f, ψ and ϕ_ψ respectively replaced by f_n, ψ_n and ϕ_{ψ_n}.*

The proof is a straightforward adaptation from the analogous result in Theorem 16.1 and is, therefore, omitted.

Consider the Cauchy problem (16.5.1) with $\overline{\rho} \in \mathbf{L^1} \cap \mathbf{BV}\left(\mathbb{R}; [0, R^*]\right)$. By the results in Sect. 5.3.1, we can approximate the initial datum $\overline{\rho}$ by means of a sequence $\overline{\rho}_n$ of piecewise constant functions with compact support, such that for all $n \in \mathbb{N}$

$$\overline{\rho}_n(\mathbb{R}) \subseteq \mathscr{M}_n, \quad \mathrm{TV}(\overline{\rho}_n) \leq \mathrm{TV}(\overline{\rho}) \quad \text{and} \quad \lim_{n \to +\infty} \left\| \overline{\rho}_n - \overline{\rho} \right\|_{\mathbf{L}^1} = 0. \qquad (16.5.3)$$

Consider the approximating Cauchy problem

$$\begin{array}{ll} \partial_t \rho + \partial_x f_n(\rho) = 0 & (t,x) \in \mathbb{R}^*_+ \times \mathbb{R} \\ \rho(0,x) = \overline{\rho}_n & x \in \mathbb{R}. \end{array} \qquad (16.5.4)$$

Each point of jump in $\overline{\rho}_n$ gives rise to a Riemann problem which we solve by means of $\mathscr{R}^n_{RC}$. Gluing these solutions, we obtain an approximate solution ρ_n defined up to the first time at which two or more discontinuities in ρ_n collide. ρ_n can be extended beyond this interaction time solving the Riemann problem arising at the interaction point. This procedure allows to extend the approximate solution to $\mathbb{R}^*_+ \times \mathbb{R}$, provided $[t \mapsto \mathrm{TV}\left(\rho_n(t)\right)]$ remains uniformly bounded. As it is standard in this context, the key difficulty is in providing a uniform bound for the total variation of ρ_n. To this aim, we use the weighted total variation introduced in Sect. 16.3. Let

$$\rho_n(t) = \sum_k \rho_n^k \, \chi_{]x_n^{k-1}, x_n^k]} \quad \text{with} \quad x_n^{k-1} < x_n^k \text{ for all } k$$

be the approximate solution at time t and define

$$\mathscr{I}_n = \left\{ k : \rho_n^k < \rho_n^{k+1} \text{ and } \rho_n^{k+1} > R \right\}$$
$$\mathrm{TV_w}(\rho_n) = \sum_{k \in \mathscr{I}_n} \left| \rho_n^{k+1} - \rho_n^k \right| + W \sum_{k \notin \mathscr{I}_n} \left| \rho_n^{k+1} - \rho_n^k \right|.$$

In other words, all entropy and non–entropic shocks with right state in the panic interval $[R, R^*[$ have weight 1, all the other entropic waves have weight W. By Proposition 16.3 the map $[t \mapsto \mathrm{TV}_{\mathrm{w}}(\rho_n(t))]$ is a non–increasing function.

As a consequence, the total variation of $\rho_n(t)$ is bounded for all $t \in \mathbb{R}_+$, and therefore the total number of interactions is finite. Thus the above wave front tracking algorithm defines a piecewise constant solution $\rho_n(t)$ to (16.5.4) for all $t \in \mathbb{R}_+^*$, which has a finite number of jumps and satisfies

$$\mathrm{TV}(\rho_n(t)) \leq \mathrm{TV}_{\mathrm{w}}(\rho_n(t)) \leq \mathrm{TV}_{\mathrm{w}}(\overline{\rho}_n) \leq W\ \mathrm{TV}(\overline{\rho}_n) \leq W\ \mathrm{TV}(\overline{\rho})\ . \qquad (16.5.5)$$

Lemma 16.3. *For every times $t, s \in \mathbb{R}_+^*$*

$$\left\|\rho_n(t) - \rho_n(s)\right\|_{\mathbf{L}^1} \leq W\ \mathrm{TV}(\overline{\rho})\ \mathscr{L}ip(f)\ |t-s|\ . \qquad (16.5.6)$$

Proof. (This proof is inspired by [9]). Let $[t_1, t_2]$ be an interval containing no interaction, and $y_k = y_k(t)$, $k = 1, 2, \ldots, N$, be the propagating fronts in ρ_n. Then,

$$\left\|\rho_n(t_2) - \rho_n(t_1)\right\|_{\mathbf{L}^1} \leq \sum_{k \geq 1} \left|\rho_n^+(y_k(t_1), t_1) - \rho_n^-(y_k(t_1), t_1)\right| \left|y_k(t_2) - y_k(t_1)\right|,$$

where ρ_n^- and ρ_n^+ are the left and right traces of ρ_n. Since each y_k' is constant in $[t_1, t_2]$, we have

$$\left|y_k(t_2) - y_k(t_1)\right| = \left|y_k'\right| (t_2 - t_1) \leq \mathscr{L}ip(q)\ (t_2 - t_1)\ .$$

By using these two estimates we find that

$$\left\|\rho_n(t_2) - \rho_n(t_1)\right\|_{\mathbf{L}^1} \leq W\ \mathrm{TV}(\overline{\rho})\ \mathscr{L}ip(q)\ (t_2 - t_1)\ .$$

Therefore, if $[s, t]$ is an interval containing no interaction, then there is nothing else to prove.

Let us now consider an interval $[s_1, s_2]$ containing one interaction at time $\xi \in]s_1, s_2[$. It is not restrictive to assume that the interaction involves exactly two incoming waves $y_k = y_k(t)$, $t \in]s_1, \xi]$, $k = 1, 2$, i.e. $y_1(\xi) = y_2(\xi)$. Let $z_j = z_j(t)$, $t \in [\xi, s_2[$, $j = 1, \ldots, h$, be the outgoing waves generated by the interaction, i.e. $z_j(\xi) = y_1(\xi)$, $j = 1, \ldots, h$. Denote by H_n the maximal number of outgoing waves after an interaction between two waves. Clearly $h \leq H_n \leq \sharp \mathscr{M}_n$. Let $y_k = y_k(t)$, $t \in]s_1, s_2[$, $k \geq 3$, be the others non–interacting waves. Then,

$$\begin{aligned}
&\left\|\rho_n(s_2) - \rho_n(s_1)\right\|_{\mathbf{L}^1} \leq \\
&\leq \sum_{j \geq 3} \left|\rho_n^+(s_1, y_j(s_1)) - \rho_n^-(s_1, y_j(s_1))\right| \left|y_j(s_2) - y_j(s_1)\right| \\
&\quad + \sum_{k=1}^{h} \sum_{j=1}^{2} \left|\rho_n^+(s_1, y_j(s_1)) - \rho_n^-(s_2, z_k(s_2))\right| \left|z_k(s_2) - y_j(s_1)\right|
\end{aligned}$$

$$+\sum_{k=1}^{h}\sum_{j=1}^{2}\left|\rho_n^-\left(s_1,y_j(s_1)\right)-\rho_n^+\left(s_2,z_k(s_2)\right)\right|\left|z_k(s_2)-y_j(s_1)\right|$$
$$\leq (5h-1)\,W\;\mathrm{TV}(\overline{\rho})\;\mathscr{L}ip(q)\;(s_2-s_1)\,,$$

since the equalities $y_j(t)=y_j'(t-\xi)+y_1(\xi)$, $z_k(t)=z_k'(t-\xi)+y_1(\xi)$ imply

$$\left|z_k(s_2)-y_j(s_1)\right|\leq\left|z_k'\right|(s_2-\xi)+\left|y_k'\right|(\xi-s_1)\leq\mathscr{L}ip(q)\;(s_2-s_1),$$

and the equalities

$$\begin{array}{ll}\rho_n^-\left(s_1,y_1(s_1)\right)=\rho_n^-\left(s_2,z_1(s_2)\right)\,, & \rho_n^+\left(s_1,y_1(s_1)\right)=\rho_n^-\left(s_1,y_2(s_1)\right)\,,\\ \rho_n^+\left(s_1,y_2(s_1)\right)=\rho_n^+\left(s_2,z_h(s_2)\right)\,, & \rho_n^+\left(s_2,z_k(s_2)\right)=\rho_n^-\left(s_2,z_{k+1}(s_2)\right)\,,\end{array}$$

for $k=1,\ldots,h-1$, imply that

$$\sum_{k=1}^{h}\sum_{j=1}^{2}\left|\rho_n^+\left(s_1,y_j(s_1)\right)-\rho_n^-\left(s_2,z_k(s_2)\right)\right|+\left|\rho_n^-\left(s_1,y_j(s_1)\right)-\rho_n^+\left(s_2,z_k(s_2)\right)\right|$$
$$=\left|\rho_n^+\left(s_1,y_1(s_1)\right)-\rho_n^-\left(s_1,y_1(s_1)\right)\right|+\sum_{k=2}^{h}\left|\rho_n^+\left(s_1,y_1(s_1)\right)-\rho_n^-\left(s_2,z_k(s_2)\right)\right|$$
$$+\sum_{k=1}^{h}\left|\rho_n^-\left(s_2,z_1(s_2)\right)-\rho_n^+\left(s_2,z_k(s_2)\right)\right|+\left|\rho_n^+\left(s_2,z_h(s_2)\right)-\rho_n^-\left(s_2,z_k(s_2)\right)\right|$$
$$+\sum_{k=1}^{h-1}\left|\rho_n^-\left(s_1,y_2(s_1)\right)-\rho_n^+\left(s_2,z_k(s_2)\right)\right|+\left|\rho_n^-\left(s_1,y_2(s_1)\right)-\rho_n^+\left(s_1,y_2(s_1)\right)\right|$$
$$\leq h\;\mathrm{TV}\left(\rho_n(s_1)\right)+2(2h-1)\;\mathrm{TV}\left(\rho_n(s_2)\right)\leq(5h-2)\,W\;\mathrm{TV}(\overline{\rho})\,.$$

Therefore, if $t>s$ and $[s,t]$ contains some interactions, then it is not restrictive to assume that at times s and t no interaction occurs, and that each interaction involves exactly two entering waves. Let t_k, $k=1,\ldots,K$, be the times corresponding to these interactions. If we denote $t_o=s$ and $t_{K+1}=t$, then, for what we proved before

$$\left\|\rho_n(t)-\rho_n(s)\right\|_{\mathbf{L}^1(\mathbb{R})}\leq$$
$$\leq\lim_{\varepsilon\searrow 0}\left[\sum_{k=0}^{K}\left\|\rho_n(t_{k+1}-\varepsilon)-\rho_n(t_k+\varepsilon)\right\|_{\mathbf{L}^1(\mathbb{R})}+\sum_{k=1}^{K}\left\|\rho_n(t_k+\varepsilon)-\rho_n(t_k-\varepsilon)\right\|_{\mathbf{L}^1(\mathbb{R})}\right]$$
$$\leq\lim_{\varepsilon\searrow 0}\left[2\varepsilon K(5H_n-1)\,W\;\mathrm{TV}(\overline{\rho})\;\mathscr{L}ip(q)+\sum_{k=0}^{K}W\;\mathrm{TV}(\overline{\rho})\;\mathscr{L}ip(q)\;(t_{k+1}-t_k-2\varepsilon)\right]$$
$$\leq W\;\mathrm{TV}(\overline{\rho})\;\mathscr{L}ip(q)\;(t-s)$$

completing the proof. □

By compactness arguments and adapting the proof of Theorem 5.1, we prove the global existence of entropy weak solutions to (16.5.1), within a class of functions with bounded variation.

Theorem 16.2. *Let* q *satisfy* ***(F.1)–(F.9)****, let* s, Δs *satisfy* (16.1.2) *and assume that there exists a constant* W *satisfying* (16.3.2)*. For any initial datum* $\overline{\rho} \in \mathbf{L^1} \cap \mathbf{BV}\left(\mathbb{R};[0,R^*]\right)$*, the Cauchy problem* (16.5.1) *admits a weak solution* $\rho = \rho(t,x)$ *generated by the Riemann solver* $\mathscr{R}_{CR}$ *and defined for all* $t \in \mathbb{R}_+^*$*. Moreover:*

$$\mathrm{TV}\left(\rho(t)\right) \leq W\ \mathrm{TV}(\overline{\rho})\,, \qquad \text{for all } t \in \mathbb{R}_+^*\,. \tag{16.5.7}$$

Proof. By (16.5.6) and the obvious estimate $\left\|\rho_n(t)\right\|_{\mathbf{L}^\infty} \leq R^*, t \in \mathbb{R}_+^*$, we can apply the Theorem 2.4 and deduce the existence of a subsequence ρ_m which converges to some function ρ in $\mathbf{L^1_{loc}}(\mathbb{R}_+^* \times \mathbb{R})$. Clearly (16.5.5) implies (16.5.7). Observing that the convergence $f_m \to q$ is uniform on the interval $[0,R^*]$, and recalling that ρ_m is an entropy weak solution to (16.5.4), we obtain

$$\begin{aligned} &\int\!\!\int \Big(|\rho - k|\ \partial_t\varphi + \big(f(\rho) - f(k)\big)\ \mathrm{sgn}(u-k)\ \partial_x\varphi\Big)\ \mathrm{d}x\ \mathrm{d}t = \\ &= \lim_{m\to\infty} \int\!\!\int \Big(|\rho_m - k|\ \partial_t\varphi + \big(f_m(\rho_m) - f_m(k)\big)\ \mathrm{sgn}(\rho_m - k)\ \partial_x\varphi\Big)\ \mathrm{d}x\ \mathrm{d}t \geq 0 \end{aligned}$$

for all $\varphi \in \mathbf{C_c^\infty}(\mathbb{R}_+ \times \mathbb{R}; \mathbb{R}_+^*)$. Finally, (16.5.3) and (16.5.6) imply that the initial condition (16.5.1b) is attained. This proves that ρ is a weak solution to (16.5.1). □

Example 16.1. Consider the specific flow represented in Fig. 16.13:

$$f(\rho) = \max\left\{\frac{\rho(7-\rho)}{6}, \frac{3(\rho-6)(2\rho-21)}{20(\rho-12)}\right\} \tag{16.5.8}$$

inspired by [8, Fig. 1, top right]. Geometrical considerations on Fig. 16.13 and elementary computations show that the assumptions in Theorem 16.2 are satisfied provided s $\in \left]0, 1/25\right]$, Δs $\in [67/10, 34/5]$ and $W \in [1.8563, 1.8599]$. □

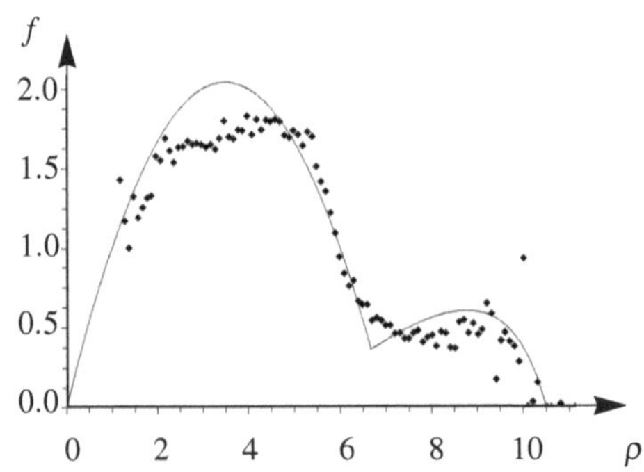

Fig. 16.13 The flow (16.5.8) and the raw data measured in [8]. ρ is on the horizontal axis ($[\rho] = 1/m^2$) and f on the vertical one ($[f] = 1/(m/s)$).

Further qualitative properties of the solutions constructed above are difficult to prove analytically. As an example we note the following straightforward consequence of the maximum principle and of the diminishing of the total variation, that hold for entropy weak solutions.

Proposition 16.5. *Fix an initial datum* $\overline{\rho} \in \mathbf{L}^1 \cap \mathbf{BV}\left(\mathbb{R};[0,R^*]\right)$. *Then, the solution* $\rho = \rho(t,x)$ *exhibited from Theorem 16.2 satisfies*

$$\rho(t,x) \le \max\left\{\|\overline{\rho}\|_{\mathbf{L}^\infty}, R_T^*\right\} \quad \textit{for all} \quad (t,x) \in \mathbb{R}_+^* \times \mathbb{R}.$$

Furthermore, if

$$\overline{\rho}(\mathbb{R}) \subseteq [0,R] \quad \textit{and} \quad \mathrm{TV}(\overline{\rho}) < \Delta s$$

then, the solution $\rho = \rho(t,x)$ *exhibited from Theorem 16.2 coincides with the entropy weak solution and does not attain values among the panic states. Hence, it is a Lipschitz function of the initial data with respect to the* $\mathbf{L}^1$*–norm.*

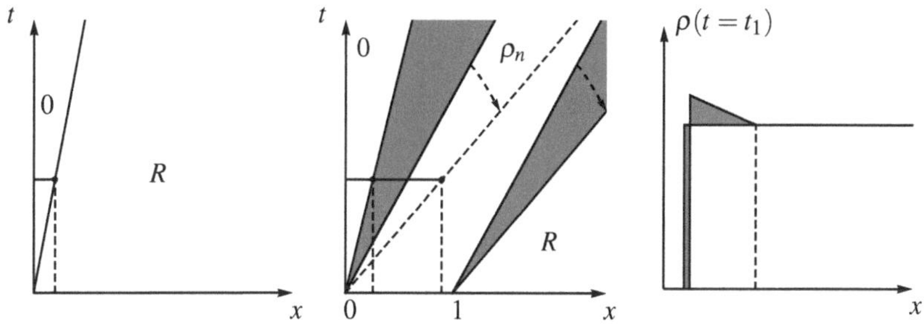

Fig. 16.14 Left: Representation in the (x,t)–plane of the solution ρ to the Riemann problem (16.5.9). It is given by a shock from 0 to R started from $x = 0$. Center: Representation in the (x,t)–plane of the solution ρ_n to the Cauchy problem (16.5.10). It is given by a non–entropic shock from 0 to $\psi(0)$, followed by a decreasing rarefaction from $\psi(0)$ to $\rho_n = R + 1/n$ both starting from $x = 0$, while from $x = 1$ starts a rarefaction from ρ_n to R. Right: Representation of ρ and of the limit ρ_∞ in the (x,ρ)–plane.

The next proposition proves that $\mathscr{R}_{CR}$ is not $\mathbf{L}^1$–continuous in $[0,R^*]^2$.

Proposition 16.6. $\mathscr{R}_{CR}$ *is not* $\mathbf{L}^1$*–continuous in* $[0,R^*]^2$.

Proof. Let $n \in \mathbb{N}$ and consider the Cauchy problems

$$\partial_t\rho + \partial_x f(\rho) = 0 \qquad \rho(0,x) = \begin{cases} 0 & \text{if } x \in \mathbb{R}_- \\ R & \text{if } x \in \mathbb{R}_+^* \end{cases} \tag{16.5.9}$$

$$\partial_t\rho_n + \partial_x f(\rho_n) = 0 \qquad \rho_n(0,x) = \begin{cases} 0 & \text{if } x \in \mathbb{R}_- \\ R + n^{-1} & \text{if } x \in [0,1[\\ R & \text{if } x \in [1,+\infty[\,. \end{cases} \tag{16.5.10}$$

Easy computations, see Fig. 16.14, show that for all $t \in \mathbb{R}_+$

$$\lim_{n\to\infty} \left\|\rho(0) - \rho_n(0)\right\|_{\mathbf{L}^1(\mathbb{R};[0,R^*])} = 0\,, \quad \lim_{n\to\infty} \left\|\rho(t) - \rho_n(t)\right\|_{\mathbf{L}^1(\mathbb{R};[0,R^*])} \neq 0$$

and this ends the proof. □

References

1. Colombo, R.M., Facchi, G., Maternini, G., Rosini, M.D.: On the continuum modeling of crowds. American Mathematical Society (AMS), Providence (2009)
2. Colombo, R.M., Goatin, P., Maternini, G., Rosini, M.D.: Macroscopic Models for Pedestrian Flows. In: Big Events and Transport: The Transportation Requirements for the Management of Large Scale Events, pp. 11–22. IUAV – TTL Research Unit (2010)
3. Colombo, R.M., Goatin, P., Rosini, M.D.: A macroscopic model for pedestrian flows in panic situations. In: Proceedings of the 4th Polish-Japanese Days. GAKUTO International Series. Mathematical Sciences and Applications, vol. 32, pp. 255–272 (2010)
4. Colombo, R.M., Goatin, P., Rosini, M.D.: Conservation laws with unilateral constraints in traffic modeling. In: Mussone, L., Crisalli, U. (eds.) Transport Management and Land-Use Effects in Presence of Unusual Demand, Atti del Convegno SIDT 2009(June 2009)
5. Colombo, R.M., Rosini, M.D.: Pedestrian flows and non-classical shocks. Math. Methods Appl. Sci. 28(13), 1553–1567 (2005)
6. Colombo, R.M., Rosini, M.D.: Existence of nonclassical solutions in a Pedestrian flow model. Nonlinear Analysis: Real World Applications 10(5), 2716–2728 (2009)
7. Dafermos, C.M.: Polygonal approximations of solutions of the initial value problem for a conservation law. J. Math. Anal. Appl. 38, 33–41 (1972)
8. Helbing, D., Johansson, A., Al-Abideen, H.Z.: Dynamics of crowd disasters: An empirical study. Physical Review E (Statistical, Nonlinear, and Soft Matter Physics) 75(4), 046109 (2007)
9. Lefloch, P.G.: Hyperbolic systems of conservation laws. The theory of classical and nonclassical shock waves. Lectures in Mathematics ETH Zürich. Birkhäuser Verlag, Basel (2002)
10. Rosini, M.D.: Nonclassical interactions portrait in a macroscopic pedestrian flow model. J. Differential Equations 246(1), 408–427 (2009)

Chapter 17
Applications

Abstract. In this chapter we apply the CR model to describe the evacuation of a corridor through an exit when panic arises. Two different situations are considered: first when the space between the pedestrians and the exit is free, then when before the exit there is a further door, through which the pedestrians have to move. According with empirical observations, the CR model confirms that placing such a door may reduce the time necessary for the evacuation. This displays the ability of the CR model to reproduce the so called Braess' paradox for pedestrian flows.

17.1 Introduction

In emergency situations, it is well known [10] that the transition to panic in the crowd approaching an exit door and the corresponding overcompression of the people seeking to exit significantly affect the door efficiency and may dramatically reduce it, making it even lower than that usually allowed by the door. To prevent this, often suitable obstacles, typically columns, are posed in front of the exit, at a suitable distance, to partially sustain the crowd pressure. In fact, the presence of an obstacle may avoid the insurgence of panic among the people, therefore keeping the door efficiency at a higher level. Paradoxically, thus, the insertion of this obstacle may reduce the evacuation time, although most individuals may have a slightly longer path to reach the exit. This remarkable behavior reminds of the Braess' paradox [3], typical of networks and is captured by the CR model, as we show in this chapter [4–8].

Let us consider a group of people that needs to evacuate a corridor (or a bridge ...) parameterized by a single coordinate $x \in [0, x_{\mathrm{D}}]$, through an exit door (or gate ...) at $x = x_{\mathrm{C}}$, with $0 \ll x_{\mathrm{C}} < x_{\mathrm{D}}$. The dynamics of the crowd exiting the corridor is described by (16.5.1) and the Riemann solver $\mathscr{R}_{RC}$ prescribed in Sect. 16.1 through **(R.1)**–**(R.4)**. More precisely, choose a flow f as in Fig. 17.1, right, which satisfies **(F.1)**–**(F.9)** given in Sect. 16.1 and let s, Δs satisfying (16.1.2). Assume that the crowd is initially uniformly distributed on $[x_{\mathrm{A}}, x_{\mathrm{B}}]$ with uniform density $\overline{\rho}$

M.D. Rosini: *Macroscopic Models for Vehicular Flows & Crowd Dynamics*, UCS, pp. 227–237.
DOI: 10.1007/978-3-319-00155-5_17

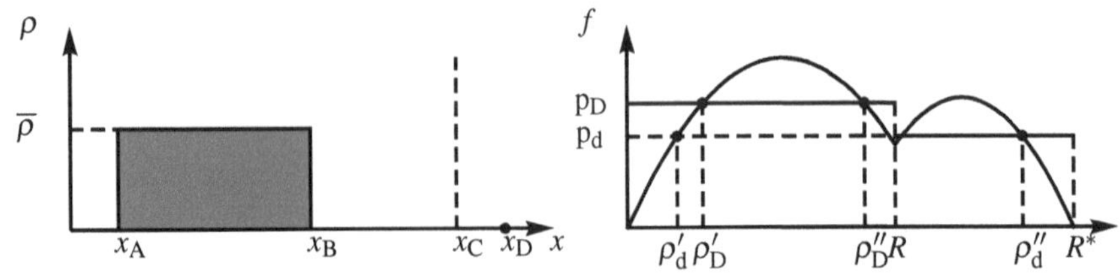

Fig. 17.1 Left: The initial datum for the constrained Cauchy problem (17.2.1). Right: The fundamental diagram, the maximal outflow through the door in standard situations, p_D, through the door in overcompressed regime, p_d, as defined in (17.1.1).

and needs to pass through the door in $x = x_C$, where $0 < x_A < x_B < x_C < x_D$ and $0 \ll \overline{\rho} < R$, see Fig. 17.1, left. The maximal possible outflow through the door at $x = x_C$ is given by $p : [0,R^*] \mapsto \mathbb{R}^*_+$. To model the drop of the door efficiency in the transition to panic situations, we choose the simple piecewise constant behavior

$$p(\rho) = \begin{cases} p_D & \text{if } \rho \in [0,R] \\ p_d & \text{if } \rho \in]R,R^*] \end{cases}, \tag{17.1.1}$$

with $0 < p_d \leq p_D$ as in Fig. 17.1, right. Here, p_D is the outflow in standard situation, while p_d denotes the same quantity in the overcompressed regime, i.e. in panic situations.

In this chapter we analytically study both the situations, with and without an obstacle, comparing the corresponding results to point out how the CR model describes the overcompression effect due to the rise of panic and the consequent fall in the outflow. These two cases are analytically described by two different constrained initial–boundary value problems. Let us recall that two different definitions of solution to initial–boundary value problems for conservation laws are found in the literature. In [2], a definition based on the vanishing viscosity limit is given. Here, we prefer to follow the other definition [9], more suitable to the present situation. Indeed, viscosity conflicts with our choice of non–entropy solutions and its physical meaning can hardly be justified. Recall that the definition [9] leads to well posed problems, see [1]. Consider the Riemann problem having as data the boundary condition and the trace of the solution on the boundary. This problem leads to waves directed towards the domain or towards the complement of the domain. According to the definition given in [9], the latter waves are neglected. Thus, there may well be a jump between the boundary and the solution along the boundary. For a further discussion of boundary conditions in conservation laws modeling traffic flows, see [11].

17.2 Evacuation without Obstacles

Let p given by (17.1.1) with $0 < p_d \leq f(R^*_M)$ and $\max\{f(R), p_d\} \leq p_D < f(R^*_M)$. With reference to Fig. 17.1, right, call ρ'_D, ρ''_D, ρ'_d, ρ''_d the densities defined by

$$\rho'_{\rm d} \le \rho'_{\rm D} < \rho''_{\rm D} < R\,, \quad R^*_M < \rho''_{\rm d}\,, \quad {\rm p_D} = f(\rho'_{\rm D}) = f(\rho''_{\rm D})\,, \quad {\rm p_d} = f(\rho'_{\rm d}) = f(\rho''_{\rm d})\,.$$

Observe that if $\rm p_d = p_D$, then $\rho'_{\rm d} = \rho'_{\rm D}$.

Remark 17.1. It is possible to consider more general values for p, but then the construction of the solution could be slightly different. □

The evacuation of a group of people, uniformly distributed in $[x_{\rm A}, x_{\rm B}]$ with density $\overline{\rho}$, from a corridor $[0, x_{\rm D}]$, through a door in $x = x_{\rm C}$ with maximal efficiency $p\,:\,[0, R^*] \to \mathbb{R}^*_+$, is modeled by the constrained Cauchy problem

$$\begin{aligned}
&\partial_t \rho + \partial_x f(\rho) = 0 && (t,x) \in \mathbb{R}_+ \times\,]0, x_{\rm D}[&& (17.2.1a)\\
&\rho(0,x) = \overline{\rho}\, \chi_{[x_{\rm A}, x_{\rm B}]}(x) && x \in [0, x_{\rm D}] && (17.2.1b)\\
&f\left(\rho(t, x_{\rm C})\right) \le p\left(\rho(t, x_{\rm C})\right) && t \in \mathbb{R}^*_+\,. && (17.2.1c)
\end{aligned}$$

The construction of the solution to (17.2.1) is made through wave front tracking method. While the overall picture of the solution to (17.2.1) is rather stable, a detailed rigorous analytical study necessarily needs to consider many slightly different cases. Below, we concentrate our attention only on the most representative situation in which panic arises. The definition of the rarefactions and the curves of discontinuity involved in the constructed solution are deferred in (17.2.4).

Assume that the door through–flow $\rm p_D$ is *"small"* and the initial crowd density $\overline{\rho}$ is *"large"*, namely

$$\begin{aligned}
&{\rm p_D} < \min\left\{f({\rm s}), f({\rm s} + \Delta{\rm s})\right\} && (17.2.2a)\\
&\overline{\rho} > \frac{x_{\rm C} - x_{\rm B}}{x_{\rm B} - x_{\rm A}} \left(\frac{\rm p_D}{f'(\rho'_{\rm D})} - \rho'_{\rm D}\right)\,. && (17.2.2b)
\end{aligned}$$

The first step to solve (17.2.1) is to consider the Riemann problems

$$\begin{aligned}
&\partial_t \rho + \partial_x f(\rho) = 0\,, \qquad \rho(0,x) = \begin{cases} 0 & \text{if } x < x_{\rm A}\\ \overline{\rho} & \text{if } x \ge x_{\rm A}\end{cases}\\
&\partial_t \rho + \partial_x f(\rho) = 0\,, \qquad \rho(0,x) = \begin{cases} \overline{\rho} & \text{if } x < x_{\rm B}\\ 0 & \text{if } x \ge x_{\rm B}\,.\end{cases}
\end{aligned} \qquad (17.2.3)$$

The former is solved by an entropic shock, whose discontinuity curve is the straight line $\mathscr{S}_1$ given by (17.2.4a), see Fig. 17.2, left; the latter is solved by a rarefaction wave, whose values are given by $\mathscr{R}_{\rm B}\,:\,\left\{(t,x)\,:\, f'(\overline{\rho})\,t < x - x_{\rm B} < f'(0)\,t\right\} \to\,]0, \overline{\rho}[$ implicitly defined by (17.2.4b), see Fig. 17.2, left.

$$\begin{aligned}
&\mathscr{S}_1\,:\, x = x_{\rm A} + \frac{f(\overline{\rho})}{\overline{\rho}}\, t && (17.2.4a)\\
&\mathscr{R}_{\rm B}\,:\, x = x_{\rm B} + f'\left(\mathscr{R}_{\rm B}(t,x)\right)\, t \quad \text{for } x \in\,]x_{\rm B} + f'(\overline{\rho})\,t, x_{\rm B} + f'(0)\,t[&& (17.2.4b)
\end{aligned}$$

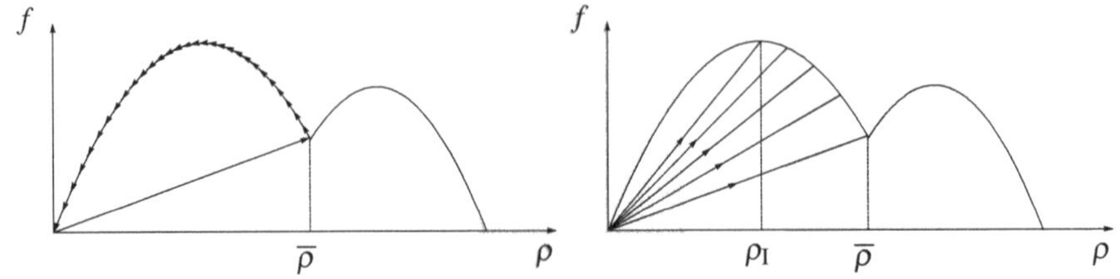

Fig. 17.2 Representation of $\mathscr{S}_1, \mathscr{R}_\mathrm{B}$, left, $\mathscr{S}_2$, right, in the (ρ, f)–plane, with $\rho_\mathrm{I} = \mathscr{R}_\mathrm{B}(t_\mathrm{I}, x_\mathrm{I})$

Therefore, for small times, the solution to (17.2.1) has the expression

$$\rho(t,x) = \begin{cases} 0 & \text{if } x \in \left]0, x_\mathrm{A} + f(\bar\rho)\,\bar\rho^{-1}\,t\right[\\ \bar\rho & \text{if } x \in \left[x_\mathrm{A} + f(\bar\rho)\,\bar\rho^{-1}\,t, x_\mathrm{B} + f'(\bar\rho)\,t\right[\\ \mathscr{R}_\mathrm{B}(t,x) & \text{if } x \in \left[x_\mathrm{B} + f'(\bar\rho)\,t, x_\mathrm{B} + f'(0)\,t\right[\\ 0 & \text{if } x \in \left[x_\mathrm{B} + f'(0)\,t, x_\mathrm{D}\right[\,, \end{cases} \tag{17.2.4c}$$

see Fig. 17.7, left. Note that pedestrians start exiting through the door at time $t_\mathrm{E} = (x_\mathrm{C} - x_\mathrm{B})/f'(0)$. At time $t_\mathrm{F} = \bar\rho\,(x_\mathrm{B} - x_\mathrm{A}) / \left(f(\bar\rho) - \bar\rho\, f'(\bar\rho)\right)$, the shock $\mathscr{S}_1$ and the rarefaction $\mathscr{R}_\mathrm{B}$ meet. Assume that $x_\mathrm{F} < x_\mathrm{C}$ or, equivalently that

$$\frac{x_\mathrm{C} - x_\mathrm{B}}{x_\mathrm{B} - x_\mathrm{A}} > \left(\frac{f(\bar\rho)}{\bar\rho\, f'(\bar\rho)} - 1\right)^{-1}.$$

Observe that the above condition is automatically satisfied for all $\bar\rho \in \left]R_M, R\right[$, indeed in this case the right hand side is negative. From F starts the shock $\mathscr{S}_2$ given once it is solved the following Cauchy problem, see Fig. 17.2, right,

$$\mathscr{S}_2 : \dot x(t) = \frac{f\left(\mathscr{R}_\mathrm{B}\left(t, x(t)\right)\right)}{\mathscr{R}_\mathrm{B}\left(t, x(t)\right)}\,, \qquad x(t_\mathrm{F}) = x_\mathrm{F}\,. \tag{17.2.4d}$$

Due to the interaction with the rarefaction $\mathscr{R}_\mathrm{B}$, the shock $\mathscr{S}_2$ accelerates, while the state to its right decreases.

If the straight line $\mathscr{R}_\mathrm{B}(t,x) = \rho'_\mathrm{D}$ does not reach the door before it is reached by $\mathscr{S}_2$, then the outflow through the door is always less than the maximal one, p_D, and panic may not arise. Therefore, we assume that this does not happen. In this case, at time $t_\mathrm{G} = (x_\mathrm{C} - x_\mathrm{B})/f'(\rho'_\mathrm{D})$, the maximal outflow through the door is reached, provided not all people already exited, i.e. provided

$$(x_\mathrm{B} - x_\mathrm{A})\,\bar\rho > \int_{t_\mathrm{E}}^{t_\mathrm{G}} f\left(\rho(t, x_\mathrm{C}-)\right)\,\mathrm{d}t = \int_{t_\mathrm{E}}^{t_\mathrm{G}} f\left(\left(f'\Big|_{[0,R_M]}\right)^{-1}\left(\frac{x_\mathrm{C} - x_\mathrm{B}}{t}\right)\right)\,\mathrm{d}t$$

$$= -(x_C - x_B) \int_0^{\rho'_D} \frac{f(\rho)\, f''(\rho)}{f'(\rho)^2}\, d\rho = (x_C - x_B) \left(\frac{p_D}{f'(\rho'_D)} - \rho'_D \right)$$

which, in turn, is ensured by (17.2.2b). A shock corresponding to a queue at the door with negative speed and support $\mathscr{S}_3$ is formed at G. It interacts with the rarefaction $\mathscr{R}_B$ and accelerates backwards, see Fig. 17.3, left, as it follows by

$$\mathscr{S}_3 : \dot{x}(t) = \sigma\left(\mathscr{R}_B\left(t, x(t)\right), \rho''_D\right) \quad , \quad x(t_G) = x_G \,, \tag{17.2.4e}$$

where $\sigma(\rho_1,\rho_2) = \left(f(\rho_1) - f(\rho_2)\right)/(\rho_1 - \rho_2)$. Along the right side of $\mathscr{S}_3$ the density is constant and equal to ρ''_D, while on its left side it increases. Assume that the

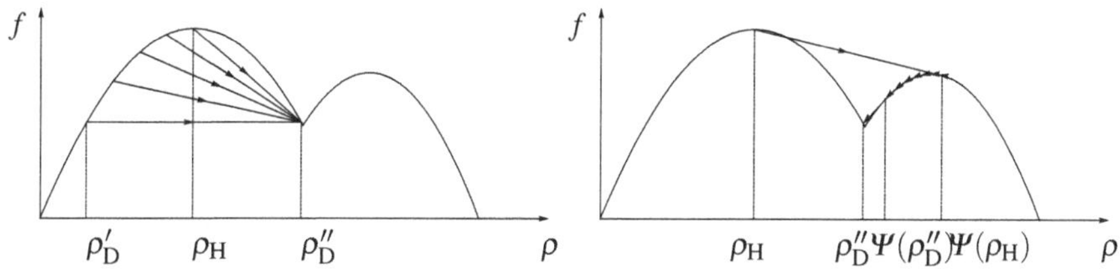

Fig. 17.3 Representation of $\mathscr{S}_3$, left, $\mathscr{N}_1$, $\mathscr{R}_H$, $\mathscr{S}_4$, right, in the (ρ, f)–plane, with $\rho_H = \rho''_D - \Delta s$

panic arises due to the fact that the line $\mathscr{R}_B(t,x) = s$ meets the shock $\mathscr{S}_3$, namely that $\rho''_D - \Delta s > s > \rho'_D$, which is ensured by (17.2.2a). Denoted by H this intersection, from it starts a non–entropic shock $\mathscr{N}_1$, followed by a rarefaction $\mathscr{R}_H$ taking values in the panic interval of densities, followed by a shock $\mathscr{S}_4$ between $\Psi(\rho''_D)$ and ρ''_D, see Fig. 17.3, right, given, respectively, by

$$\mathscr{N}_1 : \dot{x}_{\mathscr{N}_1}(t) = \sigma\left(\mathscr{R}_B\left(t, x_{\mathscr{N}_1}(t)\right), \Psi\left(\mathscr{R}_B\left(t, x_{\mathscr{N}_1}(t)\right)\right)\right) \quad , \quad x_{\mathscr{N}_1}(t_H) = x_H \tag{17.2.4f}$$

$$\mathscr{R}_H : \frac{x - x_H}{t - t_H} = f'\left(\mathscr{R}_H(t,x)\right) \quad , \quad f'\left(\Psi\left(\rho''_D - \Delta s\right)\right) < \frac{x - x_H}{t - t_H} < f'\left(\Psi\left(\rho''_D\right)\right) \tag{17.2.4g}$$

$$\mathscr{S}_4 : \frac{x - x_H}{t - t_H} = \sigma\left(\Psi\left(\rho''_D\right), \rho''_D\right) . \tag{17.2.4h}$$

$\mathscr{S}_4$ has positive speed because by assumption $p_D < f(R^*_M)$. Due to the interaction with the rarefaction $\mathscr{R}_B$, the non–entropic shock $\mathscr{N}_1$ accelerates backwards and rarefaction waves could be generated on its right; in this case they are given by

$$\mathscr{R}_{\mathscr{N}_1} : \mathscr{R}_{\mathscr{N}_1}(t,x) = \Psi\left(\mathscr{R}_B\left(t_o, x_{\mathscr{N}_1}(t_o)\right)\right) \quad , \quad \frac{x - x_{\mathscr{N}_1}(t_o)}{t - t_o} = \dot{x}_{\mathscr{N}_1}(t_o) \,. \tag{17.2.4i}$$

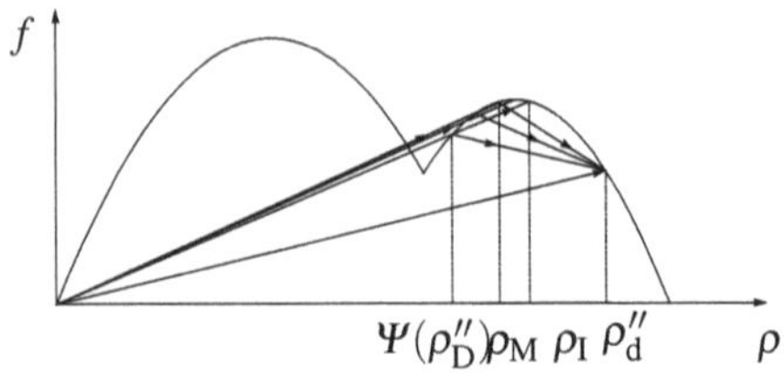

Fig. 17.4 Representation of $\mathscr{S}_5$, $\mathscr{S}_6$ and $\mathscr{S}_7$ in the (ρ, f)–plane, where $\rho_{\mathrm{M}} = \mathscr{R}_{\mathrm{H}}(\mathrm{M})$ and $\rho_{\mathrm{I}} = \mathscr{R}_{\mathrm{H}}(\mathrm{I})$

$\mathscr{N}_1$ meets the shock $\mathscr{S}_2$ at I and starts the discontinuity $\mathscr{S}_5$ given by

$$\mathscr{S}_5 : \dot{x}_{\mathscr{S}_5}(t) = \sigma\left(\mathscr{R}_{\mathscr{N}_1,\mathrm{H}}\left(t, x_{\mathscr{S}_5}(t)\right), 0\right) \quad , \quad x_{\mathscr{S}_5}(t_{\mathrm{I}}) = x_{\mathrm{I}} \,. \tag{17.2.4j}$$

$\mathscr{S}_5$ interacts with the rarefaction $\mathscr{R}_{\mathrm{H}}$ and consequently accelerates while the state to its right decreases, see Fig. 17.4. Depending on the situation at hand, $\mathscr{S}_5$ can be an entropic shock or a non–entropic one. Assume that $\mathscr{S}_4$ reaches the door before it is reached by $\mathscr{S}_5$. Then, at time t_{L} the panic reach the door and this causes the fall down of the efficiency of the door to $\mathrm{p_d}$. From L starts a shock $\mathscr{S}_6$ given by

$$\mathscr{S}_6 : \dot{x}_{\mathscr{S}_6}(t) = \sigma\left(\mathscr{R}_{\mathscr{N}_1,\mathrm{H}}\left(t, x_{\mathscr{S}_6}(t)\right), \rho''_{\mathrm{d}}\right) \quad , \quad x_{\mathscr{S}_6}(t_{\mathrm{L}}) = x_{\mathrm{L}} \,, \tag{17.2.4k}$$

that interacts with the rarefaction $\mathscr{R}_{\mathscr{N}_1,\mathrm{H}}$ and accelerates backwards, see Fig. 17.4. $\mathscr{S}_5$ meets $\mathscr{S}_6$ in M and then starts $\mathscr{S}_7$ given by

$$\mathscr{S}_7 : x = x_{\mathrm{M}} + \frac{f(\rho''_{\mathrm{d}})}{\rho''_{\mathrm{d}}} \, (t - t_{\mathrm{M}}) \,, \tag{17.2.4l}$$

that can be a non–entropic shock or an entropic one, depending on whether the segment $s(0, \rho''_{\mathrm{d}})$ intersects the graph $f = f(\rho)$ or not. Finally $\mathscr{S}_7$ reaches the door and the evacuation time results to be t_{N} given in (17.4.1a).

The solution of (17.2.1) resulting from the above construction is represented in Fig. 17.7, left, in the (x,t)–plane and in Fig. 17.8, up, in the (x,ρ)–plane.

17.3 Evacuation with an Obstacle

Keep now the same situation as in Sect. 17.2, but insert another door / obstacle at a point $x_{\mathrm{O}} \in \left]x_{\mathrm{B}}, x_{\mathrm{C}}\right[$, see Fig. 17.6. The resulting problem is

$$\partial_t \rho + \partial_x f(\rho) = 0 \qquad (t,x) \in \mathbb{R}_+ \times \left]0, x_{\mathrm{D}}\right[\tag{17.3.1a}$$

$$\rho(0,x) = \bar{\rho}\, \chi_{[x_{\mathrm{A}}, x_{\mathrm{B}}]}(x) \qquad x \in [0, x_{\mathrm{D}}] \tag{17.3.1b}$$

$$f\left(\rho(t, x_{\mathrm{O}})\right) \le \mathrm{p_O} \qquad t \in \mathbb{R}_+^* \tag{17.3.1c}$$

$$f\left(\rho(t, x_{\mathrm{C}})\right) \le p\left(\rho(t, x_{\mathrm{C}})\right) \qquad t \in \mathbb{R}_+^* \,. \tag{17.3.1d}$$

Here, p_O is the given maximal possible flow at the obstacle location. With reference to Fig. 17.5, right, we assume that the obstacle through–flow p_O and the corresponding densities ρ'_O and ρ''_O are such that

$$\rho'_D < \rho'_O < \rho''_O < \rho''_D\,, \qquad f(\Delta s + s) < p_O = f(\rho'_O) = f(\rho''_O) < f(s)\,, \quad (17.3.2a)$$

and the initial crowd density satisfies (17.2.2b) together with the condition

$$\overline{\rho} > \frac{x_O - x_B}{x_B - x_A}\left(\frac{p_O}{f'(\rho'_O)} - \rho'_O\right). \qquad (17.3.2b)$$

As for the problem (17.2.1), to solve (17.3.1) we first consider the two Riemann problems (17.2.3) obtaining (17.2.4c) as solution for small times.

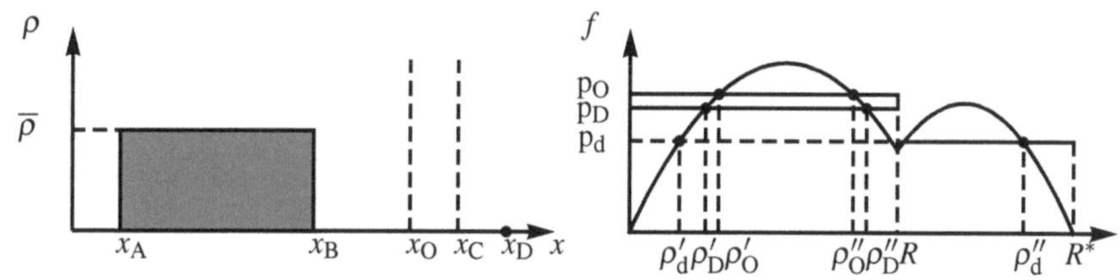

Fig. 17.5 Left: The initial datum for the constrained Cauchy problem (17.3.1). Right: The fundamental diagram, the maximal outflow through the door in standard situations, p_D, through the door in overcompressed regime, p_d, as defined in (17.1.1), and through the obstacle, p_O, as defined in (17.3.2a).

Fig. 17.6 A uniformly distributed crowd along $[x_A, x_B]$ in a corridor $[0, x_D]$ with an obstacle x_O before the exit x_C.

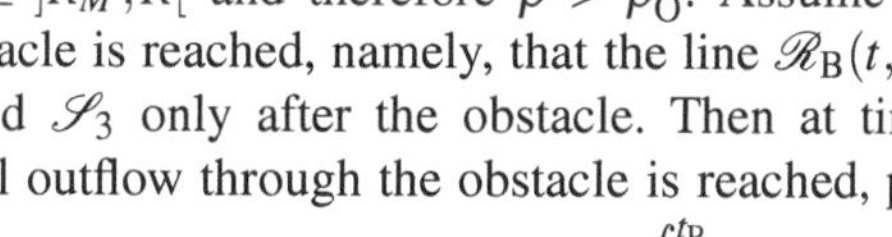

Note that pedestrians start passing through the obstacle at time $t_U = (x_O - x_B)/f'(0)$. By assumption $\overline{\rho} \in]R_M, R[$ and therefore $\overline{\rho} > \rho'_O$. Assume that the maximal flow through the obstacle is reached, namely, that the line $\mathscr{R}_B(t,x) = \rho'_O$ meets both the shocks $\mathscr{S}_2$ and $\mathscr{S}_3$ only after the obstacle. Then at time $t_P = (x_O - x_B)/f'(\rho'_O)$, the maximal outflow through the obstacle is reached, provided not all people already passed it, i.e. provided $(x_B - x_A)\,\overline{\rho} > \int_{t_U}^{t_P} f\left(\rho(t, x_O-)\right)\, dt$ which, in turn, is ensured by (17.3.2b).

Assume also that the line $\mathscr{R}_B(t,x) = \rho'_O$ meets $\mathscr{S}_3$ before $\mathscr{S}_2$. Then from this intersection starts a shock $\mathscr{S}_{10}$ between the values ρ'_O and ρ''_D. Assume that $\mathscr{S}_{10}$ reaches the obstacle before $\mathscr{S}_2$ and call Q the intersection point. Then from Q starts the shock $\mathscr{S}_9$ between ρ''_O and ρ''_D. A shock $\mathscr{S}_8$ arises from P, where the line

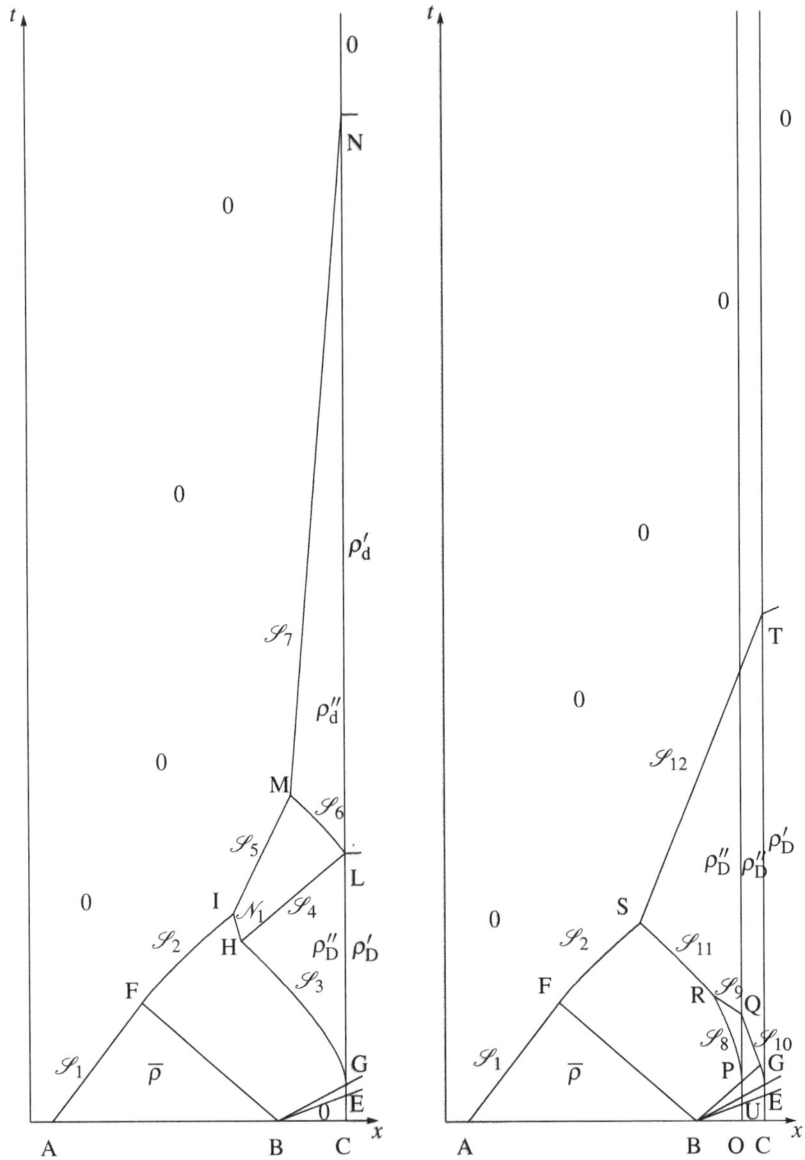

Fig. 17.7 Left: The solution to (17.2.1). Right: The solution to (17.3.1).

$\mathscr{R}_B(t,x) = \rho'_O$ reaches the obstacle. Along the right side of $\mathscr{S}_8$ the density is constant and equal to ρ''_O. Due to the interaction with the rarefaction $\mathscr{R}_B$, the shock $\mathscr{S}_8$ accelerates backwards, while the state to its left increases. Assume that the shock $\mathscr{S}_9$ reaches the shock $\mathscr{S}_8$ before the shock $\mathscr{S}_2$. Then, at time t_R, the shocks $\mathscr{S}_8$ and $\mathscr{S}_9$ meet and the result of their interaction is the shock $\mathscr{S}_{11}$. Analogously to $\mathscr{S}_8$, also $\mathscr{S}_{11}$ interacts with the rarefaction $\mathscr{R}_B$ and accelerates backward reaching $\mathscr{S}_2$ in S. Along the right side of $\mathscr{S}_{11}$, the density is constant and equal to ρ''_D, while on the left side it increases. Note that by hypothesis (17.3.2a), $s > \rho''_O - \Delta s$. We also assume that $\mathscr{S}_2$ meets $\mathscr{S}_{11}$ before the line $\mathscr{R}_B(t,x) = s$. Then, from S starts the shock $\mathscr{S}_{12}$ that reaches the door at time t_T, which corresponds to the time necessary to evacuate the corridor.

The solution of (17.3.1) resulting from the above construction is represented in Fig. 17.7, right, in the (x,t)–plane and in Fig. 17.8, down, in the (x,ρ)–plane.

Remark that depending on the situation at hand, the evacuation time for (17.2.1) could be higher than that for (17.3.1), as for the case described in Fig. 17.7.

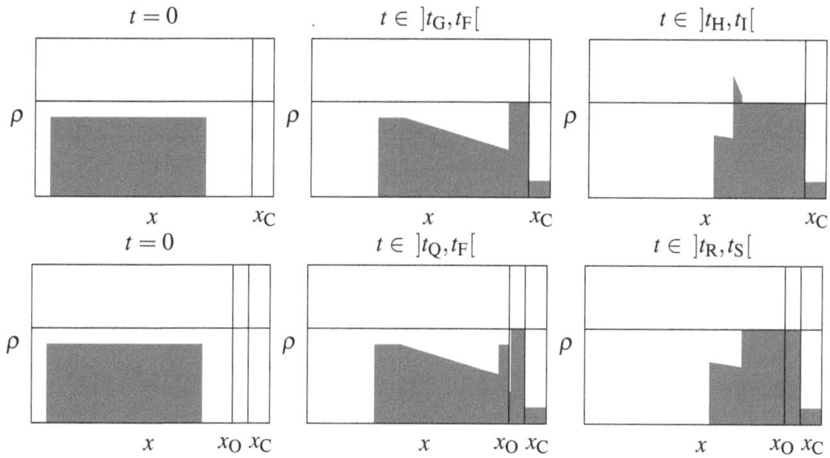

Fig. 17.8 Numerical integrations of (17.2.1), up, and of (17.3.1), down, using the wave front tracking method. The vertical segments denote the positions of the obstacle, x_O, and of the exit, x_C. The horizontal segment denotes the value $\rho = R$.

17.4 Evacuation Time

The evacuation time T is particularly relevant and can be computed integrating (17.2.1) or (17.3.1) numerically. Having a simple initial datum, i.e. uniformly distributed on a given segment, an analytical study is also possible. Indeed, the wave front tracking technique applied to (17.2.1) yields Fig. 17.7, left, while applied to (17.3.1) yields Fig. 17.7, right.

Proposition 17.1. *The evacuation times for the cases analyzed in Sect. 17.2 and Sect. 17.3 are respectively*

$$t_N = \frac{(x_B - x_A)\,\overline{\rho} + (x_C - x_B)\,\rho'_D}{p_d} - t_L\left(\frac{p_D}{p_d} - 1\right) \qquad (17.4.1a)$$

$$t_{\mathrm{T}} = \frac{(x_{\mathrm{B}} - x_{\mathrm{A}})\, \overline{\rho} + (x_{\mathrm{C}} - x_{\mathrm{B}})\, \rho'_{\mathrm{D}}}{\mathrm{p}_{\mathrm{D}}}\,. \tag{17.4.1b}$$

Proof. Consider the case studied in Sect. 17.2. In this case

$$\begin{aligned}(x_{\mathrm{B}} - x_{\mathrm{A}})\, \overline{\rho} &= \int_{t_{\mathrm{E}}}^{t_{\mathrm{G}}} f\left(\rho(t, x_{\mathrm{C}}-)\right)\, \mathrm{d}t + (t_{\mathrm{L}} - t_{\mathrm{G}})\, \mathrm{p}_{\mathrm{D}} + (t_{\mathrm{N}} - t_{\mathrm{L}})\, \mathrm{p}_{\mathrm{d}} \\ &= (x_{\mathrm{C}} - x_{\mathrm{B}}) \left(\frac{\mathrm{p}_{\mathrm{D}}}{f'(\rho'_{\mathrm{D}})} - \rho'_{\mathrm{D}} \right) + \left(t_{\mathrm{L}} - \frac{x_{\mathrm{C}} - x_{\mathrm{B}}}{f'(\rho'_{\mathrm{D}})} \right) \mathrm{p}_{\mathrm{D}} + (t_{\mathrm{N}} - t_{\mathrm{L}})\, \mathrm{p}_{\mathrm{d}}\end{aligned}$$

and therefore (17.4.1a) holds true. The proof of (17.4.1b) is analogous. □

Remarkably, in this particular situation, the evacuation time with no obstacle is larger than the evacuation time with the obstacle. Note that the presence of the obstacle avoids the density to reach the high values $\rho \in\ ]R, R^*]$, thus allowing for a faster evacuation from the corridor. In particular, we get the diagram in Fig. 17.9 for the evacuation time, T, as a function of the position of the obstacle, $x_{\mathrm{O}} \in\ ]x_{\mathrm{B}}, x_{\mathrm{C}}[$. Note that the presence of an obstacle too close to the exit is negligible. Indeed,

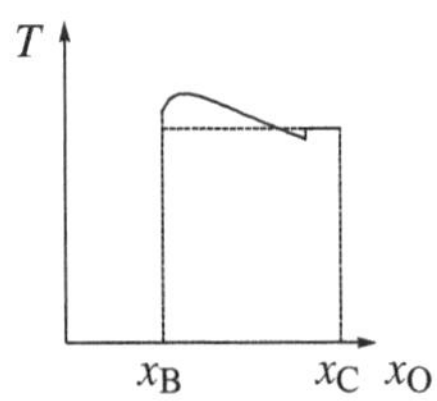

Fig. 17.9 The dotted horizontal line is the evacuation time with no obstacle. The solid line is the evacuation time, T, as a function of the position of the obstacle, x_{O}.

the efficiency of the obstacle is primarily dependent on avoiding the insurgence of panic. We also underline that the expressions for the evacuation times computed in Proposition 17.1 are not general since they deeply depend on the situations at hand.

References

1. Amadori, D., Colombo, R.M.: Continuous dependence for 2 × 2 conservation laws with boundary. J. Differential Equations 138(2), 229–266 (1997)
2. Bardos, C., le Roux, A.Y., Nédélec, J.C.: First order quasilinear equations with boundary conditions. Comm. Partial Differential Equations 4(9), 1017–1034 (1979)
3. Braess, D., Nagurney, A., Wakolbinger, T.: On a paradox of traffic planning. Transportation Science 39(4), 446–450 (2005)
4. Colombo, R.M., Facchi, G., Maternini, G., Rosini, M.D.: On the continuum modeling of crowds. American Mathematical Society (AMS), Providence (2009)
5. Colombo, R.M., Goatin, P., Maternini, G., Rosini, M.D.: Using conservation Laws in Pedestrian Modeling, pp. 73–79 (2009)
6. Colombo, R.M., Goatin, P., Maternini, G., Rosini, M.D.: Macroscopic Models for Pedestrian Flows. In: Big Events and Transport: The Transportation Requirements for the Management of Large Scale Events, pp. 11–22. IUAV – TTL Research Unit (2010)

7. Colombo, R.M., Goatin, P., Rosini, M.D.: A macroscopic model for pedestrian flows in panic situations. In: Proceedings of the 4th Polish-Japanese Days. GAKUTO International Series. Mathematical Sciences and Applications, vol. 32, pp. 255–272 (2010)
8. Colombo, R.M., Goatin, P., Rosini, M.D.: Conservation laws with unilateral constraints in traffic modeling. In: Mussone, L., Crisalli, U. (eds.) Transport Management and Land-Use Effects in Presence of Unusual Demand, Atti del Convegno SIDT 2009 (June 2009)
9. Dubois, F., LeFloch, P.G.: Boundary conditions for nonlinear hyperbolic systems of conservation laws. J. Differential Equations 71(1), 93–122 (1988)
10. Farkas, I., Helbing, D., Vicsek, T.: Simulating dynamical features of escape panic. Nature 407(6803), 487–490 (2003); Randomness and complexity (Eilat, 2003)
11. Lebacque, J.P.: Intersection Modeling, Application to Macroscopic Network Traffic Flow Models and Traffic Management. In: Hoogendoorn, S.P., Luding, S., Bovy, P.H.L., Schreckenberg, M., Wolf, D.E. (eds.) Traffic and Granular Flow 2003, pp. 261–278. Springer, Heidelberg (2005)

Index

MIX
Papier aus verantwortungsvollen Quellen
Paper from responsible sources
FSC® C105338

If you have any concerns about our products,
you can contact us on
ProductSafety@springernature.com

In case Publisher is established outside the EU,
the EU authorized representative is:
Springer Nature Customer Service Center GmbH
Europaplatz 3, 69115 Heidelberg, Germany

Printed by Libri Plureos GmbH
in Hamburg, Germany